The Book of Immortality

ALSO BY ADAM LEITH GOLLNER

The Fruit Hunters:
A Story of Nature, Adventure, Commerce and Obsession

The Book of Immortality

The Science, Belief, and Magic Behind
Living Forever

Adam Leith Gollner

Doubleday Canada

Library and Archives of Canada Cataloguing in Publication is available upon request

ISBN: 978-0-385-66730-2

Jacket illustration by Janet Hansen
Jacket design by Tal Goretsky and Janet Hansen
Printed and bound in the USA

Published in Canada by Doubleday Canada,
a division of Random House of Canada Limited,
A Penguin Random House Company

www.randomhouse.ca

10 9 8 7 6 5 4 3 2 1

To my mother and father

Host of the 1994 Miss USA competition, to Miss Alabama: "If you could live forever, would you want to, and why?"

Miss Alabama: "I would not live forever, because we should not live forever, because if we were supposed to live forever, then we would live forever, but we cannot live forever, which is why I would not live forever."

I haven't any clear idea what I'm saying when I'm saying "I don't cease to exist." . . . If you say to me—"Do you cease to exist?"—I should be bewildered, and would not know what exactly this is to mean . . . and this is all there is to it—except further muddles.

—Ludwig Wittgenstein, *Lectures on Religious Belief*

Contents

Part 3: Science

The Book of Immortality

Prologue

On Finitude and Infinity

The only secret people keep
Is Immortality.

—Emily Dickinson, poem number 1748

My dear colleagues: good bad, religion poetry, spirit skepticism,
definition definition,
that's why you're all going to die,
and you *will* die, I promise you.
The great mystery is a secret, but it's known to a few people.

—Tristan Tzara, *Seven Dada Manifestos*

WE'RE FOREVER DREAMERS. Humans have always believed in immortality. In search of longevity, if not eternal youth, we've tried elixirs, hormones, prayers, pills, spells, stem cells. The Hungarian countess Erzsébet Báthory bathed in the blood of murdered virgins. Throughout the Middle Ages, old men tried to hot-wire faded energy levels with veinloads of fresh blood, often resulting in gruesome transfusion mishaps (as when three boys died draining themselves for Pope Innocent VIII). Seventeenth-century Englishmen guzzled buzzard stones and pulverized boar pizzles hoping to solve the puzzle of aging. In the 1960s, booster shots of fetal lamb cells became a trend, with Swiss tissue clinicians administering embryonic injections to the likes of Noël Coward and Somerset Maugham. Modern-day gene-regenerating creams are made with baby human foreskin fibroblasts. Some Jamaican men still grate dried tortoise scrotum into bowls of soup as an antiaging tactic. If it won't bestow never-ending life, at the very least, they tell each other (and curious reporters), it's like Parmesan for the erectile soul.

Where haven't we gone? Elderly and hopeful we've traveled to backwater Romania for procaine hydrochloride treatments of Gerovital-H3, to Tibet in pursuit of pure lama urine, to the South Pacific seeking rainwater cures. In the 1990s, the abundance of centenarians in the Caucasus region led to speculation that kefir extends life; but in 1998, a 121-year-old Azerbaijani divulged his secret to investigators: he *never* ate yogurt. We don't care; just tell us again and again that there are hot spots, hidden valleys, and other blue zones where people live extraordinarily long, fulfilled lives. And then sell us ways of incorporating their secrets into our daily grind so that we, too, can hum forever.

How confused can we get? Immortality is as oxymoronic and straightforward as surviving death. After all, doesn't radical life extension just lead to eternal life? Heaven's Gaters convinced themselves they could reach the comet of paradise through cyanide-laced applesauce. The poet Charles Baudelaire's suicide note (from a failed 1845 attempt) explained, "I'm killing myself because I believe I am immortal." The pre-Socratic philosopher Empedocles leapt into an active volcano to prove that immortality is real. He was never seen again, but his name lives on in perpetuity. Eternal life is twisted like that, a molten knot, a Möbius striptease, a pretzel made of mirrors.

We die to live forever; and we use immortality to keep dead people alive. Decades after their deaths, the preserved bodies of Chairman Mao, Ho Chi Minh, and Lenin remain on public view. "Lenin, even now, is more alive than all the living," declared Vladimir Mayakovsky, at the great leader's funeral. "Lenin's death is not death," clarified the suprematist painter Kazimir Malevich. "He is alive and eternal."

How weird have we been? In our desperation, we've eaten Egyptian mummies. Entwined and embalmed, preserved for millennia, they seemed connected to the beyond. For hundreds of years, until World War II, scraps and powders of shredded or ground mummified corpses were prescribed as medicine (*Mumia vera aegyptiaca*) by European medics. Sixteenth-century physicians claimed that our vitality is nothing more than "a certain embalsamed Mumia," a self-generated healing balm that prevents us from rotting alive. Early chemists described it as "the liquor of an interior salt most carefully and naturally preserving its body from corruption." This corporeal potion could also be manufactured chemically, they argued, with mercury, or as a saline solution incorporating the smoked flesh of dead youths mixed with myrrh, agarwood resin, turpentine, and other distillates. Alas.

Still, ancient Egyptians were capable of building pyramids; surely they knew a thing or two about extending lives? When the four-thousand-year-old Edwin Smith Papyrus first resurfaced, it seemed to contain ancient secrets of rejuvenation. The scroll commences with a tantalizing promise: "The beginning of the book for making an old man into a youth . . ." Once the hieratic scribbles were fully decrypted, however, the directives turned out to be a base recipe for fenugreek oil—used to mask liver spots and as a hair restorative for balding men.

The truth is bald: we all get old and die, even if we wish there were a shampoo of eternal life. In late medieval times, experts thought the answer lay in usnea, a moss or lichen from the skull of a hanged man. Some of us still keep tufts of dead loved ones' hair in hopes of one day cellularly regenerating them. After the poet Milton died, his grave was pillaged for mementos. His hair didn't even need to be cut off—the follicles were no longer stuck to the scalp, and bunches came off in handfuls. These, alongside sundry other bits, made their way into *collections d'élite*. Poets called the locks of Milton's hair in their possession "lovely things that conquered death."

Half in love with the impossible, we've always wanted to conquer death. In 1854, the medical specialist Léopold Turck published a groundbreaking work (*De la Vieillesse Etudiée Comme Maladie*) characterizing old age as a curable illness. He believed electrical-shock treatments could revitalize and rejuvenate the elderly and the infirm. He was wrong. Aging is a fact of life, not a disease.

Or is it? We're bombarded with media reports on the inevitability of living forever: all we need to do is lengthen telomeres, target sirtuins, or activate CREB1, the brain's latest "longevity molecule." Cover stories in the *New York Times Magazine* tell us there are immortal jellyfish, but then it turns out the jellyfish aren't immortal at all. In reality, the immortal jellyfish is extremely weak, easily killed, and often eaten by slugs.

We want immortality so badly that we're always ready to be swept away into unthinkingness. Pitchmen claim that injecting stem cells into our skin will make us young forever, but then women who've undergone pioneering surgical stem-cell face-lifts discover tiny bones growing in their eyelids. (Each blink sounds "like a tiny castanet snapping shut.")

In 1971, longevity researchers declared that science would unravel all the mysteries of aging within five years. Five years later, the *San Francisco Chronicle* reported that "human life could be extended to 800 years." That same year, an outfit called Microwave Instrument Co. in Del Mar, California, said they'd have immortality drugs on the market within three years. Here we are, decades later, still croaking. The expression *hope springs eternal* itself first appeared in a 1733 poem mocking the foolish desire to become godlike through science. That vain hope is eternally ours.

In 1220 CE, the undefeatable Genghis Khan summoned a cave-dwelling Taoist monk called Qiu Chuji to his court. The Great Khan felt sure that Qiu Chuji had penetrated the essence of the Tao. Here, finally, was someone who knew the secret to infinity. "Communicate to me the means of preserving life," the Khan implored in his missive. "Say only one word to me and I shall be happy." The wise old hermit obliged, traveling thousands of miles over Mongolia and the Tian Shan mountains, across Kazakhstan and Kyrgyzstan, into Afghanistan, finally arriving at the warlord's encampment in the Hindu Kush. "Sainted man, you have come from a great distance," announced Genghis Khan, welcoming him. "Have you a medicine of immortality?"

Qiu Chuji shook his head: "There are no medicines for immortality."

Today, according to the market-research firm Global Industry Analysts, we spend somewhere between $80 billion and $114 billion per year on antiaging products and other modern medicines for immortality. And none of them work. All we can really do is eat our vegetables and exercise. Even then, no matter what we sacrifice, how disciplined we may be, whether we chug wheatgrass or not, we can still get stuck in a tsunami, pulled into drawn-out illness, trampled by an elephant. We all have to go the way Genghis did, whether dying in battle, tumbling off a horse, succumbing to pneumonia, or being shivved by a lover. Maybe one day we just don't wake up. However it happens, we enter the mystery.

Introduction

The Nature of Immortality

It is apparent that there is no death.
But what does that signify?

—Edna St. Vincent Millay, "Spring"

Because they believed in nothing, they were ready to believe anything.

—Lucian Boia, *Forever Young*

IMMORTALITY DOESN'T ACTUALLY EXIST. It's not something tangible we can point to, see, or demonstrate. It resides in thought but not in reality. Immortality is an abstract concept that helps us make sense of death. The idea emerged from our fear of dying, from the sense that life *must* go on in some way.

Immortality means nonmortality, undeath, never-ending existence in this world or some other. It is the permanent absence of death. It entails evading or outliving the end. But that can't be done, or at least we can't prove that it can be done. No examples of anything immortal have ever been found by science. There are just visions, tales, hopes, fears, and maybe some inferential cognizers.

In most definitions, immortality occurs *after* death. The unending perseverance of a mind or a soul following the decay of the physical body is spiritual immortality. The basic premise of this cosmology is simple: we die but our soul (or some other part) doesn't. Just as our flesh must necessarily decay, our spirit or intellect or entelechy returns to the primordial source. An energy or force within us outlives its mortal container, ending up in the afterlife or hurled back into rebirth.

Spiritual immortality is a narrative of numberless incarnations, from eternal sanctification to damnation to reincarnation. The very word *immortality* conceals infinite possibilities. It's a one-word poem. It can mean whatever we want it to mean, whatever we believe it to mean. In recent years, the idea of the indefinite persistence of an undying material body has captivated us. But physical immortality is also a mythology. It, too, helps followers cope with an uncertain world, just as a Christian uses the idea of redemption.

We tend to imagine that these are secular times. The facts suggest otherwise. Belief in posthumous immortality is very much alive today. Data collected by the General Social Surveys show that 80 percent of Americans believe in life after death. The figures are around 70 percent in Canada, 65 percent in Australia, 60 percent in the UK, and above 50 percent throughout much of Europe. According to the World Values Survey, close to 100 percent of those surveyed in parts of the Middle East believe in the afterlife. Not exactly a faithless world.

For those of us who don't believe in immortality, we can either dismiss it or contemplate it. Either way, it's not something we can resolve. Immortality is a matter of belief, not fact. Like death, immortality is something we dance with. But there's no denying the existence of death. We can *believe* we won't die, but dying is ineluctable, devastating, *real*.

Every single day around two hundred thousand people die worldwide. There are two deaths every second. Six people just died. Make that eight. Ten. It can happen to anybody at any time, and yet exposure to death profoundly bothers our mind precisely because we can't understand it. Death is what's called a "meaning threat." When confronted with the incomprehensible—such as losing a loved one—our mind scrambles to find another pattern that alleviates the confusion. For some, it's enough to say, "They're gone." Others have such an urgent need to escape feelings of meaninglessness that they create alternate, more coherent plausibilities, such as myths about immortality.

Grief forces us to have an opinion about the end. Imagining that everlasting life exists is a common reaction. We tell ourselves the loved one is somewhere else now, somewhere better. To make sense of insensateness, we wrap ourselves in beliefs. We're all apprentice magicians trying to master the trick that transforms loss into understanding.

Thoughts of eternal life shuttle between the terminals of knowledge and belief. There are things we can know and things we can't know. The knowables are gathered into knowledge. We deal with everything else through belief. Science is our means of exploring all that can be known; belief is how we approach that which cannot be known. Beliefs allow the brain to assert truths when lacking material evidence. Death tells us nothing knowable, only that we are currently alive and that our bodies won't last forever.

As a result, psychologists claim we're all frightened of dying, but it isn't simply anticipatory worry; it's the not knowing that bothers us, the lack of control. What we want is something that doesn't exist: resolution.

Because patternlessness cannot be borne, the brain represses thoughts of its eventual extinction. It's impossible to understand what it would be like to have no more thoughts. We have a central incapacity, a bug built into the operating system: our consciousness cannot imagine a lack of consciousness. Trying to imagine our own death is like trying to think thought. We cannot do it. "Try to fill your consciousness with the representation of no-consciousness, and you will see the impossibility of it," wrote Unamuno. "The effort to comprehend it causes the most tormenting dizziness. We cannot conceive ourselves as not existing."

Nonexistence is nonconceivable. The brain conceptualizes things as being somehow similar to other conceivable things, so we compare death to life, minus the body—which is why we imagine that our consciousness (whatever that is) will outlive us. All our fantasies of stymieing the inevitable stem from an inability to grasp the fact of finality.

Consider what would happen if certain species did *not* die. They would simply keep on breeding and accumulating. In the time it takes to read this sentence, several hundred million ants will have been born across the planet. It would take a single tiny bacterium mere hours to generate a mass equivalent to that of a human child—and there are countless billions of bacteria within a hundred-foot radius of everyone. Imagine if they could live forever? "In less than two days, the entire surface of the earth would be covered in great smelly dunes of prettily colored bacteria," explains zoologist Lyall Watson. "Left similarly unhindered, a protozoan could achieve the same end in forty days; a house fly would need four years; a rat eight years; a clover plant eleven years; and it would take almost a century for us to be overwhelmed by elephants." We can thank death for the fact that our atmosphere isn't clogged with hedgehogs all the way to the ozone layer. Like everything else in nature, we're all terminal cases.

The oldest person who ever lived whose true age could officially be verified died at 122 in 1997. (She only gave up smoking at 119.) From the dawn of the *Homo* genus up to the 1800s, the majority could expect to live for approximately twenty-five to forty years. Largely due to basic realizations about hygiene, life expectancy has increased significantly over the past century and a half. Some demographers argue that life spans have attained their utmost and are starting to decrease slightly. Others disagree, suggesting that 125 is a reasonable target for baby boomers. Most scientists maintain that human life has a maximum expiry date, but immortalists speak of Plastic Omega (*omega* being the end of life, and *plastic* being malleable). As of 2013, all parties can anticipate living somewhere between seventy to ninety years unless an accident, disease, or disaster strikes—or immortality becomes reality.

Intriguing, genuine discoveries are being made in the field of gerontological studies. Scientists have dramatically increased the life spans of simple organisms such as yeast, roundworms, fruit flies, even mice. So far, those breakthroughs haven't yielded human applications. But even if we learned to cure every major disease, to resolve every cause listed on death certificates, some biologists argue, we'd still only add ten or fifteen years to human life expectancy. Research being made on a genetic level could eventually prove beneficial to our health span. Even so, death will become us.

There's an important distinction between medicine and miracle work: miracles prevent death; medicine counteracts illness. The main aim of mainstream medicine is prolonging health, and contemporary doctors know more about keeping people alive than ever before. That doesn't mean we can make people stop dying. The average Westerner only gets eighty years, not eighty trillion. And there's a price to pay for extended life: the degenerative diseases of senescence. Our bodies aren't supposed to live forever, which is why they have built-in obsolescence mechanisms.

Longevity is starkly different from immortality, yet somehow the two have fused in public consciousness. Most of us will live longer than our ancestors did, but all of us can still die at any moment. Spurred on by our gains in life expectancy, a pandemic of magical thinking about science's unlimited capabilities has led to a wider discussion about the possibility of eternal life.

Prime-time TV specials with titles like "Can We Live Forever?"

fuel the mass delusion. Every year more conferences pop up purporting to reveal the latest means of attaining eternity through technology. "Immortality Only 20 Years Away," blare newspaper headlines. Philanthropic organizations (the Immortality Institute, the Methuselah Foundation, the Fuck Death Foundation) are joining together to eliminate death. "There are people living today who may extend their life spans indefinitely," declare salesmen, triangulating faith, biology, and magic into a unified worldview.

This confusion has led to an alarming increase in the availability of untested antiaging remedies over the past two decades. Countless products are presently being sold as having life-extending qualities, even though there aren't any demonstrable means of increasing human life. "No treatments have been proven to slow or reverse the aging process," announced the National Institute on Aging in 2009, trying to staunch the hype. Contrasting the claims made by life-extension companies with the genuine science of aging, fifty-two scientists signed a "Position Statement" on longevity that clarified the situation explicitly: no currently marketed intervention—none—has yet been able to stop or even affect human aging. "The prospect of humans living forever is as unlikely today as it has always been," they wrote, "and discussions of such an impossible scenario have no place in a scientific discourse."

The dominant mythology of triumphalist scientism is the idea of progress. For the most part, we don't question the idea that everything is constantly getting better and better and better. It's just the way things are, we tell ourselves. We're so close to perfection. And progress necessarily leads somewhere: to a world in which we're all immortal.

A solid belief system is one we don't realize is a belief system. Because science's veritable achievements are so impressive, almost everybody today believes in the unidirectional march of progress. Technology is unceasingly propelling us forward, and science has become synonymous with progress, so it becomes easy to imagine that life everlasting is around the corner. We take it for granted that suffering can be eliminated, that poverty will ultimately be eradicated, that we should never be sick again, that science will soon make everybody never die. The illusion of continual betterment is a pervasive enough mythology that it can overlook the environmental crises, the scale of warfare, and the fact that over a billion people live on less than $1 per day.

Is progress even real? Microchips certainly get smaller and processing speeds faster, but not everything has progressed over the past centuries. Have our emotions changed since Shakespeare's time? Since Sophocles's time? Are we moving toward a time of universal happiness? Genocide is not an anachronism. Neither is inequality. Is progress a law of history, or is it a story we tell ourselves?

In 1869, the avant-garde writer Comte de Lautréamont published *Les Chants de Maldoror,* a book exploring "the spiritual crisis brought on by scientific progress." In it, he characterized immortality as "the terrifying problem that humanity has not yet solved." A century and a half later, we're still nowhere close to solving it.

Although our congenital belief in progress means we're more ready than ever before to believe in physical immortality, misinformed life-extension stories have been around for millennia. There's nothing new about bearded hustlers such as Aubrey de Grey vowing to help us live forever and cryonicists who claim to have "cured death itself." They're all tapping into a longing that has always been with us.

Yellowing medical journals are filled with stories about how "the great alchemical dream, the 'Elixir of Life,' seems almost ready to be bottled." Following World War II, the personal goal of attaining immortality moved from religious aspiration to "actual possibility." In 1966, biophysicists at the California Institute of Technology wrote, "We know of no intrinsic limits to the life span." In the 1970s, a group of molecular biologists and gerontologists mobilized as "the Immortalist Underground." In 2010, a special issue of *Time* magazine about longevity announced that "elixirs of youth sound fanciful, but the first crude anti-aging drugs may not be so far away."

We're drowning in misinformation. How-to books such as *Why Die? A Beginner's Guide to Living Forever* and *Young Again! How to Reverse the Aging Process* and *Physical Immortality: The Science of Everlasting Life* each outline various ways to defeat reality by harnessing miracles of technology. Finding such miracles is abundantly easy, especially online. Searching "immortality device wanted" leads to a site called www.achieveimmortality.com that claims to own US patent number 5,989,178 for "the most imporatnt [*sic*] invention in human history," a gear-based magnetic pinkie ring that "ALLOWS HUMANS TO STAY PHYSICALLY YOUNG FOREVER."

Entire subcultures of enthusiasts are dedicated to deathlessness. There are bloggers with "a passion to create an environment where all sickness, aging and death are eliminated." There are amateur philosophers who argue that everyone is bodily immortal until proven otherwise. Facebook Transhumanists list their religious views as "the abolition of suffering." They end posts with the movement's abbreviation: H+, as in "human plus." Superlongevist manifestos confidently assert that we can all live for hundreds of years, that the "eventuality" of a "modern Fountain of Youth" is nigh.

It is utterly ordinary to not want to die. But, as Dame Edith Sitwell once wrote, ordinariness carried to a high degree of perfection is precisely the definition of eccentricity. In her view, eccentricity entails "some rigid, and even splendid, attitude of Death, some exaggeration of the attitudes common to Life." As she concluded, there's something askew about people who don't understand they will die—or that they are actually dead, in the case of cryonicists buried upside down in frozen thermoses, five per container, wrapped in sleeping bags, awaiting reanimation.

Eccentrics really, *really* don't want to die. Caloric restrictors, sun gazers, nightwalkers, potion peddlers, cybernetic Nostradamus-types, and outright charlatans: the immortality community boasts a plethora of unorthodox individuals. Most noneccentrics consider physical immortality a nonsensical fantasy, but physical immortalists don't care. They're convinced that science will soon unlock the codes that regulate aging. In their hunger to live forever, or at least to confirm their bias that science can solve all problems, they're so willing to put their critical faculties on hold.

I am not personally interested in living forever, but I *am* interested in writing about people obsessed with the impossible. Nonfiction writers are characterists—we look for characters living real stories. Perhaps not unexpectedly, people who want to never die may appear unsavory: a hint of the corpse hangs about them, a whiff of swamp yawn. Those seeking endless physical existence are undoubtedly peculiar, but there's also something profoundly human about them. *More than human,* they would say. *Human after all* might be more accurate.

We all believe. A belief is a relationship, something we fall into or grow up with; we can cherish it, desert it, stay loyal or cheat. A peculiarity of all belief systems is that those in the throes of belief do not see themselves as believers, but rather as those who know the Truth—even though the Truth cannot be known. We can view ourselves as believers or as nonbelievers. It's a personal choice. Either way, we all don't understand death. So whatever anyone sees death as is what it is.

Building palisades of belief is what we do when we can't understand something that has no provable explanation. When it is impossible to know something with any certitude, we turn to belief to feel like we know. Beliefs are mirages that provide the illusion of certainty. Unlike a fact, a belief can persist even when disproved.

To this day, belief precedes knowledge. Before we can know whether something is true or not, first we have to perceive or experience it, and then believe or disbelieve it. All scientific tests begin as beliefs before becoming testable hypotheses—but once we come up against the limits of the knowable, we either turn back to rational ground or take a leap into faith.

Examining our own base assumptions invariably means realizing that some things we take as knowable facts are simply beliefs. We may find it hard to distinguish between what we believe and what we know. We've classified ourselves as the species that knows: *sapiens*. But there is so much we don't know, that we can't know. We believe that we know; that's all. *Homo credulis* would be more accurate.

We all believe, and we all need mythologies. As Einstein brokered it, "Science without religion is lame, religion without science is blind." In his view, nonbelievers don't realize how many unattainable secrets surround us. Einstein tended to be more critical toward debunkers than those of faith. "The fanatical atheists," he wrote, "are like slaves who are still feeling the weight of their chains which they have thrown off after hard struggle." He espoused an approach of humility, rather than one of hubris. Just as religious zealots deserve to acknowledge the truths of science, rabid unbelievers could benefit from recognizing the limitlessness of the unknown.

We are pattern seekers. And so-called nonbelievers, like the devout of any denomination, are simply doing what humans have always done: they are looking for meaning. The conviction that disbelief is a preferable alternative to religious belief has, paradoxically, transformed atheism into a religion. It isn't very organized, but it *is* a system of thought based on a relationship with the unknowable. And any story purporting to explain death is an indication of faith.

Approximately 5 to 10 percent of Americans don't believe in God. Somewhere between 0.7 and 2 percent of Americans define themselves as atheists. The central tenets of atheism are that humans have no soul, that

God doesn't exist, and that nothing happens after death. None of these are provable or disprovable—they're matters of belief. Calling atheism a "belief system" is anathema to atheists, who insist that their position is one of no beliefs whatsoever. But they do believe. They believe they know what death means, just like others. And they also have key texts, prophets, myths. They attend atheist gatherings, where they can feel that sense of belonging and community others find in traditional churches, temples, or places of meditation. As with traditional believers, they insist things would improve if everyone adhered to their view.

Merely talking about belief can be particularly emotional for those who think they "understand" that the way for the world to be perfect is to dispense with belief. This position stems from a fundamental misapprehension of what belief actually is, and is itself a means of imposing one's own belief system on others. I tell such people that I respect all belief systems (including theirs) as long as nobody's getting hurt. Then again, I hasten to add, I'm the sort of centrist who believes that intelligence and faith can coexist—and who also believes that conflict is never-ending. Such conversations are akin to discussing abortion rights with evangelical Christians. But just because something is difficult to talk about doesn't mean we *shouldn't* talk about it. Quite the opposite.

Atheists aren't the only nonbelievers. A large percentage of that unbelieving 5 to 10 percent identify themselves as nothing in particular. Others are agnostics, or undecided. Agnosticism was defined by Thomas Henry Huxley in 1869 as a way to "neither affirm nor deny the immortality of man." But being noncommittal does not exempt a person from the need to believe. Even those agnostics who would remove themselves from any position are nonetheless eventually forced into a belief-officiated relationship with death.

Whether we take out billboard advertisements calling for religions to be dispensed with or whether we protest the teaching of evolution in schools, it's as erroneous to assume scientific theories are the literal Truth as it is to imagine that a religious text contains accurate history. Both methodologies are thought games, tools that allow us to contend with a universe whose ultimate nature will always elude us.

An ingrained certainty about eternal life helps many people function, including physical immortalists, even though thanatologists (the technical term for those who advocate an acceptance of the inevitable) think thanatophobes are delusional. Conversely, prolongevists deride "deathists" as pessimistic pushovers. "As a physician," Carl Gustav Jung once said, "I am convinced that it is hygienic to discover in death a goal toward which one can strive; and that shrinking away from it is something unhealthy and abnormal." But physical immortalists are convinced that human beings think *too much* about death rather than too little; that we have all been too accepting of death—and that such a viewpoint is self-defeating, serving to perpetuate death. Their fondest hope is to render thanatology obsolete.

Attempts to do justice to both sides of the argument rapidly spiral into meaninglessness, which is as it should be: beliefs are illogical. Demonstrable evidence has no place in belief. Regarding death, the only certain thing is uncertainty.

My own inquiries into the topic of immortality began on a fog-enshrouded fall night in London. There to research the history of labyrinths, I had spent the afternoon walking Hampton Court's seventeenth-century hedged maze. The train back to town passed a dilapidated brick brewery with a story-high sign saying TAKE COURAGE. In my hotel room, jotting down some observations about the silvery mirror I'd come across at the center of the labyrinth, I fell asleep on the couch, an overstuffed, royal-blue antique.

Shortly before dawn, I was startled from my slumber by an unsettling, beauteous dream of the fountain at the source of all life. The liquid gushing forth resembled water mixed with mercury. Its droplets crystallized into a blindingly radiant sea of diamonds. The fountain's location wasn't specified, and the rest of the dream is hazy, but the image of a crescendo of water bursting into the sky stayed with me.

Turgenev spoke of his stories beginning as visions hovering before his eyes, soliciting him. The apparitions that inspired his writing, he said, often seemed to embody the complexities of existence, intricacies he couldn't yet understand, subjects he could only arrive at by completing his story. I didn't know what the fountain was, let alone what it represented, but the dream had an urgency, as though it were a demand, more than a clue, something to pursue.

In the end, the dream became the start of a story. Philip Roth, when beginning a book, would always ask himself, "If this book were a dream, it would be a dream of what?" I felt the reverse. "If this dream were a book," I wondered, "it would be a book of what?" There was only one answer: a book of immortality.

The common impulse toward making one's life worthwhile stems from our ambivalence about the meaning of death. We've been granted the opportunity of a life's time, so we try to be significant, notable—but why? Beyond the necessities of earning a living or of achieving status, every answer points toward our being terrified of death. The accomplishments we pursue, whether shooting for athletic excellence, expressing ourselves creatively, having children, striving to build a legacy that helps our fellow humans, or volunteering in the service of a charitable or spiritual cause, are all undertaken in the hopes of transcending mortality, of garnering a dash of salvation. We do and make to deal with oblivion.

The urge to be a writer taps into this well as well. Poets "write their poems to ward off dying," explains Harold Bloom. Horace felt that his contributions to the poetic arts ensured his immortality: "I shall not altogether die." Anyone worried about living and dying in obscurity, wrote Chekhov, "reflectively snatches up a pencil and hastens to write his name on the first thing that comes handy."

Writing is an attempt to find a way out of a situation from which there is no way out, like Houdini escaping from a water-torture chamber, or a soul escaping a body. In John Fante's view, writers try to "endow posterity with something like a monument" to their days upon this earth. But so do sculptors. And architects. And painters. And so on. It's not just artists who are like this. In all our deepest reveries, immortality is always only a few synapse bursts away.

Marketers long ago figured out how to tap into this basic desire to escape death. Gucci's clothes are "what lasts forever." The Canadian cookware company Paderno offers "pots for eternity." Wine critics call the 1961 Jaboulet Hermitage "truly immortal," but it'll taste like vinegar in a few thousand years. Grocery stores stock utopia: in their gleaming aisles, everything is always in season and you can eat anything you want whenever you want. Even the meat there doesn't seem to have come from actual living beasts. Nothing ever dies under those neon lights. Aeterna is the name of a flashy funeral home near my studio. Aeterna: a place to help you live on after death. Forever—for a fee.

The rich, powerful, and important have always tried to circumvent destiny by throwing money at it. But we cannot bribe our way out. PayPal founder and Facebook board member Peter Thiel has invested heavily in foundations dedicated to ending aging. In 2009, he sank millions into a Silicon Valley nanotechnology start-up called Halcyon Molecular. Its founder, William Andregg, told TechCrunch.com he plans to live for "millions, billions, hundreds of billions of years." Halcyon Molecular quietly went out of business in the summer of 2012. Other technocrats are still trying. Google's Sergey Brin has donated hundreds of thousands of dollars to the Ray Kurzweil–affiliated Singularity University. And Larry Ellison, CEO of Oracle software, has vowed to defeat mortality. As his biographer notes, Ellison sees death as "just another kind of corporate opponent he can outfox." But in this case, the house always wins.

We may daydream of becoming godlike, but distressingly human we remain. Julius Caesar was deified, despite accomplishing the one thing no divinity could: he died, as did his heir Augustus, the self-designated Son of God. Subsequent emperors fantasized about making it to forever, but Tiberius was smothered, Caligula stabbed, Claudius poisoned, Commodus strangled. Killed in battle. Drowned. Struck by lightning. Their profiles were permanently sutured into coins, but they spent themselves. Starvation. Decapitation. Apoplexy. The plague. The end.

Yet, we continually long for something greater. This yearning to go beyond, to transcend our current state, is built into our DNA, because that's precisely what DNA does. It takes on new forms of complexity. Its impetus is to overreach, to overextend, to mutate on up, and the miracle (or reality) of life is precisely that it has managed to express itself in such an infinity of shapes and sizes. There is something in life that wants to replicate variations of itself, something vital and revitalizing, something as creative as creation, something that opens a passage through inertia and pulsates into being: call it the Life Force.

"Your own fate you can't escape," wrote Turgenev. My fate, as that dream fountain decreed, became researching eternity.

At the outset, I had no idea what immortality even meant. I started by reading the oldest stories ever written, the newest discoveries in molecular biology, the densest philosophy, the soaringest poems, the freakiest online message boards. Nothing dealing with immortality offers any closure. Some thinkers out there have spent decades attempting to prove or disprove eternal life. Their treatises invariably conclude with lines about how the belief in personal immortality "inevitably and inescapably involves unresolved and perhaps unresolvable tensions." Perhaps. Grappling with such books can be a thrilling speculative workout, but the mental gymnastics inevitably and inescapably drop you where you were at the start, only more confused. "He is convulsed," wrote Byron. "This is to be a mortal and seek the things beyond mortality." Finding certitude in the world of immortality is like trying to paint with air.

Immortality doesn't want to be understood. It wants to be believed in. And trying to comprehend how the believing mind works often left me feeling autistic. Medieval alchemists deemed investigations into the nature of eternity to be voyages into the *sylva sylvarum*—the forest of forests. In no time, the topic possessed me. I lived under its canopy. The forest dictated the rules, the timetable, even some of the words. I'd find myself writing sentences that made no sense. But then, months later, I'd understand what I'd been getting at. The forest guided me.

Part of the magic of exploration is its unpredictability. False starts, painful detours, dissolving paths: mistakes are how stories, and lives, take shape. The word *error* derives from the Latin term *errare,* which has a dual meaning: to be wrong and to wander, two verbs about being human. We are built for the hunt, for losing our way, and in hopes of finding the unfindable we can only continue searching, always moving forward, never arriving, until we actually die. We're never really out of the forest.

As I familiarized myself with the literature, I started speaking with geneticists and gerontologists about the tumult surrounding scientific promises of indefinite life extension, with classicists and mythologists about the fountain of youth in history, with balneologists and homeopaths about the powers of healing waters, with mystagogues about the meaning of rebirth and regeneration. I embarked on countless concentric conversations plumbing their, and my, innermost values. The leaders of different religious denominations told me about immortality in their respective faiths. Quacks demonstrated their products, and philosophers of science shared their insights.

I wholeheartedly loved learning about never-ending life, but the subject often felt so endless that I despaired of ever completing a manuscript. I hung a poster of a skull over my desk and tried to keep calm and carry on. I joined a yoga studio, bought bottles of burgundy (wine being a means of realizing mysteries), and attempted to understand the ways we can't understand.

I had entered the dark woods. Groping my way out took five years.

That's normal. Descending into the underworld isn't complicated, as the Cumaean Sibyl informs Aeneas: "But to return, and view the cheerful skies, / In this the task and mighty labor lies." I chipped away at a never-ending book. I grew older. I broke some bones. I took up whittling. I always only saw infinity. Whether or not I believed in free will, I couldn't not continue. Life's inextinguishable current had zapped me in a dream.

Throughout the process, my dreams overflowed with fountains, bodies of water, magical liquids. I dreamed that I was in Water School, that I was chewing on the ocean's blue flesh, that I'd found the formula to explain water's mysterious link to the symbolic realm. I dreamed that a bird made of water flew into my room and told me I could ask it one question. "Are you really here?" I asked. It didn't respond. I dreamed I was crawling up a mountain, deliriously stopping at every puddle in case it contained the fountain of youth. On the atoll of Apollo, I dreamed I saw Dionysus with my own eyes. Sacred light shone down, filling me with awe and terror.

Waking from these dreams, I struggled to relax. There is no need to stay in the narrative, I'd remind myself. There is only this moment, unconnected to what came before, to any others, there is only now. I learned to breathe deeply.

In hopeful moments, I'd envisage the book's unfinishedness as a stay of execution. If some chthonian dream deity had wanted this project to be undertaken, then—simply by following through—I was protected unto completion by that subterranean force majeure. In bleaker climes, however, I felt haunted for not moving fast enough, and I often thought the book would kill me if I didn't kill it first. Such is the believing mind, or at least the writing mind. I can vouch for Margaret Atwood's formulation; she describes writing a book as an act of "negotiating with the dead."

One afternoon in my studio, surrounded by heaps of paper, on the verge of giving up entirely, I got a call from an unfamiliar number.

"Infinity of Manhattan," said the voice on the other end of the line.

I assumed it was a friend messing with me, or maybe a telemarketing scam. "Okay. *Infinity?* What is that?"

"We're returning your call," the man explained. He sounded serious.

"Excuse me?"

"You left us a message. Your inquiries weren't totally clear, so I'm calling back."

"*I* made some inquiries to *Infinity*?" I asked incredulously.

"Yes, you did, or maybe someone else did and they're using your number. You wanted to know about long-distance endurance?"

If this was a prank, it was pretty cosmic in nature.

"It can't have been me," I stammered. "I didn't realize I could call Infinity."

"This must be a wrong number—we'll take note of it. Sorry for the confusion."

Googling the number on my call display, I learned that it was a car dealership, for the luxury-vehicle brand Infiniti ("Accelerating the Future"). Regardless, the message sank in. Long-distance endurance. Perseverance. I pressed onward.

An ancient tale about Alexander the Great describes his encounter with a wise old man in the Land of Darkness. The sage asks Alexander why he wants to venture further into the uncertainty. "I have heard that therein is the Fountain of Life," replies Alexander, "and I desire greatly to go forth and see if, of a truth, it is there."

Humans are *alphestes,* Homer wrote; *searchers*. Searching for immortality is an age-old impulse; that doesn't mean it can be found. Still, an impossible quest is a good quest. "Sometimes it doesn't help to know what it is you are really hunting," explains the shaman Martín Prechtel, "because the beauty that the hunter becomes and creates through his willingness to fail in pursuit of what he deeply longs for and doesn't yet understand can cause the incomprehensible thing to show its divine face."

The cuneiform tablets of Nineveh, among the earliest written documents, tell of King Gilgamesh, whose best friend dies. He is stricken with grief. But he is the omnipotent king of Uruk, the one who has gazed into the depths—the one who slayed the Bull of Heaven!—surely he's almighty enough to bring dead loved ones back to life. Mute with sorrow and pride, he buries his friend beneath a river and sets out to find eternal life. The scorpion people, whose knowledge is fathomless and whose glance is death, warn him about dangers ahead. A lady of the vines tries to console him, telling him that love is the closest mortals can come to immortality. Crossing the Waters of Death, he discovers a marvelous underwater plant that contains the secret of perpetual youth—the watercress of immortality, as the clay etchings call it, or the "never-grow-old"—but, alas, a serpent promptly steals it away. History's prototypical protagonist fails, yet his story ends the only way it can: with acceptance of reality. Of mortality.

Any story about immortality is really a story about death, the greatest mystery, the stumping question, the undiscover'd country from whose bourn no traveler returns. Because we can't comprehend it, we have a need for stories about continuing—or to see death as a complete stop. Physical immortalists live a mythology in which dying is avoidable; their story isn't far off from the story of floating up to heaven. The atheist notion that death simply means the end of the "neurons that store the informational patterns of our bodies, our memories, and our personalities" is also unverifiable. But myths don't need to be real to be true. Stories provide something stronger than facts. They satisfy our craving for cohesion, for a thread in the maze of experience, for a beginning, middle, and end. (And possibly a sequel.)

The deaths of others bring with them a faint hope that the void will explain itself, that we'll grasp mortality's incomprehensibility, but surviving grief cannot reveal what, if anything, lies beyond. That will be disclosed only when we pass into something else or nothingness. The language of the dead is off-limits to the living. The only way to become fluent is to die.

Until then: life, the sea. As the waves sweep by, part of the journey from innocence to maturity, from magic to reality, entails coming to a deeper understanding of human nature, of our vulnerabilities and limitations. "Charity," as Saint Paul put it, "believeth all." Being charitable means being aware of others' beliefs. It means accepting that others are like us (even if their beliefs differ). This requires seeing ourselves as one of them. Charity also involves recognizing the importance of belief in creating that most precious substance: meaning. To be human is to hunt for meaning. Death makes no sense. But imagining that we'll live forever—whether physically or spiritually—is an elemental solace. Immortality renders death meaningful.

"The question of immortality is so urgent, so immediate, and also so ineradicable that we must make an effort to form some sort of view about it," wrote Jung. "But how? My hypothesis is that we can do so with the aid of hints sent to us from the unconscious—in dreams, for example." But our dreams can lead us astray. The Persian king Khosrow Anushirvan reigned from 531 to 579 CE, at which point he vanished into a mysterious fountain that he'd been told, in a dream, would unite him with the Creator. According to one account, Khosrow plunged into the water and was never seen again: "And not a trace was left behind, not a dimple on the wave." Before his death, Khosrow had sent physicians to India in search of the secret of eternal life. All they found were stories.

Our beliefs have magical powers. They can prevent us from seeing reality; they can also allow us to accept reality, a double yellow line we're forced to cross whenever we're confronted with death. For the fortunate few, decades can go by without a funeral. But sooner or later, what Heidegger called a "sudden inflashing"—the realization that one day we're going to die—forces us to contemplate our own encroaching mortality. A gaping gate opens. Losing someone is a reminder, the most potent memento mori. And the bereavement it brings has a way of reacquainting us with ourselves, with the need to believe, of making us hold on to something, anything, as we fall.

Part 1

Belief

There is only one supreme idea on earth—the idea of the immortality of the human soul.

—Dostoyevsky, *Diary of a Writer*

1

We Bereave, We Believe

Can I learn to suffer
Without saying something ironic or funny
On suffering? I never suspected the way of truth
Was a way of silence

—W. H. Auden, *The Sea and the Mirror*

We have to die, we have to leave life presently. Injustice and greed would be the real thing if we lived for ever. As it is, we must hold to other things, because Death is coming. I love death—not morbidly, but because He explains. . . . Behind the coffins and the skeletons that stay the vulgar mind lies something so immense that all that is great in us responds to it.

—E. M. Forster, *Howards End*

EVERYONE IN the family called her Auntie Tiny. She'd always been minuscule—"of stature elegantly wee," as our Hungarian relatives put it. She spoke in a helium-pitched, young-girl voice, even when discussing serious matters. We kids considered her one of us. She was an emissary from the grown-up realms, a benevolent Old World pixie who kept shrinking into her dotage. We worried she'd get smaller and smaller, so teensy she'd eventually vanish.

I first heard Auntie Tiny's real name when the funeral notice appeared:

With deep sorrow, but in acquiescence to Divine will, we inform those who knew and loved Ilona Köver Göllner that, in the 96th year of her life, she returned home to her Lord and Saviour.

Childless and widowed, our childlike Auntie Tiny had been closer to a grandmother than a great-aunt. My father's mother died when I was very young. And the last time I tried to visit my maternal grandmother, a British antiquarian with hoarding tendencies, she asked me not to come. "I'm in an absolute muddle," she sighed, engulfed by four stories of belongings.

Auntie Tiny loved visitors. Her one-room apartment was on the third floor of a sooty, bullet-ridden building in Budapest. She'd creak open the door, greeting my brothers and me with that witchy falsetto. Kissing her powdered cheeks felt like kissing marshmallows. *"Entrez, entrez,"* she'd trill. The noble bearing wasn't affected: when she was born, the dual monarchy still reigned and her family belonged to Austria-Hungary's landowning gentry. Following communism, all that remained was this somber tenement overlooking the Danube.

Inside it smelled of dust and paprika. Her décor was strictly Habsburg Empire: embroidered lace, illuminated vellum, ikons, dried flowers. We'd sit at a table next to the doll-size bed and eat bowls of chilled sour-cherry soup. She baked little cylindrical biscuits called *pogácsa,* telling us how one should never set off on a journey without a knapsack full of them.

She herself embarked on a mini-pilgrimage every morning. After lifting her hair into an onionlike beehive and lacing up her booties, she headed out to various crosses around town, beaming thoughts of eternity. Then, without fail, she attended church service.

The last time we saw each other, I'd become an adult. Her milky eyes had clouded into near blindness. A vein cluster near her right iris was tangled in the distinct shape of a heart. She could no longer read her beloved books, but she seemed capable of seeing into the next world. Despite suffering more than I could imagine, she said it was all a blessing. She loved life, but was not afraid of death.

What about those who don't know what they believe in? I wondered. As though sensing my thoughts, she gave me a small cross and said she prayed for me.

Her funeral invitation included a biblical passage: "Whoever believes in me will live, even though they die; whoever lives and believes in me shall never die. Do you believe this?"

Not able to answer, but unwilling to lose her forever, I located the quote in the Gospel of John, where God is defined as love and faith as living water: "Whoever drinks of the water that I shall give him will never thirst; the water that I shall give him will become in him a spring of water welling up to eternal life."

We imagine ourselves impervious, but at some point in every life, grief overpowers us. When it descends, we feel physically ill. We drift asleep and fall awake with it. We catch ourselves asking unanswerables. We call that person's name. We repeatedly anticipate being reunited and then have to re-remember reality, painfully. The depth of our reaction can come as a shock, but the slow grind of mourning work is how we accept an absence where a presence had been. Bereavement brings with it a sense of having embarked upon a lengthy search for something that can't be found. Gnawing reminders. Wringing rearrangements. Memory folders on our mental desktop get filed away in forlorn subcompartments. Eventually, what's gone becomes a part of us.

The modern attitude toward death has been described as "a hedonistic avoidance of the issue." In the past, we dealt with it socially. Wakes, cremation ceremonies, and other burial rites entailed community participation. Today, being bereaved is often seen as interfering with productivity, a sign of weakness, an unpleasantry best done in private. Should we tweet about it? Do we press *like* when someone else does? Doctors dole out dolorifuging pills. Even in death we pump ourselves full of chemicals to fit in, with morticians embalming bodies to give them a waxy semblance of life. But no matter how much a cadaver appears like a napping person, the irrevocable can't be airbrushed away. Everyone ends up sleeping the same sleep.

My friend Elena was in the hospital with her comatose mother. In the middle of the night, her mother started shaking violently. She appeared to be trying to pull her clothes off. Elena, realizing she couldn't calm her unconscious mother, decided to help her undress. Once naked, she quieted down. The following day, she passed on in peaceful repose. Elena felt sure her mother wanted to exit life the way she had entered it. "There was something greater than us in the hospital room that night," she concluded.

When someone close to us dies, we escape into beliefs. It's not unusual to find religion in loss. At the age of twenty-five, another friend of mine lost her best friend and her cousin within a few weeks of each other. The grief was so all-consuming, she told me, that "the only way I could stay alive was by starting to believe in the afterlife."

Our mind ceaselessly churns out plausible interpretations of unexplainables in attempting to reconcile itself to death's implications. The idea of "person permanence"—that our dead relatives, or parts of them, such as a soul, are still floating around somewhere—is a venerable way of negotiating the question. The recovering brain prefers to imagine them continuing to exist in some undetermined afterlife. Social psychologists involved in a discipline called terror-management theory explain that envisaging others' postmortem continuance has an added benefit: if their spirits are kicking around the starry skies, it follows that we'll join them out there, too, when our time comes. In this ostensibly mythless age, person permanence remains the mind's preferred means of handling the destabilizing possibility of its own demise. Such illusions can be survival mechanisms. At least until we die.

If dealing with the deaths of others is so hard, then mustn't actually dying be brutal agony? Not necessarily, it turns out. Elisabeth Kübler-Ross, who famously developed the theory that grief passes through five distinct stages, from denial to acceptance, found that those on the brink of dying often experience feelings of peaceful contentment. Peter Pan was right: "To die will be an awfully big adventure." It'll also be different from what anyone supposes—and luckier—wrote Walt Whitman. As survivors of life-threatening injuries attest, the anticipatory worry is far worse than the actuality. Gallup polls of patients who've experienced clinical death and were then revived show that "confronting and undergoing death frequently seems more pleasant than life itself."

Our neuroprocessors may prevent consciousness from actually experiencing its own annihilation. On death's doorstep, the mind produces narcotic tranquilizers to protect itself. In his 1892 study of mountain-climbing accidents, *Remarks on Fatal Falls,* Albert von St. Gallen Heim interviewed people who'd fallen from Alpine heights (and lived to tell the tale). Many reported feeling a calm lucidity as they drifted through the sky. They plummeted not in stabbing terror, but rather like lemon pips sinking into tea.

Of course, death can also be painful or violent, but for the most part, we're pretty out of the loop as far as the whole "dying experience" is concerned. Because the brain naturally suppresses thoughts of its eventual extinction, much of our thinking about death takes place on an inaccessible, subconscious level. Modern neuroscience has demonstrated that the mind consists of multiple layers of cerebral functioning. Within the vast interlocking choreography of transmitters, axons, peptides, circuits, and spanned gaps, many processes that resemble conscious activity are actually performed entirely without our rational mind realizing it. We can be totally unaware of our feelings about death—until we lose someone, triggering a realization of our own mortality.

A part of us accepts that we will eventually disintegrate; other neural subdivisions harbor furtive aspirations. Rationally, we know the end will come. Irrationally, we hope to get around it somehow. As we face finality, complicated feelings arise, as do hopes of evading the unavoidable. Inexorable though the situation may appear, we are infinitely creative when it comes to concocting alternative scenarios. We convince ourselves that if we search long enough, we just might stumble upon a loophole, a VIP pass, a get-out-of-jail-free card. Imagination is an essential existential consolation.

Ernest Becker's 1973 book, *The Denial of Death,* argued that undertaking heroic acts is a way of challenging the loamy unknowability of death. We use our imaginative powers to concoct "immortality projects" that will allow our name to outlive our mortal transience. Many creative types are fueled by this instinct, but so are people in every other walk of life. Donors to public institutions receive commemorative plaques in their honor. "Achieve Immortality! (We're not kidding)" is the tagline for advertisements by the New York Community Trust that encourage benefactors "to leave a charitable legacy that will make gifts in your name forever." Scientists hope that their life's work, while it may not ever explain the mystery of mysteries, will grant them a kind of immortality after their passing. Einstein, after all, is remembered as much as van Gogh. Our pursuit of impressive acts or deeds is known as "achievement immortality."

During an 1841 breakdown, Abraham Lincoln confided to a friend that he had "done nothing to make any human being remember that he had lived." He found the idea of dying intolerable, yet started seriously considering killing himself. His only solace, he explained to his friend, was the idea of surviving in others' memories. What kept him alive was the ambition that, by accomplishing deeds that would link his name to monumental exploits, he would attain immortality. And, in a way, he did.

The phenomenon of occupying memories after we die is called social or vicarious immortality. Even though we may lose people, they exist in our minds when we imagine them. A person's sense of identity depends on the knowledge that he or she is in others' thoughts. This is a normal strangeness that can become warped under duress. Suicide cases may start to reason (unconsciously) that by becoming dead, they, too, will posthumously inhabit others' minds, therefore becoming immortal. And those who've been kidnapped, or imprisoned in solitary confinement, or who've faced an extended period of almost-certain death, can find themselves starting to write down the names of every person who loved them, everyone who will remember them. It's a curious thing to do, and it's a form of consolatory immortality. Nobody wants to be forgotten.

After we die, our corporeal remains can be cremated, carried off by carrion eaters, or chewed upon by worms and microorganisms. My friend Melanie, vice president of horticulture at the Brooklyn Botanic Garden, doesn't believe in linear religions. Based on her work with plants, she prefers the idea of the nutrient cycle. "When I'm buried, my corpse will gradually be dismantled and embodied by millions of bacteria, roots, bugs, amoebae, fungi. There's a massive congregation of lifeforms in the soil. Whenever something dies, countless lives are enriched. Why would it be any different for us? If you're buried properly—not in some impenetrable coffin that prevents nature from its due—you'll get to escape your body, to go out there into all those new organisms." As right as she is, Mel is also talking about continuity, about escaping the body, about parts of us becoming something else. We all have semantic approaches to death's incommensurability. Mel's version fits into a category called cosmic immortality, the nontheistic notion of a person coming from the elements and returning back to them.

Intimations of immortality surround us. Molecules in the dead animal we eat become part of our cellular makeup. When a pollinated flower wilts, it becomes a fruit that dies into ripeness, containing within its spent flesh a seed that becomes a tree. Quantum Immortality (or QI, for those who frequent speculative-physics chat rooms) holds that there are many universes in which each of us lives parallel lives, and even if we die in this world, we'll survive in faraway galaxies forever. This is about as verifiable as the precise geographical coordinates of nirvana, but it's an interesting, if excessively technical, example of scientific religiosity.

Heredity, that stream of acquired traits, is a more generally accepted form of immortality. Linking eons, DNA is a means of encoding and preserving information that is transferred from generation to generation.

Posterity immortality is the phenomenon of genes living on through one's children. "What is mortal tries, to the best of its ability, to be everlasting and immortal," wrote Plato. The most obvious way of doing this, he added, is by making babies—by leaving behind replicas. The notion of trumping death through having progeny also got the biblical stamp of approval. The Old Testament isn't concerned with the afterlife as much as it is obsessed with generations of descendants. Having kids who then have their own children is the path. Parents may die, but parts of them persist. The selfish-gene theory suggests that we value our offspring because we're *in* them. They are made from us, just as we consist of those who came earlier. We all wear the faces of our dead ancestors. The desire to bring life into the world is an attempt to leave something behind that will outlast us, that will overcome our demise. As the Hindu Dharmasutra of Āpastamba tells us, "You beget children, and that's your immortality, O mortal."

Some scientists consider DNA "immortal" in the sense that it flows on even though we die. But if the entire species gets wiped out, then the DNA also dies out. And if the planet explodes—all its gene pools will disappear along with it. DNA, alas, is not really immortal. But neither is anything else, unless—as Auntie Tiny's funeral notice stated—we believe.

My parents weren't religious, and neither were my two brothers and I. We were baptized but never went to church. My paternal ancestors were Hungarian Protestants, but my father—nonpracticing yet still a spiritual man—raised us without imposing any organized religion on us.

My mother's family had an even more tenuous relationship to faith. Her father had been an Irish Communist. He later renounced his membership, becoming an executive for a multinational pharmaceutical company and an atheist. He died when I was eight years old. My mother and I flew to London for the funeral. While there, trying to make sense of my relatives' mourning, I started writing notes about "piles of people in their naked emotions." Two decades later, in the same city, I dreamed of the fountain.

As kids, the closest my brothers and I came to some higher level of reality was when our parents opened junk mail on Sunday mornings and pretended to read messages between the lines. These transmissions were written in invisible ink, they said, and had been sent from outer space.

The arrival of bills fluttering through our suburban mail slot caused giddy delight.

Throughout childhood, most of my life revolved around one block. I trained each morning at dawn in hopes of becoming an Olympic swimmer. School was across the street from the pool, which was connected by a walkway to the library I studied at when classes ended. The first stoplight around the corner overlooked a flat cemetery. Twice a day, the bus paused there, and gazing out onto the tombstones, I'd imagine all the rib cages, tibiae, and vertebrae prostrated in the frosty soil below. The vision was both unsettling and comforting; whatever life holds in store, there's one certainty awaiting us. The light changed, we rolled forward, crystals glittered in their sunlit blanket of snow.

Water, books, and buried bones were the holy trinity of my formative years. After high school, I'd spend days off in graveyards, reading or taking photos of the headstones. My favorite one simply said HURRY on it. Friends occasionally joined me; some were more uptight than others. Fear of mortality is forever, one of them told me. She put it on her grad school application under "motivations." To her, death was a perpetually present shadow, an inchoate horror residing in the mind. Rather than scared, I felt curious. Perhaps the anxiety was buried so deeply in my subconscious I wasn't even aware of it.

Despite being raised irreligious, I spent much of my twenties studying both mythology *and* the sciences. I learned to view all theologies with equal fascination and detachment. The particularities of doctrinaire activity interested me less than the overarching fact that our minds utilize belief. Every religious text, every idea relating to immortality, every fountain-of-youth saga, I came to believe, is interesting and valid—whether as a window into the inexplicable or a way of understanding ourselves.

Those ideas came flooding back to me as I read Auntie Tiny's funeral notice. The image of a spring welling up to eternal life recalled that dream I'd had in London. A fountain bursting from the sea of infinity. What did it mean? Why do religions link immortality to water? I needed to speak with someone who knew.

2

Journey into Remoteness

With mountain winds, and babbling springs,
And moonlight seas, that are the voice
Of these inexplicable things,
Thou didst hold commune, and rejoice
When they did answer thee

—Percy Bysshe Shelley, "To Coleridge"

There is another world, but it is in this one.

—Paul Éluard (attributed)

THE FIRST priest I tried was Father Emmett Johns. Known locally as Pops, he'd founded a charitable organization that dispenses food, shelter, and aid to homeless kids. I'd interviewed him before, around the time a newspaper voted him "Montrealer closest to sainthood." We hadn't probed any metaphysical topics during our first conversation, but he had a straight-ahead, no-nonsense demeanor, and I knew he would tell me what he could. I left a message for him at the Chez Pops day center and headed out for a coffee.

Among the newspapers at Café Souvenir, I came across a brochure for a conference examining the human psyche's posthumous survival. According to the organizers, we are all immortal sparks who, after dying, end up in "the invisible portion of the cosmic sea of the great Uncreated Light." Many of the talks focused on communicating with spirits floating out there in the invisible sea.

One of the workshops featured Raymond Moody, the American scholar who coined the term *near-death experiences,* or NDEs. He'd spent decades documenting and investigating case studies of people pronounced dead and subsequently resuscitated. His 1975 bestseller, *Life After Life,* explored the uncanny commonalities in NDEs, such as a sensation of floating above one's own body, traveling down a tunnel toward a brilliant light, or being reunited with long-deceased relatives. Skeptics considered his research unscientific, while others interpreted the testimonials collected by Moody as confirming the existence of life after death. He endorsed neither side. Even if his findings suggested something more to death than impenetrable darkness, he didn't consider his data capable of *proving* anything. Moreover, he gravitated toward the consolatory nature of his findings. He'd gone on to become a grief counselor and had written extensively about mourning and how the wish to reconnect with those we've lost is among the most universal and deep-seated of human desires.

As I looked over the program guide, I received a call from an administrative manager at Pops's charity. Unfortunately Father John wouldn't be available for an interview. "He just doesn't have the energy to do it anymore," she apologized. "The last time I brought someone to him was six months ago. He can't do publicity anymore. He received a major award last week and decided not to attend the ceremony. He normally loves an audience, and the fact that he didn't come says a lot about where he's at these days. He's had several bypasses and needs a lot of rest. I'm sorry."

Minutes after thanking her for the call, I'd already signed up for the all-day session with Raymond Moody, "expert on the unknown."

Around fifty people sat rapt in a ballroom beneath the Delta Hotel in Old Montreal as a short, elderly woman with dyed-red hair and wraparound sunglasses stood at the podium introducing the event. I recognized her as Marilyn Rossner, a local medium who'd been a television personality during my childhood. She often appeared on a CBC program called *Beyond Reason,* a kind of psychic game show with a panel of intuitives sitting in soundproof parallelograms trying to guess the identity of a mystery guest.

She had a power far greater than her four feet nine inches would suggest. In her usual thrift-store ball gown, pink knee-high socks, and sixties bangs, Rossner looked like an ancient child. But unlike Auntie Tiny with her unwavering kindliness, Rossner seemed equally in touch with the light and dark. She belonged to another time. Her actual age didn't matter. If someone inquired, she'd say she is "as young as a blade of grass and as old as eternity."

At her international appearances, she's billed as one of the world's most gifted sensitives (*una psíquica canadiense considerada como la mejor medium del mundo*). Here at home, she runs the Spiritual Science Fellowship, a center for learning about channeling. Onstage, she told the audience that our speaker today, Raymond Moody, would be covering a variety of apparitional experiences that prove dying isn't the end. She firmly believed in spiritual immortality. Her conference had one supreme aim: to spread the truth of spiritual life after death.

Moody took to the stage. Bald, clean-shaven, and bright-eyed, he had the genial disposition of a doctor in his seventies. His natural charisma put the crowd at ease immediately. Not only did he come across as compassionate, learned, and generous, he also had that rare ability to simplify complicated matters. He clearly knew a great deal about his subject, and many others as well, but he kept reminding the audience of how little we can really know. "I am a great fan of science," he said, "so I am in complete agreement with skeptics—these experiences don't prove life after death. Reason as it is can't tell us anything certain about the afterlife." But the audience members, many of whom could be described as New Agers, wanted proof about immortality. What Moody had learned researching NDEs, however, is that nothing is straightforward when it comes to death.

Born in rural Georgia and based in Choccolocco, Alabama, he spoke with a singsong, Southern twang, articulating every syllable, emphasizing key words such as *su-per re-ali-teee,* and ending them on a mild vibrato. He started discussing the impossibility of accurately conveying what transpires in NDEs. *"In-de-scri-ba-bili-teee,"* he pronounced, with that tremulous voice. "It's something near-death experiences have in common the world over. An *in-eff-a-bili-teee*. The experiences are clear; but there are no words to explain what happened. It just can't be properly described."

What also cannot be explained is the meaning of NDEs. Are they simply hallucinations? Or do they provide some insights into what occurs at death? Can they tell us anything about human prospects for immortality? Whatever the interpretation, one thing is certain: NDEs are a real phenomenon. According to the International Association for Near-Death Studies, between 4 percent and 15 percent of humans have had one. Many recount having a life review, seeing their whole life flash before their eyes. It isn't uncommon to have out-of-body experiences: patients rise up and look down upon their own bodies in a dramatic transformation of perspective. In their detached aloofness, they may even hear the doctor pronouncing them dead. Conversely, they can be overwhelmed by white noise, a loud ringing or buzzing sound. Some feel that they have actually died; others feel fearlessness or unreality. It can be a peaceful, painless sensation—or a harrowing overdose of despair and guilt. Many come back unafraid of death.

A travel narrative is frequently invoked. A barrier or threshold between our world and the beyond is crossed. We fly through the ceiling or into walls. We traverse a tunnel or narrow passageway. At the end is a light. Thought accelerates, space wrinkles, and we enter a see-through energy field. "At death a sort of portal opens up, a pathway," Moody explained. "To what or where we don't know."

Numerous accounts describe something scholars call the vestibule effect. It's as though the front door to a mansion opens, and we find ourselves in the foyer of another dimension not subject to everyday laws. There's a conjunction of opposites. Some compare it to being in a little room that's big. The furniture comes toward us while moving away at the same time. We stand in place and simultaneously go forward. Our thoughts turn into people.

Moody spoke of how the light may brighten in a room when someone dies. "I have seen people's faces light up and transform as they pass on," he said. "You might think this is crazy, but when my own mother died, I remember the shape of the room changed. There was a narrowing in the middle, a broadening at the top, as though in an hourglass. I saw a funnel above us."

He asked how many people in the audience had undergone an NDE. A smattering of hands went up. One woman near me described going through a tunnel as a child. Beings tried to pull her into the darkness. Right before crossing over into the light, something stopped her and explained it wasn't her time yet. "How would you interpret that?" she asked him.

"I don't know." Moody shook his head sincerely.

A skeptical lady chimed in, "It was a bad dream, that's all."

Seated in the back row of the room, I found myself lifting my hand with the other NDEers.

It happened to me as a teenager, in Hungary. I'd been playing soccer in a schoolyard next to home. A neighborhood kid brought a BB gun to the field. We all took turns firing it—unloaded, of course. Then we went to play a game. My team lost. Returning to the bench, I picked up the gun lying there. In mock-despondency at our defeat, I placed it in my mouth and pulled the trigger. Unfortunately, someone had loaded the chamber during the match and left it armed and unattended.

The bullet tunneled through the length of my tongue, a fortuity that slowed it sufficiently to not actually perforate my spinal cord. The round steel pellet, flattened into a D shape by the resistance of its fleshly trajectory, came to rest inside the tissue around the cervical vertebrae behind my throat, mere millimeters from terminal damage.

The moment the shot rang out, I hovered out into midair, two stories up, as though on wings. I looked down benevolently as a terrific volume of blood poured from my mouth. It felt like watching a dramatic reenactment, peering in on someone else's traumatic accident. Decentered, disassociated, a part of me floated there, a kite in the breeze. Time stopped, or bent itself. For a spell, I dangled there languidly, taking everything in from an Archimedean vantage point. I could see the white, wooden goalposts, my sweater draped over the fence, an apricot tree, the hill sloping down to the shrubs that bordered our home. Minutes seemed to fly by—but the whole thing couldn't have lasted more than a few seconds. In that warped temporality, I watched it all slowly for a fast instant, then leapt back into my usual point of view to see the crimson torrent gushing forth onto the field.

It never hurt. The wound's intensity must have overloaded my neuronal signals, short-circuiting any sensation of pain. Two friends took me to the hospital in a taxi. We weren't sure where the bullet had ended up. The entry hole suggested it might be inside my tongue. X-rays revealed a white dot lodged in my esophagus. The doctor's hands trembled as she picked up the forceps and attempted to disinter the pellet without pushing it deeper. She succeeded. After several days of observation, I went home. It never hurt. But it also didn't convince me that death is—or is not—the end.

"Nobody can contradict you if you've had a near-death experience," Moody explained to the crowd. "It cannot be challenged." The experience is so powerful it can convince us that there is life after death, he continued, the same way some people find certainty through faith or religion. "But is there a rational way of answering the question of whether life continues after death? No, there isn't."

"But in your work, you use the scientific method," interjected a man in the crowd who seemed sure that NDEs offer proof of the afterlife.

"No, I don't," Moody answered. "There's no control group. Even in systematic parapsychological studies of NDEs, there's no way of establishing any proof. But reason and conceptual analysis tell us there are other avenues for rational inquiry. Consider this premise: the only method we have of finding truth and establishing knowledge is science. It's the premise of scientism. But how can we demonstrate whether that premise is true? Not through the scientific method. So reason tells us that we have to acknowledge that there are other ways of finding truth and establishing knowledge."

Moody spoke eloquently about how all the wonderful accomplishments of science make it so easy to slip into the belief system of scientism. "Scientism is a doctrine subscribed to by those who don't know science. Really good scientists don't *worship* science. They aren't *deh-vo-teees.* They're just as baffled by the world as the rest of us."

During the break, I walked over to a table covered in books, CDs, and pamphlets on the far side of the banquet room. Photographs showed Marilyn Rossner: holding hands with Pope John Paul II, communing with the Dalai Lama, playing with children at Mother Teresa's convent in Calcutta. I picked up a memoir written by her husband, John Rossner, an Anglican priest.* *The Priest and the Medium: Heaven Is Closer Than You Think!* told the story of their life and love. It described Marilyn as a diminutive, colorful whirlwind, a grown-up flower child in sundresses and pigtails. When they first met, in 1972, he actually thought she was a child, "a little doll-like character who just stepped out of a children's illustrated story book."

The book chronicled the strange experiences they'd had together. They fraternized with Tibetan lamas, Jain acharyas, and Hindu mahasiddhas. They strolled over burning coals with swamis. They cured their cocker spaniel, Blackie, diagnosed with terminal leukemia, by praying

---

* In a past life, he also served as a priest in ancient Egypt. In this life, he died doing yoga on August 27, 2012.

over a fragment of fabric snipped from the stole of Saint Seraphim of Sarov. John wrote lovingly of the way Marilyn once restored twenty-twenty vision in a blind Native American girl by a laying on of the hands while channeling the Himalayan patron saint of the blind.

Marilyn's telepathic powers emerged in childhood. Teachers were aghast when she predicted a classmate's death to the day. Asked how she'd known it would happen, she explained that she'd heard the date in her mind while noticing the swirling gray color surrounding the girl. Marilyn was surprised to learn that not everybody could see auras.

She started wearing sunglasses to protect her eyes from the clouds of blue, pink, and green billowing out of people. Perhaps her strongest (or strangest) ability is seeing dead people in the moments just after they die, whether it be three children murdered in an icy snowdrift or her own grandmother. On the night her mother died, she sent John a fax from Spain: "Dear Dad,* Please tell me what is going on there! I have just had a vision last night, of my mother. My mother said to me, 'Marilyn, I have just died, but I am all right!'"

According to one of the pamphlets on the table, Marilyn and her circle offered free supernormal-percipience services on Sunday evenings in the basement of the Days Inn in downtown Montreal. Each week, they gathered to channel the spiritual world and relay extraphysical messages. I folded the pamphlet into the copy of the book I purchased.

In Pew Trust surveys, 29 percent of Americans believe, based on personal experience, that it's possible to contact the dead. Visionaries of all nations and all times have evoked immaterial beings, but the nineteenth century was the heyday for mediumship in the Western world. Psychics performed to packed houses. Séances were conducted in the White House. Spiritualism attracted millions of followers.

The sheer scale of paranormal claims led to the formation of Britain's Society for Psychical Research, established in 1882 to investigate the paranormal "in the same spirit of exact and unimpassioned enquiry which has enabled Science to solve so many problems." The society's two founders, Henry Sidgwick and Frederic Myers, attempted to devise a means of demonstrating whether the soul can survive death. Alas, as

---

* Marilyn called her husband Dad.

Moody pointed out, studies concerning the hereafter remain outside the scientific method's remit.

Still, as a psychologist, Frederic Myers played a foundational role in exploring our understanding of the unconscious mind. He forwarded the by-now accepted idea of subliminal consciousness—which illuminated everything from hypnotic suggestion to dream imagery. But he also believed our subliminal self could contact beings that existed in a parallel realm he called the "metetherial world." He coined the term *telepathy* as a way of referring to "the communication of impressions of any kind from one mind to another independently of the recognized channels of sense." Myers's final work, *Human Personality and Its Survival of Bodily Death,* shows that he believed to the very end in the possibility of posthumous immortality. He exerted a profound influence on William James, whose famous 1898 lecture on human immortality argues that "we need only suppose the continuity of our consciousness with a mother sea, to allow for exceptional waves occasionally pouring over the dam."

James and Myers made a pact that whoever died first would communicate a sign—if at all possible—from beyond the grave. As Myers passed on in Rome, James sat outside Myers's room with "his note-book on his knees, pen in hand, ready to take down the message with his usual methodical exactitude." None came.

When James's turn came, he wasn't as uncooperative—at least in John Rossner's account. At one public séance Rossner attended, in which the host medium went into a trance and passed out, a semitransparent entity walked onto the stage and identified itself as the materialized spirit of William James.

James's ghost walked around the room and then plunged his hand into the solar plexus of a young man seated beside John Rossner, who looked on in amazement. "Professor James's spirit hand progressively disappeared up past the wrist as he pushed it into my friend's midriff until it was gone from sight. In a few seconds he began to draw it out again and in it was a white, cloudy, luminous substance which came at first in a lump, but still attached to my friend somehow." Because of the ectoplasm's volatility, James kept making a patty-cake motion with his hands to help the energy stay in place, as though using a spoon to prevent a poaching egg from disintegrating into wisps.

Then something went wrong. James started quivering. He asked for the lights to be dimmed. Being in this dimension for so long was weakening him, he said, and the vibrations from one particular red light were messing with his ability to stay materialized. The wattage was also eating away at the eight feet of ectoplasmic goo seeping from Rossner's friend's aural field all over the floor.

A momentary panic struck when it became apparent that nobody knew where the switch was—except the medium, who, as Professor James's conduit, lay comatose. Nobody wanted to flip the wrong switch and turn the fluorescent lights on, pulverizing all the fragile ectoplasm everywhere. But then John realized he could cover the bulb with his cupped hands if he stood upon a chair. Crisis averted, the spectral professor crammed the misty pillar of etheric energy back into the volunteer and then took a bow. "Finally he began to disappear in full view of all," explained Rossner's book. "First his feet and legs, then with the usual characteristic *poof* of the dematerialization process the rest of him was gone."

Part of the beauty of this subject, I was starting to realize, is the way our desire for immortality manifests itself in the most exceptional stories.

The remainder of Raymond Moody's presentation concerned his rediscovery of a twenty-five-hundred-year-old technology that he claimed allows people to see and speak with departed loved ones. He himself had seen it work on numerous occasions. All one needs to do, he said, is construct a simple apparatus using materials already available in most homes.

"You can set up the apparition chamber in a walk-in closet," he explained. Any small, windowless room is fine. Fabric needs to be placed over the entryway so no cracks of light bleed through. The next step is finding a mirror. "The dimensions are irrelevant," Moody said. "Four feet by four feet is what I use. Position it just above head level so you don't see yourself, or anything else in the reflection. Once the room is in complete darkness, fetch a candle or a very dim light. Enough to make out the mirror a little bit. Because nothing is reflected in the mirror's surface, the effect is akin to gazing into infinity. Just look into the space with no agenda. It's like you are in outer space, seeing the depth of the infinite."

He'd been inspired to explore the lost art of crystallomancy, a form of scrying also known as mirror gazing, based on his research into the Pawnee Indian tradition of sacrificing badgers. After draining the badger's blood into a bowl, they'd see mirror visions in the blood's reflective surface. This practice synced up with discoveries made at archaeological digs of ancient Greek psychomanteums. Some scholars feel that these sanctuaries, also known as "oracles of the dead," were places where the spirits of the dead could be called upon. Based on his own exegeses of papyrus scrolls, Moody decided that the bronze cauldrons found in the torch-lit Greek apparition hallways used to be filled with olive oil, creating a mirrorlike screen upon which to commune with shades.

In 1990, Moody opened a psychomanteum research laboratory in a rural Alabama gristmill. He named it the John Dee Memorial Theater of the Mind, in honor of Queen Elizabeth I's official scryer. The chamber itself is covered in floor-to-ceiling curtains of black velvet suspended by a framework of plastic tubing painted black. Fifty percent of those who've mirror-gazed at his facilities, he said, get a visitation from a dead relative. "You seem to see loved ones there in the mirror," he said. "You see full-color images as though in a screen. They may stay in the Middle Realm, that in-between space. Or they sometimes come out of the mirror and stand in front of you. Or you yourself go through the mirror and meet your relatives on the other side."

The notion of darkened mirrors as portals to the otherworld had such widespread appeal in classical times that Pythagoras cautioned against looking in mirrors next to lamps. Leaving Moody's lecture, I decided to heed Pythagoras's warning. But I also resolved to check out one of Marilyn Rossner's message circles. Perhaps a spirit—or at the very least Rossner's belief in them—would shed more light on the meaning of spiritual immortality. Maybe I'd even hear from Auntie Tiny.

Montreal boasts many places of worship in unlikely settings: charismatic churches in movie theaters, Orthodox synagogues in former breakfast restaurants, Hindu temples in repurposed auto body shops. None are quite as site-nonspecific as the midsize conference hall in the basement of the downtown Days Inn,* Sunday-evening gathering place for the Spiritual Science Fellowship.

When I arrived, just in time for the 7:30 p.m. liturgy, a forty-person-

* Subsequently rebranded as Hotel Espresso.

strong congregation had already assembled. For some reason, I'd expected a celebratory mood, but most attendees sat solemnly. The five clairvoyants onstage guided us through a meditation. We were asked to clap arrhythmically, possibly to scare maleficent entities away, or perhaps to heighten the shamanistic vibe. The mediums placed themselves in an unfocused state of mind, the better to let the spirit world in.

The dispatches they passed on were, for the most part, nurturing messages of reassurance, along the lines of "your deceased sibling/spouse/parent/child/relative wants you to know they are doing just fine in the afterlife." At other times, the speakers appeared to simply be tapping into random transmissions from their subconscious minds. "Ballerina shoes!" exclaimed one of them, pointing at a dimpled, middle-aged woman in a stiff dress with a high, black-and-gold collar. "You are going to be walking on tiptoes—remember: ballerina shoes."

While I didn't believe that actual dead spirits were communicating through them, at no point did the actions of the psychics come across as staged or contrived. The tidings they brought were all too peculiar, too personal and exotic, like time-lapsed flowers blooming from their intuitive selves. The messages emanated with enough velocity and uniqueness to preclude memorization or conscious thought. When receiving, the diviners also spoke differently, their antennae tuned in to some murmuring frequency within—or perhaps without?

"I see your face off your head," exclaimed a female medium, addressing a bearded man in the first row. "Work on that."

"You won't believe this, but I see a pink, fluffy slip in a hotel bed," said the psychic next to her, shrugging her shoulders. "I see a roll of toilet paper with lips on it, unfurling into the distance. Just rolling away."

On it went, a ticker tape of *pensées automatiques* unraveling through the subterranean conference room. It called to mind André Breton's *Surrealist Manifesto,** with its emphasis on psychic automatism, on thoughts unfettered by the constraints of reason or rationality. In the same way that Surrealist writers transcribed nonsensical, dreamlike phrases that bubbled up from the depths, these clairvoyants sought to articulate them live. Whereas Breton advocated automatic writing, they were automatic speakers.

While I didn't perceive anything indicative of occult powers at work

---

*Breton acknowledged the movement's indebtedness to Frederic Myers, of Britain's Society for Psychical Research.

in the room, I did find the experience entertaining. At one point, fading the psychobabble out, I peered into the rug's foliage design, trying to convince myself I could make out the image of a cyclops, a baby with an old face, and a bodiless cherub with wings. Perhaps our imaginations are made to run away with us. Seeking patterns where there may not be any, we love finding meaning in unrelated coincidence.

The sensitives onstage made sure to single out everyone in the crowd at least once, whether sharing visions of the dead or of bathroom tissue. Toward the end of service, a giggling medium in tinted glasses singled me out. "I see books all around you," he said, looking through me. "There is important information in these books. You will use this information to make a document of your own. That's it." He bowed and sat back down. I nodded, impressed.

When the messages were done, the organizers passed around a basket for donations. At the end, they sprinkled their hands over the basket as though stirring energy into a receptacle. On the way out, I overheard the seer who'd pegged me as a writer telling a couple behind me that those seeking a more profound messaging experience could become members of the SSF and attend séances in the organization's headquarters. Occasionally Marilyn Rossner herself guided the sessions.

There's little outside indication that 1974 De Maisonneuve West is a school for mediums. Inside the graystone town house, I caught a glimpse of "the little one" striding purposefully down the hall, like the doyenne of a dollhouse. An attendant directed me toward the second floor. At the top of the peeling-paint stairwell, in an octagonal room with purple walls and gold moldings, I found a dozen people sitting there, meditating. It smelled of antique furniture, Oriental rugs, and dry tears. An exceptionally mellow, long-haired graybeard started the session by playing plaintive tugboat-foghorn drones on a trombone. The sounds abetted the conduits' ability to enter a receptive state. Unfortunately, Rossner did not join us.

"I see a vortex," said one of the clairvoyants, her eyelids fluttering. "A vortex. At the center all is calm. You can let the whirlpool whirl you around, or sit observing. The choice is yours." She brought her palms together, kissing her thumbs as her fingertips grazed a burgundy dye-job hairsprayed into place. The severe sideways parting of her hair glowed ghostly white, an arrow pointing to the recesses of her mind.

"I see several people dancing in a rainbow," announced another medium. Perhaps someone's dead relatives? She described them vaguely and looked us over. No one in the circle recognized the deceased. She wasn't discouraged. "They want us to know they are happily eating all sorts of meat pies and drinking pilsners in the afterlife."

A squinting, myopic South Asian man in a lumpy suit pointed at me. "Can I come to you?"

I nodded.

Concentrating hard, his scrunching face swallowed his eyes. It seemed as though he were peering so far into his own infolded head he could see into another continuum. A communiqué came through. "I see a young man carrying a beam of light," he declared, tugging on his jagged-geometric tie. "Spirit says you can share this radiance with many people. You can touch them. You are voyaging to sacred destinations, recording your impressions in a notebook. Spirit wants you to know you can do whatever you want the way you want."

Afterward, I spent a moment in the building's lobby perusing announcements for upcoming events. The walls, painted in primary colors, were plastered with prohibitive signs. DO NOT *EVER* TURN THIS OFF. NEVER PUT ANYTHING IN THIS FURNACE. NO RECORDINGS OF ANY KIND! I figured that documenting the events in a notebook wouldn't count as a recording, seeing as Spirit had green-lighted my quest. On a door adjacent to the exit, a thumbtacked piece of paper said SILENCE PLEASE, HEALING IN PROGRESS. I loitered in front of it for a moment, hoping to overhear a snippet from the beyond. But then the blue door opened, and Marilyn Rossner herself stared out at me. "Well, are you joining us? We're about to begin."

The room's large, brown table had a dozen or so disciples sitting around it. The boardroom feel was offset by trippy artwork and the sounds of interstellar pan flutes playing some sort of outer-space Chinese ragas.

I grabbed the one available chair, two seats away from Marilyn. A couple of twentysomethings paused to look me over inquiringly. As they started talking again, it became clear that they were studying mediumship. A suburban couple, both of whom wore mournful countenances, sat across from me. They were discussing how the term *bless you* derives from the ancient belief that when you sneeze, your soul can escape your bodily form. A pregnant woman sitting beside them joined in, saying that she'd heard that gravestones began as a way of keeping spirits in the ground.

The white-haired woman at my left spoke with Marilyn about the caregiving trip to Africa they'd just been on. It took a moment for me to realize the woman was a nun. Ending the chatter, Marilyn adjusted her sunglasses and began by explaining how the antimalarials she'd been taking were making her feel out of sorts. Any discomfort paled, she said, in comparison with the rewards her team had felt ministering to impoverished, abandoned, and infirm children in Soweto. She said she'd go around the table, counterclockwise, and tell each of us what the heavenworld had in store for us on that day.

She commenced with the nun next to me. Full of tenderness and gratitude, Marilyn spoke of the special rewards for people as devoted to the spirits as the nun, of how she would be having the most incredible visions on the insides of her eyelids.

My turn came next. Marilyn started off innocuously enough, talking about how I'd gone through a great trauma. "You will soon be relieved as you will receive a great blessing," she said, starting to cough and shiver. She wrapped herself up in a blanket, looked up at the ceiling, and grew perturbed. "You are in great danger, do you realize that? Drugs! You must not take any drugs if they are offered to you!" Her breaths sounded strained, as though blocked by gravel.

"Your mind is *sensitive,* do you understand?" she shouted at me, someone else speaking through her. Before I could react, she leapt from her chair and cast off her blanket, yelling about blood, hearts, and "that which isn't which is." Startled, I wanted to calm her down, to reassure her that I wouldn't shoot up or pop pills or whatever it was she was seeing. Raving, she stood up and frantically pawed the countertop at the rear of the room. She found the sink and turned on the faucet. I watched, aghast, as she started throwing tap water at me.

Was she baptizing me? "We are in the kingdom," she growled. She rushed over to a nearby watercooler, filled her palms, and hurled their contents at my head. The people around the table were all staring at me, mouths gaping, as I tried to not look too mortified. She kept coughing painfully and prophesying my violent death.

Moving away from the cooler, she fell back into her chair, shuddering. Her head was twitching. *She's possessed,* I told myself. *This is about her, not you. It's the antimalarials, some semipsychotic side effects.* Bolting upright, she looked at me and began speaking in a hoarse voice: "The time has come for you to do the work you were put on this earth to do. In the beginning was the word and the word was God, and you too shall come to understand the meaning of your name: *Ah-dahm*."

How does she know my name? I wondered, as she collapsed forward onto the table.

I never went back to the SSF.

After the class, a young medium-in-training followed me to an Italian restaurant around the corner. I hadn't invited him to join me but he did anyway. Over a plate of *supplì al telefono,* he suggested I return to investigate the energy that had "mounted" Rossner. I politely declined. It wasn't always this apocalyptic, he insisted. Indeed, following our baptismal pas de deux, Marilyn Rossner had regained her composure and showered everyone else in the room with positive, easygoing messages of light. The young man said he hoped to see me there again. I remarked that some chemistries are better left unexplored. Before leaving, he informed me that I would come to love sunrises more than sunsets.

3

The Valley of Astonishment

There is another reality, the genuine one, which we lose sight of. This other reality is always sending us hints, which, without art, we can't receive. . . . This essence reveals, and then conceals itself. When it goes away, it leaves us again in doubt. But we never seem to lose our connection with the depths from which these glimpses come.

—Saul Bellow, Nobel Lecture, 1976

Fishes, asking what water was, went to a wise fish. He said that it was all around them, yet they still thought that they were thirsty.

—Azīz-i Nasafī, *The Most Sublime Goal*

"CAN I help you?" an elderly Middle Eastern man in a taqiyah asked.

I explained that I'd come for the open house, that I was hoping to speak with someone about immortality in Islam.

The newspaper that morning carried a notice that the Muslim Council of Montreal had organized an awareness program called "Montreal Open Door Mosques." Thinking about it, it dawned on me that Islamic spiritual leaders would likely be able to shed light on the watery shadow world of immortality. I decided to visit the temple around the corner from my studio, the Al Sunnah Al Nabawiah Mosque.* When I showed

* This was several months before a WikiLeaked Guantánamo Bay document deemed it one of the nine "top Al Qaeda recruiting zones" in the world, allegations the mosque (where over two thousand faithful show up for Friday sermons) dismissed as defamatory.

up, on a Saturday afternoon, the building was nearly deserted. A few men were milling around in a room near the entrance, and I joined them in what turned out to be the mosque's library.

"No one is here to help you right now," the man said, sternly, responding to my query about the idea of immortality in Islam. "I am only the librarian. But perhaps I can get you something which will answer your questions." He turned to the stacks of books and pulled down two volumes.

The first one offered a general outline of Islamic principles: devotion to Allah; praying five times a day; the commitment to make the hajj to Mecca, if possible, during one's lifetime; fasting during Ramadan. It spoke of the importance of Jesus to Muslims, the major prophet antedating Muhammad. The religion emphasizes peace and mercy; any act of terrorism, the book clarified repeatedly, is a sin.

The second book, a beige hardcover, offered a detailed exploration of Muslim immortality beliefs. Following death, souls end up either in the paradisiacal gardens of heaven or in the reeking abysses of al-Nar, where sinners have their bowels dissolved by boiling, festering water while wearing garments of flame and eating meals of fire through burnt-off lips. For dessert, the damned are fed torment-intensifying Zaqqum fruit, covered in thorns and shaped like devil's heads. As they're digested, they release innard-scalding oils.

While I read, the librarian came over to ask whether I was planning on speaking to any other Islamic experts.

"I'm not sure, perhaps some Sufis?" I answered.

The librarian shook his head and frowned. "Sufism is not Islam," he explained, in weary tones. "Us Sunnis speak about what is True. All that is True is in the Qur'an and the words of the prophet. Everything else is human, confused. The Sufis are like other religions. A blend of beliefs. In Islam, you need proof. And it's all provable. You are seeking the Truth and the Truth has to be provable."

He reached into a drawer of his desk and pulled out a thin booklet called *A Brief Illustrated Guide to Understanding Islam.* "It's a good book because it gives you proofs," he explained, handing it to me for perusal.

The booklet was definitely proof-oriented. The first chapter outlined the ways Muhammad's explanations of natural phenomena have been substantiated by science. For example, the Qur'anic description of mountains having deep "roots" is a concept now confirmed by modern geologists. Additionally, according to the booklet, modern meteorologists explain cloud formation the same way the Qur'an did. And the Islamic contention that the universe at one point consisted solely of smoke has also apparently been validated by astrophysics, with contemporary observational cosmologists characterizing the cosmos's beginnings as an opaque, dense, and hot gaseous composition.

One of the most unusual chapters of the book focused on child growth in utero. The Qur'an posits a period of fetal development called the *mudghah* (or "chewed substance") stage. X-rays prove that human embryos do in fact resemble a chewed substance, a claim the booklet substantiates by printing an image of a used piece of Bubblicious chewing gum next to an image of a twenty-eight-day-old embryo.

"Take the book, it contains further proofs," the librarian concluded. "Just don't speak to Sufis. They'll mix you up. They're not the Truth."

Soon after my trip to the mosque, I bumped into a couple of Sufis in long robes coming out of the alley behind my apartment. One of them—red bearded, pale skinned, wearing a kufi—I recognized as Husseyn, the owner of a nearby restaurant called Rumi.

We spoke for a few minutes, as neighbors do. I mentioned that I'd started researching immortality. Husseyn's friend, also in a kufi, recounted a story about a friend of theirs whose cat had died that week. "The thing about immortality is that most people are completely detached from the reality of dying and decrepitude," he added.

"Yes, and that's what's so interesting about belief," I answered. "It helps people contend with that reality."

"We can believe what we want," said the friend, firmly. "Then there's tasting."

I'd read enough to know he was referring to a famous Sufi dictum: "He who tastes, knows." Tasting what? The truth.

Historically, Sufism emerged synchronously with Islam in the seventh century CE. The sect is considered to be Islam's mystical kernel, its inner reality. Sufism is Islam, insiders say, and it is not Islam. Adherents believe that all religions share an essence, that a single, underlying truth resides within everything we consider spiritual. This makes them open to other denominations, dubious toward the exclusivism of institutional paths, and prone to esoteric utterances. "Sufism is the milk," they say, "and religion is the butter, after it has been churned."

Sufi teachings form an energizing chain of transmission they call the

Sufic stream. They describe it as a yeast that leavens one's inner bread, permitting spiritual potentials to rise. Sufi means "clothed in wool," symbolizing the renunciation of material concerns. The aim of the Sufi is to realize and participate in one's own divine nature.

Sufism's take on immortality is not written down in books because it cannot be. Attempts at articulating it have never succeeded because, in their words, "those who know do not need it; those who do not know cannot gain it without a bridge."

A teacher is a bridge, but one does not simply select a Sufi guide; nor can one even be found; a true master chooses to adopt the disciple. Poetry is also considered a bridge; poets have been the Sufi path's primary disseminators since its inception. And while poems convey tenets of their faith, the canonical versifiers always insisted that what they were really getting at is beyond language. Their couplets are a bridge to a bridge, a fuel that lights the torch carried into the darkened silence of the real journey.

In Sufism, the Barzakh is a threshold zone between life and death where souls end up after leaving the body. The Barzakh is both a barrier *and* a bridge, a liminal midway point that we cannot attain until we die—unless we let go of our conscious mind and tap into it through a mystical experience. Beyond that paradoxical barrier-bridge lies an even greater ultimate reality that cannot be described. In the words of Nobel-winning British author Doris Lessing, a prominent Sufi, "all our associations with the word *mysticism* are wrong or limited." As her teacher Idries Shah noted, Sufism's mechanisms take over at the point where words leave off. According to the sufi Al-Ghazali, there's a limit to how much one can communicate about mystic truth: "The rest can not be learnt by study or by speech."

The off-kilter detachment inherent to Sufism is radically different from the stability that accompanies knowing something. When I tried to copy and paste one Sufi treatise into an e-mail, my browser translated the word *knows* as *!wows*. It felt meaningful. In Sufism, to know is to attain an exclamation point of pluralistic recognition.

Sufis, calling themselves the Lovers, follow the Creed of Love. They greet each other with the Arabic word *ishq,* which means "love." The analogy of love gets close to the heart of Sufism, but the love they speak of both is and is not familiar love. For them love *is* God. In Sufism, the feeling of love can, when pursued beyond the ordinary, bring with it ecstatic glimpses of the beyond, resulting in reassurances about one's own postmortem prospects, a certainty that, though we presently belong to the world of dimension, we come from and return to nondimension. Rather than ignore death, the Sufi way is to constantly stare death in the face, bringing about a state of indifference toward living or dying. Advanced masters, asked whether they are mortal or immortal, will respond thusly: "I know nothing, I understand nothing, I am unaware of myself, I am in love but with whom I do not know." Something about the Sufic version of love yields a deeper awareness of our interrelation with nature. Just as inert matter is taken up by bacterial organisms, which feed plants, which in turn die nourishing animals, we, too, die as humans, only to rise again in another form.

Their texts describe us all as fractions of the divine. In this view, the soul enters a body the same way rain falls upon the ground; then, at death, it evaporates back to the Universal Main. "Let us but see the Fount from which we flow," concluded the poet Attar, "and, seeing, lose ourselves therein!"

Empirical and nondemonstrable, the love that Sufis worship is a love that floods borders and washes words away. It is a mystery beyond understanding, they assure us, understood only by dying. Their literature speaks of a water that flows down from paradise, sparkling with grace. Its source is the Fountain of all purities, the Fount of Immortality. This overflowing water—the Water of Life, as poets calls it—can only be tasted through love, the mystical love that can't be known through reason. Tasting it, we learn that supreme reality is inside one's own heart. Each person's heart is God's throne in the water. To experience the fountain's outpouring is to taste. And to taste is to know.

On a Friday around noon, shortly before prayer service, I strolled over to Montreal's Naqshbandi Sufi Center, a mosque within a converted brick storefront several blocks from my home. The sign said ZAWIYA, meaning a lodge for Sufi assemblies. Inside the window display were books about purifying the heart, several thriving houseplants, and a framed poster listing Allah's ninety-nine names. I walked in. Two dozen sets of eyes swiveled in my direction.

The midday Jumu'ah hadn't yet started, and most of the congregation, hailing mainly from North Africa (Tunisia, Algeria, and Morocco), sat against the far wall, their backs propped up by decorative pillows. Floral incense perfumed the air. The large, open room was blanketed with Persian rugs. On the wall hung calligraphic inscriptions, framed images of holy sites, embroidered fabrics, a long scroll. Barefoot kids bounced around playfully. Five women in hijabs were speaking together in one corner. A turban-wearing, gray-bearded contemplative in an ankle-length, wool tunic smiled in my direction. I smiled back and took off my shoes. He nodded slowly.

A younger man, in his early forties, greeted me. He was the imam, Omar Koné, the community's leader. Originally from the Sahara in Mali, Koné cut a dashingly ascetic figure, his orange scarf draped over a dark, flowing, wide-sleeved robe. I asked him if we could speak about immortality in Sufism. Stroking his black beard, he said we'd be able to chat until prayer service commenced. His eyes—wise, serious, kind—looked out beneath a large green turban, an indication that he'd undertaken the pilgrimage to Mecca. In Sufism, I'd soon learn, the color green also represents eternal life.

Koné gently cut short my jumbled attempt at explaining my book. "You can't know what the questions are. The search is like that. Anyway, I can sense you want to be here." He welcomed me as though I were treading the path myself, a *murid*-in-waiting rather than just a spiritual tourist.

"Okay—so the concept of immortality," he commenced, as we sat down, cross-legged, on sheepskin rugs. "The Sufi point of view is the Islamic point of view. And according to Islam, physical immortality doesn't exist. It's incontestable. The body cannot live forever. The immortality of the soul, however, is obvious. The soul is created for infinity. It has a beginning and no end. The soul is a most divine favor. Everything else in creation will perish."

As we spoke, he adjusted his ring, a lozenge-shaped, turquoise gemstone set in a silver band. "Our actions, too, are eternal," he continued. "Everything we see, hear, say, act, think, intend, is registered for eternity and affects us in the hereafter. Eternity is granted to souls either in the heavens or in hells. That stay in hell can be temporary. And to this we must add, *And God knows best*. We pray that no one will ever enter hell, that it's only a pedagogy. There's already enough pain on earth."

Here he looked toward Mecca and uttered a silent prayer.

"It is said that every human being has a fixed amount of time to live." Koné turned to me. "When humans reach the end of their life, no matter what age, within what we call the seven last breaths, time is stopped, a space is opened, and humans are taken on another track to live and complete their life until they reach the age of 137. This is a fixed standard for human life spans. Even a child who passes away will be granted 137 years to live their entire life."

I wasn't sure what he meant by an "other track." "Where are those remaining years lived out?" I asked.

"In a part of life that can't be seen by humans."

The belief in a supramundane unity beyond all material perceptions has been with humanity since prehistorical times. It's not a "primitive" belief; it is the "perennial philosophy." Every cultural tradition refers to another plane of existence alongside our own from the time documenting began. This other world can't be apprehended by the instruments of rationality. There's no way to demonstrate the veracity of something apart from "this world." It'll never be proven, yet, in some form or another, it's a crucial fact of life for the majority of humans. The sheer quantity of testimonials by mystics who've participated in the phenomenon is the only indication that it exists.

Mysticism is the ability to expand regular perception and attain what's defined as "a direct experience of ultimate reality." All religious systems pivot on the idea of a unity behind all appearances, a single suprasensible Reality. It takes a certain equipoise, or binocular vision, to even accept the notion there is more than one plane of being. Whatever's beyond the sphere of logic cannot be reasoned about; it is, after all, outside the reach of the human mind. Still, most people consider it an essential truth that part of us is connected to (and derives from) an indestructible world. We cannot discern this spiritual plane with our eyes, let alone apprehend it through everyday thinking.

The essence of mysticism is that we *can* approach it by cultivating a state of inner receptivity. By love God may be gotten and holden, the Sufis say, but by logical understanding, never. "Sufism is about attunement with true reality," Koné told me. "It is a discipline, the science of the fight of the ego. The aim is extinguishing the self, removing the ego, which is an impediment to understanding."

The mind and the senses cannot approach ultimate reality. Only our mystical faculties can. The very word *mystic* stems from a Greek term meaning "to close the eyes or the mouth." Mysticism is inner sight, blinding insight, seeing through shut eyelids. The Jesuit visionary Ignatius of Loyola spoke of his interior eyes, of the eyes of his mind, of eyes that can envisage the unseen.

All the Romantic poets versified their half-seen intimations of a something beyond mutability. Wordsworth used something he called an "obscure sense of possible sublimity." Obscure though it may be, it's a sense that comes as naturally to us, he claimed, as the gift of reason—which it has nothing in common with. Keats told of the "sensual ear" picking up unheard melodies, of seeing with awaken'd eyes, of sight in blindness. Fusing senses, Coleridge invoked a kind of aural-vision, a composite seeing-hearing, a light made of sound, a soundlike power within light. His couplets repeatedly attempted to convey "the attainment through the visible world of an insight into the invisible world beyond." The eternal speaks, Coleridge wrote, whether we can hear it or not, whether droplets of incommunicability melt enough to trickle into our imaginations, or whether locked up in frozen muteness. Such is the secret ministry of frost.

Because Sufic beliefs are deeply mystical and cannot easily or completely be explicated, *murshids* developed a system of imparting wisdom by means of something called the "scatter" dissemination. They use stories, anecdotes, poems, and jokes to help readers see from different perspectives while also letting them fill in the blanks. In scatter, the aim is to bombard the unconscious mind with multiple impacts, seeding it with new growth.

Many Sufi tales revolve around a trickster character named Nasruddin, who shares complex truths. In one famous example, a beatific-looking monk walks into a teahouse where Nasruddin is hanging out with some friends. "There is nothing which cannot be answered by means of my doctrine," announces the monk, dreamily.

"And yet just a short time ago," replies Nasruddin, "I was challenged by a scholar with an unanswerable question."

"If only I had been there!" the monk says. "Tell it to me, and I shall answer it."

"Very well. The scholar said, 'Why are you trying to get into my house by night?'"

End of story. This koanlike tale conceals layers of implications. First off, the notion of a house at night refers to the subconscious parts of the mind, those not accessible to the powers of cognition. It is upon these nonrational regions that Sufism operates. Their teachings can't be apprehended rationally; they can only take hold in the unthinking stillness.

The "joke" also mocks the idea that any system of thought can possess all the answers. To Sufis, the assumption that everything is comprehensible rationally is a flawed viewpoint hindering the possibility of spiritual growth. "Man's ignorance of his own ignorance is the real enemy," writes Idries Shah. "Man has been taught that he can understand everything by the same process, the process of logic. This teaching has undermined him." Only by getting away from thinking that we understand can we begin to understand. For Sufis, understanding immortality requires accepting that there are limits to what can be known. It requires faith, practice, and devotion.

Positioning themselves on prayer rugs, the Montreal Sufis knelt and prostrated themselves toward a semicircular *mihrab* at the far end of the room, indicating the direction of Mecca. As they prayed, I sat in a green leather armchair taking notes and wondering what life would be like were I to become a Sufi. Not that it would happen. I didn't have the commitment, or the inclination, to align myself with any faith. Maybe I was too lazy? Or maybe I was just a writer working on a book, happily lost in the forest.

When the Jumu'ah was over, Omar Koné invited me to stay for lunch with the brotherhood. They spread a thin, scarlet sheet over the rugs and placed tagines and platters of couscous on the sheet. We sat down and started conversing about Koné's *tariqa,* or spiritual order. The Naqshbandi are known variously as the Silent, the Masters of the Diagram, or the Painters. They practice sober Sufism. "In our order, you have few visions," Koné clarified. "We are not about entrancement and the opening of spiritual states."

"But other orders are?"

"Yes. Between the divine presence and humans there are seventy thousand veils. Other orders attempt to lift the veils from the human side heading toward the divine presence. In this order, the veils are removed from the divine presence's side. The thickest veil is the one on our side. That's why we aren't so much about unveilments. In other orders, when a veil is removed there are experiences."

"Ecstatic?" I asked.

He nodded. "In every unveilment, there's a danger of becoming lost in one's own illusions. Once you start seeing things, you think you've become something. Being in the delirium of a mystical experience is lifting a veil, which means one is still one or seventy thousand veils away. On that path, states will occur, it can happen that you enter into a realm of ultraperception, but we will not speculate on that."

For all Sufis, visionary inebriation is only a developmental stage, not the final aim. The ecstatic breakthrough is a bud that hopefully blossoms into an experience of "the hidden dimension beyond the metaphorical drunkenness." The goal is not sacred intoxication, but rather tasting True Reality, being transformed by it, and then, instead of staying absorbed in the effulgence of nonexistence (called *'adam*), returning to tell others about the truth of immortal life.

Despite sharing a name with nonexistence, I'd never experienced it myself, but something about the idea of True Reality clicked with me. Whatever we can't prove, we can believe in. And just because something can't be nailed down doesn't mean it isn't there. One of the last things Koné said to me was that a skeptic is like an unrequited lover who assures himself that love doesn't exist.

My time among the Sufis left me burning to speak with a priest about the biblical connections between water and eternal life. While pondering whom to contact, I remembered my former professor Father Marc Gervais. A Jesuit priest, he taught film-studies courses at Concordia University in Montreal. I admired him so much, I enrolled in six different Gervais courses, even becoming one of his teaching assistants. We hadn't seen each other in a decade or so, but I still had Father Gervais's number. It led straight to an unidentified voice-mail box. I left a message after the beep. Having a mysterious answering machine wouldn't be uncharacteristic of him.

4

Lessons of the Teachings

The most that any one of us can seem to do is to fashion something—an object or ourselves—and drop it into the confusion, make an offering to it, so to speak, to the life force.

—Ernest Becker, *The Denial of Death*

A teacher affects eternity.

—Henry Adams, *The Education of Henry Adams*

IN UNIVERSITY, my classmates and I thought of Professor Gervais as an avant-garde cinema scholar rather than a Catholic priest. We were less interested in any personal encounters he may have had with the Lord than those he'd had with film gods such as Jean-Luc Godard and François Truffaut (whom he was occasionally mistaken for). Based on those meetings, Gervais wrote a thousand-page dissertation about the French New Wave for his doctorate at the Sorbonne. He'd also published the first-ever book about Pier Paolo Pasolini, an Italian, communist, homosexual director whose films were as out there, and seemingly unpriestly, as possible.* Sure, Gervais said mass and heard confessions,

*At the 1968 Venice film festival, Gervais headed a jury that awarded Pasolini's *Teorema* the International Catholic Cinema Office's grand prize. His decision created an uproar. The film, despite its religious overtones, clearly conflated sexuality with divine truths. The Vatican formally denounced it, the Italian government charged Pasolini with obscenity, and Pope Paul VI rescinded the award. Long after the scandal, Gervais defended the film's spiritual profundity. "I've always felt that it really says everything," he once told me. "Everything that can't be said."

but we students knew nothing about that catechist side. Each week, he screened art films in the basement of a stone chapel on school grounds.

A beloved professor, Father Gervais was also a recognized figure in the international cinematic milieu. He attended Cannes each spring, and his final grades invariably arrived late because he was busy hobnobbing on the Croisette. Film cognoscenti knew him simply as Le Jésuite. Directors considered him an éminence grise and hired him to consult on such movies as *The Mission*. Father Gervais was also an authority on the work of Ingmar Bergman. The reclusive Swede had even asked Gervais—our teacher!—to do the audio commentary tracks for his DVDs.

Gervais's one-room cell in the Jesuit residence on campus was filled with books and films. "I live not so much in a cell as I do in a celluloid world," he liked to say. With that wave of silver hair and his star-studded extracurricular activities, he was as cosmopolitan as a man of the cloth could be. The other Jesuit teachers at school were uniformly stern, and while Gervais had a grave side, he was often humorous, speaking more about love than God. "The soul that beholds beauty becomes beautiful," he'd repeat almost every class, quoting some Neoplatonist we'd never heard of.

I liked him from the start, even though Father Gervais and I weren't exactly cut from the same swath. He lectured in tweed suits and ascots; I took my fashion cues from the Ramones. During those first years of university, I spent far more time and energy on my punk group than on my studies. My bandmates and I formed a small church of like-minded obscurantists who listened exclusively to garage music recorded between 1967 and 1978 and tried to ape that trebly sound in our own analogue recordings. Clear boundaries were demarcated at our all-night jam sessions between the very, *very* few bands we deemed cool and everything anybody else had ever heard. It was rock 'n' roll fundamentalism.

Where we were naïve, Gervais was idealistic. He believed that art's main role is to connect us with the unseen. For him, watching films was the greatest form of prayer. "Through art," he'd say, "one comes across the question of the sacred, of the ineffable, paradoxically expressing itself in a most dialectical and ambiguous manner." Cinema helps us know ourselves, he'd say. But for him, knowing oneself meant knowing one's connection to the divine. This mystical truth cannot be grasped by the mind, Gervais assured us in class. It enters our lives like a spring gushing from the earth. As he spoke about it, I didn't even understand that I couldn't understand.

After class one day, I asked him to clarify things. His closet-size office was decorated with stills from Murnau's *Sunrise* and *Das Kabinett des Dr. Caligari*.

"You're a musician, right?" he began, nodding as I nodded. Bringing up Stravinsky's contention that one can never understand music, he spoke about how songs don't need to be lyrically cogent to strike a chord. I could relate to the idea that music had transcendental powers, as I'd often felt them when improvising with my bandmates. Films, too, he continued, have a pulse and a rhythm that can take us beyond thought, into emotion, other levels of existence, other soul states. The sacred, he emphasized, cannot be properly expressed, but it *can* be felt.

I felt confused.

He put his palms together in a kind of horizontal prayer and pointed them in my direction. "Making art can be a religious activity," he explained, winding his arms forward. "But it isn't easy." His serpentining hands came to a stop as he aimed them directly at my heart.

They fluttered apart, and he started speaking about how life requires striving: "There's meaning in the struggle. And there's a name for the energy that impels us forward, that allows us to survive, to keep living, no matter how difficult it gets. It's called the life force."

"Yes, but you believe in God," I countered.

"Being a Jesuit is my fountain, my well, but the life force applies equally to nonbelievers. You feel it through inner turmoil. It comes from not giving up, from confronting your suffering, from knowing you still have a ways to go. It's a continual push. In Latin, this is called *magis*. It doesn't end. And it's not supposed to be easy. Just think of love; the joy and anguish. Love is an experience of the life force. You understand when you love."

I'd heard that line in class before. Even though, as a priest, he'd vowed to renounce ever being in love, Gervais often stressed the importance of love in his lectures. "You can see the universal life force at work in romantic comedies like *Breakfast at Tiffany's*," he told me, "when it starts raining and the protagonists kiss for the first time." Water symbolized the idea of a grace, of an emanation from the beyond entering this world of mortals.

He described the "gratuitousness" of grace, its flouting of reasonableness, the way a charge of renovating vitality can surge through us. Grace is an intangible booster-shot of spiritual energy, a sustaining substance, a form of assistance, un-asked-for, unmerited, unannounced, unpredictable. It's a profound, life-restoring breakthrough. It gives us the strength to keep on living, no matter the shipwreck. Its regenerating intrusion carries us from darkness to daylight.

Even though much of it went over my head, that talk with Gervais had an effect. Before then, I hadn't ever thought of immortality, or the life force, or the meaning of art, but for the rest of my university years, I wrote essay after essay grappling with those ideas, trying to understand something I'd never experienced: the divine.

Now, ten years later, I needed help navigating that paradoxical ambiguity. Gervais would know how to steer me right. But a week went by without any response. I dialed the number again and left another message on that nameless answering machine.

The department secretary didn't know how to reach him, but suggested I try the university's chaplaincy. I sent an e-mail. The response was disheartening: "I am sorry to be the one to tell you, but Marc is unavailable. About 18 months ago Marc was transferred to the Jesuit Infirmary and his health continues to decline. He is, at this time, incapable of any consultation or collaboration. It is highly probable that he may not remember you at all. At last report a transfer to the 'full-care' section was imminent. He was not in great shape the last time I visited."

Filled with sadness, I asked the chaplaincy if they knew any Jesuits who'd be willing to chat about Gervais's teachings on the life force. Nobody responded.

"Good luck talking to a Jesuit!" scoffed one of Gervais's friends and former colleagues at Concordia. "That's like talking to the CIA. Marc was different. He carried a lot of magic in him. I'd often forget he was a priest. Sure, he performed baptisms and heard confessions, but he was also such a little boy, kooky and out there. It's so sad the way they just shipped him off to that infirmary."

"What happened to him?" I asked.

"Dementia. Alzheimer's. The last time we spoke, he described it as a dark force descending on his mind—'a terrible blackness.' I've thought of going out to see him, but it's just too depressing."

The university announced a new Marc Gervais prize for undergraduates. "He'd put on accents and do Cary Grant impressions in class," explained the brochure. "He was a colleague, a mentor, a friend and confidant."

He *was*. Everyone spoke of him in the past tense.

It felt wrong. Even though the university told me not to contact him, I couldn't imagine he'd be bothered by a visit from a former TA. After all, he did love talking about the life force.

The only infirmary for Jesuits in eastern Canada is located in Pickering, Ontario, not far from Toronto. When I called, a nurse named Felicitas picked up. I told her that Father Gervais had been my professor. Was he staying there? She put me on hold. A few moments later, Gervais himself picked up the phone.

"Helloooo?" he said, undeniably cheery.

I hadn't expected to get ahold of him so suddenly, or to hear him in such high spirits. I introduced myself as a pupil who'd taken six of his classes.

"Poor devil! And you called to brag that you survived somehow?"

"The only way I survived university was by taking the courses you offered," I said, laughing. "And I was wondering, do you ever have visitors?"

Infrequently, he said, but he enjoyed seeing friends.

Then maybe I'd stop in to say hello, I suggested. I almost said something about paying my respects, a phrasing usually used for wakes, but caught myself. I hadn't described my project to him, yet he sensed some ulterior reason for my call.

"Well. This is a journey you are on of enormous importance—so we must make it possible. I'm about five hours from Montreal, but outside of that, it's reasonably simple. Anytime is good. I'll be here, as long as I don't take a huge trip to . . . the next chapter in my existence."

"That's precisely what I wanted to come and speak with you about."

"Aha. Presuming that nothing—how should I put it?—*dramatic* happens, to either of us, you can come whenever is convenient for you. I look forward to this, shall we say, extremely, um . . . oh, I can't find the words. Anyways, barring the much . . . I don't know what . . . I'm here."

5

To Sea and Hear

Do you think the holy ghost
has all the answers? What
does he know about love?
That great ghoul of a butterfly
living for the eternal moment

—Vladimir Mayakovsky, "A Cloud in Pants"

It is tangible yet immeasurable.

—Ingmar Bergman, *Fanny and Alexander*

THE TRAIN rushed past a spot on the riverbank where I once stood, years earlier, watching ice floes break apart after visiting a family member who'd been hospitalized for months. Toward the end of his institutionalization, I spent three days at the hospital learning about his illness and how to help him adjust upon release. The ice floes, white rectangles crackling in the blue rapids, mirrored his own return to health, the melting away of his frozen state. His recovery depended on finding faith in a higher power. Now I was returning to the patch of earth that helped heal him, searching for my own beliefs.

When I booked the trip, I made sure to tell Father Gervais as well as the nursing staff about the time of my arrival, in case he'd forget. They recommended I take a bus from the train station. It dropped me at the corner of Finch and Liverpool Streets. I followed their directions to the end of the road, where a sign said NO EXIT. For a moment, all the existentialist confusion of his classes came rushing back.

A smaller signpost affixed to the closed fence read VISITORS ARE ASKED TO RESPECT THE PEACE AND QUIET OF THIS HOLY GROUND. Behind the gate was a spiritual retreat center for Jesuits. To my right, I noticed an arrow and AMBULANCE ENTRANCE HERE.

I signed in at the infirmary and entered a large room with sofas arranged around a fireplace. At the reception desk, Feli, the nurse I'd spoken to on the phone, greeted me warmly. Daily mass had just ended, she informed me, and everyone was eating lunch. She went off to find Father Gervais. As she walked away, I peered into the chapel.

Light was cascading through a stained-glass skylight. The walls were pure white, with a few pews facing the pulpit. Just across from the doorway, a crystal vial of water was perched within a recess in the wall. I moved in for a closer look.

"Loyola water," said a woman standing behind me. She had long, stringy white hair and wore a raincoat. The water had been taken from a source in St. Ignatius's hometown, she continued. It had miraculous powers, such as making barren women pregnant.

"Do you work here?" I asked.

"No, no." She laughed. "I come here for mass, and to walk around the property. This is holy land. You can really feel it out there." Her name was Vie, the French word for "life." She told me a story about God being a raindrop she kept in her freezer's ice-cube tray. One sip and you'd never thirst again.

"What madness is this?" Father Gervais asked, standing up, arms open for a hug. He looked the same, down to the suit and tie, but he clearly didn't recall our phone calls. He hadn't been expecting me and I wasn't sure if he remembered me, though he did seem to realize that I was a former student of his. "What brought you out here—to the moon?"

"I wanted to see you again." I laughed, emotionally. "I heard you were out here and wanted to speak with you."

"You hear that?" he said, turning to the two elderly Jesuits at the table beside him. "This lout says he wanted to see me! Right. You came to cadge a free lunch, didn't you?"

Each of the cafeteria's half dozen tables was encircled with aged priests. Many of them craned around, curious about the unexpected visitor. It was just like my grandfather's old retirement home, from the vinyl-covered tabletops to the overboiled vegetable medley on each plate. But this building radiated a distinct intensity. It felt like a spaceship bound for infinity.

"So," I said, sitting down, "what do you guys do out here on the moon?"

"We circle the sun," Gervais replied, immediately. His two dining companions chuckled to themselves. One of them, wheelchair-bound and serious-faced, had been a high school principal. "He was known as the Strap," said Gervais, cupping his hand over one side of his mouth.

"I only used a strap on very rare occasions," he protested earnestly.

The other priest, with bristly white hair and sparkling eyes, told me he was a former math teacher from Nova Scotia. "He mighta been a math professor," added the Strap, "but he doesn't know his own age."

"Bullshit!" said the math teacher, waving his fist threateningly. "I'm ninety-three."

"He's quite intelligent," added Gervais. "Or at least *he* thinks so."

He turned his fist toward Gervais.

"Oh-oh, if you don't watch it, you'll get his vocabulary," said the Strap.

"Or his math," I ventured, trying to participate in the repartee.

"Yes! I'll use math on you." The former teacher laughed, dropping his fist and adjusting his wool sweater.

The Strap asked what I was doing with my life. When I said that I'd become a writer, Gervais demanded to know how I paid the bills. As an act of charity, he offered to give me a copy of his book about Bergman, which I hadn't yet read. A young nurse came over to ask what I wanted for lunch. I said I'd take whatever Father Gervais was eating. He told the nurse I was one of the worst students he'd ever had.

"Oh, we like Father Gervais, don't we?" She smiled at me. "It's very nice of you to come visit your teacher."

"I don't like these people," he stage-whispered. "They beat me every day."

Some friends show their affection by insulting one another constantly, but I hadn't expected this level of irony in a nursing home for deteriorating priests.

"I'm starting to see a pattern here," I said.

"You're slow, by Jove, but you're getting there." He laughed.

Over lunch, the ribbing continued, but we also managed to speak seriously about the films he'd shown in class: Fellini, Hitchcock, John Ford. At one point, he remembered that I'd become his teaching assistant for the final couple of years. We discussed Pasolini's *Teorema,* how it shows the sacred entering people's lives. "What our world needs so desperately is to be reminded that there is something beyond," Gervais said, "and *Teorema* is filled with that message. It really goes after the depths." Pasolini, an atheist, felt that we're detached from the divine—but that we have the capacity to reconnect with it.

"You see, I showed them films full of Christian iconography," Gervais said, turning to his fellow Jesuits. "All those films have shots of the cross, in window frames or telephone poles. And the hero walks up the hill. Sacrifice and all that. I kept trying to rescue a few of these poor students. It didn't always work. I ruined you, right?"

"You made me think, a lot," I answered.

"Impossible," he declared, jutting his chin out comically.

The other priests had never seen any of the films Gervais loved. As we spoke about them, they grew uncomfortable. "You wouldn't appreciate them anyway," Gervais muttered. They nodded in assent.

When lunch ended, the cafeteria emptied out. Gervais and I stayed behind, chatting openly. "This is the best place I've ever lived," he confided. "There's a very nice spirit here. People like people. I'm the youngster, eighty-one years old, and I'm so happy here. You'd think—a bunch of crotchety old priests, everybody must be so cranky—but, no, we all like each other so much. The word for how everybody treats one another is *love*."

I smiled, unsure what to say.

"So tell me," he ventured. "What is it you came here to speak about?"

"I want to ask you about immortality. And about the life force."

"Mmmmm, yes, you've come to the right place," he said, spreading his arms. "That's my area of expertise. And it brings up an interesting thought. As a writer, you might wonder what the question of belief is like for a priest. You know, in films, priests are always *nice* priests. . . . Now, how did we get to this? What are we talking about again?"

Alzheimer's. He'd forgotten his thought midsentence.

"You were speaking about priests in films."

"Of course, yes, yes. Nice priests. They're always in a parish, they smile benevolently, and they sprinkle holy water on your head—but are they ever shown as being *afraid*? The spiritual dimension of their lives, the problematic of it, is something writers today never get into. It's incredible, given the stats of people who still believe, whatever that means, that there aren't more stories dealing with this question."

"So belief is a problematic?"

"Well, let's get into it, shall we? But not here."

We headed out of the cafeteria. In the corridor, Gervais pointed to a drawing of various Jesuit leaders on the wall and asked me to identify each of them. I had no idea who they were, and he shook his head. "Did you know skunks are religious?"

"How so?"

"When something scares them, they say, 'Let's spray.'"

I looked at him blankly.

"So they pray," he said, marching on and giggling at his own joke.

A nurse stopped to ask where we were going.

"To my room," Gervais said. "Where else?"

"Your room is the other way," she answered.

"It is? How silly of me." This was why he needed intensive care. The nurse guided us down the hall.

"This is a former pupil of mine," Gervais said, putting a hand on my shoulder. "Perhaps the most brilliant I ever taught. . . . That's not true, by the way. He is actually a retarded student, as you can see."

The nurse smiled. "Father loves being a troublemaker."

"See how they bully me?" Gervais said, stopping in the reception area. A vacuum cleaner with a large, rectangular intake port lay on the floor. "Hey—anybody need to brush their teeth?"

We couldn't help laughing. He grabbed a Coca-Cola from the fridge near the fireplace and mentioned that he was having a bit too much fun. He then headed off in the wrong direction again.

The nurse pointed the way to his room and let me walk him there. "The nurses here are so sweet," he said, in a low voice. "When one of us dies, it's so hard for them. It isn't hard for us Jesuits, of course, but when we go into the next world, they get so sad. And we do depart quite often. I don't mean we die every two weeks, but there's at least one death a year over here."

"Why isn't it hard for the Jesuits?"

"Well, because we have this nutty little concept of eternal life," he deadpanned. "And I'm so grateful to have it. Obviously there's no way you can prove the existence of heaven. On the other hand, you look around and see all the tribes, all the nations, all of today's sophisticated people, and they all still need to believe. The idea of immortality is still such a huge thing."

His cell was roomy enough for a bed, a large television, a desk, and a couple of bookshelves filled with tomes on cinematic history, deconstructionism, and aesthetics. An array of suits hung neatly in the closet. He grabbed a copy of his Bergman book from the top shelf, cracked its spine, and sat down to inscribe it for me. His armchair faced a large window looking over a forested ravine. I walked around the room, peering at the framed photographs on the wall. One was of François Truffaut, on set, looking young and handsome. "Can you tell that's me?" Gervais joked.

"Who's that?" I asked, indicating a black-and-white shot of him deep in conversation with Bergman, on Fårö.

"He was nice to me," Gervais said, putting down the book. "At the end, you know, Bergman went Christian. He didn't want it in the papers, but when he died, he had a Lutheran priest there. And I was pleased to hear that. He didn't want people to know, just as he didn't want any religious affirmations there in his last films. After all, a man has to preserve his dignity. But Christians would say it was a good ending."

Gervais turned to an image of John the Baptist on the wall. "Were you ever baptized?" he asked me.

"Yes, at a Hungarian church in Montreal, actually. My dad's parents insisted, and my mother didn't object."

"Lutheran, hmmm? Like Bergman. And are you practicing?"

I shook my head.

"Talk about an idiot!" he cried. "What are you doing here, then? *Maids! Take him away!*"

"But how does one even become a practicing believer in this day and age?"

He looked at me warily, then said he didn't know.

"So why did you become a Jesuit?"

"I was twenty, I was drunk, I didn't know what I was doing." He took a gulp of Coca-Cola.

"Come on, I'm being serious, I want to know."

"That babe"—he pointed at a photo on the wall—"was my mother. Silvia Mullens, more Catholic than the pope. So I grew up with the religious imagination. As a child, I'd go for walks in the evening, in the countryside, when I stayed with my grandparents. Out in the fields, I'd look up at the stars in the night sky, or when the sun was just setting, and I'd say to myself, 'That's the truth.' On those walks, when I looked up, I knew I'd become a priest. . . . But how did we get onto this topic?" he asked, suddenly confused, pushing his silver, side-parted hair behind his ear. He seemed perplexed that I was in the room. "I must confess, I often forget things. My gray cells aren't what they used to be. Soft pillows, you know."

"I asked you about becoming a believer."

"Oh, yes, *that*. For me, it was the sky at night. Did I already say that? Well, the sky, the stars. That was enough. I knew my vocation was to be a priest. Later on, girls appeared on the horizon, but I weathered through. Now here I am, eighty-one and ready to go. As far as practicing goes, the best time is always when I'm seeing a good movie. Or listening to Beethoven. I always felt the immensity of God when listening to him."

The Jesuit motto, *ad majorem Dei gloriam,* means "to the greater glory of God." It refers to the idea that anything done with pure intentions can be meritorious for spiritual life, even activities not of a directly religious nature. Watching films, for example, or writing, even making food or cleaning the apartment. Gervais's version of God didn't seem like the usual Old Prude in the Sky. Then again, Gervais wasn't a typical Jesuit. And Jesuits themselves aren't the most straight-ahead Catholics.

The Society of Jesus can be traced back to a vision experienced in 1522 by Saint Ignatius of Loyola. While sitting on the banks of Spain's Cardener River, gazing at the water's cavorting shapes and sinuous foamings, the eyes of his mind were opened. All our surroundings, he realized, down to the smallest details, are emanations of the divine, "streams of water leaping from the spring well." Loyola wrote *The Spiritual Exercises* to help others attain a level of receptivity in which God can be perceived in everything. But if I'd learned anything in film class, the means of perceiving God isn't at all rational.

"As Saint Gregory put it," Gervais continued, "'Then only is there truth in what we know concerning God when we are made sensible that we cannot know anything concerning Him.'"

"But doesn't the life force have something to do with it?"

"It's not my word, it's their word." He shrugged. "I use it to not alienate nonbelievers. But the life force is more than something to believe in. It's something felt, something quite concrete—even though it's something we can't calculate or capture."

"Is the life force the same thing as grace?"

"Perhaps. It could be. I use the term *life force* for lack of a better word."

That might be true, or perhaps Gervais knew not to speak too much about that which cannot be spoken about. In his list of "Rules for Thinking with the Church," Ignatius insisted on the need to "be very cautious" when discussing grace with others. Rule seventeen explicitly prohibits talking of grace at length and with any insistence.

Gervais walked over to the shelf and pulled down an old tome called *Mystical Theology*. He opened it and put his finger at the beginning of a paragraph and told me to read. *These things are not to be disclosed to the uninitiated, by whom I mean those attached to the objects of human thought, and who believe there is no superessential Reality beyond, and who imagine that by their own understanding they know Him who has made Darkness His secret place.*

"So . . . we can't talk about this?" I asked.

"Nobody can talk about this," he said, lying down in bed, still wearing his suit. "There are no words for 'this.' All we can say of the infinite is indefinite. Keep reading."

The mystic is plunged into the Darkness of Unknowing, whence all perfection of understanding is excluded, and he is enwrapped in that which is altogether intangible and noumenal . . . and through the inactivity of all his reasoning powers is united by his highest faculty to Him who is wholly unknowable; thus by knowing nothing he knows That which is beyond his knowledge.

"Therein lies the problematic," Gervais intoned, as I transcribed the excerpt in my notebook. "'A saintly, rather pathetic old man,'" he suggested I write, to describe him. "'And then he smiled. . . .' Look, you came here and asked a simple question about immortality. That'll teach you! Never speak to a Jesuit."

Mysticism is the quest for the divine source, to tap into its infinite impulsion, to feel "truth flowing into our soul from its fountainhead like an active force," as Henri Bergson described it. We do so intuitively. Bergson defined intuition as an inner fluidity made from the essence of life and therefore ready, at any time, to be reabsorbed into those living waters without. The vague nebulosity of intuition forms an indistinct fringe surrounding our rational mind. Only through this encircling fluidity can we fuse into the vital impetus, *l'élan vital,* or the life force.

The term *intuition* is apt. Intuitive activity takes place on a mental—but not conscious—level. Picking up where intelligence falls off, intuition is how the nonthinking mind approaches the metaphysical. It leads to the within of things. Intuition is a spark that plugs us into the current of life itself. Our mind can't wrap itself around the life force, but intuition taps into that wave of unceasing creation. The aim of mysticism is to realize our identity with this energy, to realize that we, too, are a continuation of this outpouring.

"We're dealing here with deep, deep things," Gervais reiterated. "Spontaneities."

"Spontaneities?"

"Things we can only understand through intuition, or the imagination. Spontaneities are things embedded in the human matrix, things all cultures have always spontaneously resonated to—like love, and the life force."

He was trying to speak about something that teases us out of thought, something we can only know by not knowing. Whether it's called God, or the Absolute, Eternity, or Truth, it's a heightened reality that cannot be analyzed, understood, or taken in. It is inaccessible. It is in the hiddenness.

The life force is a way of speaking about it. This vitalist hypothesis has been forwarded by everyone from Heracleitus (who saw the life force as celestial lightning transforming into water) to Schopenhauer (who viewed "the will to live" as in irrational, unintelligible urge). It's a fundamental principle in spiritual philosophy—whether or not we agree with it.

According to Bergson, life bears within itself an explosive, elemental force that propels growth. This innate momentum is why plants stretch their tendrils to the sun. It's what causes an embryo to grow. It's the breath behind the sails of evolution. It's the reason eggs, flowers, and minds came to be. No one can prove it, but it can be felt in love, in visionary moments. It's the *aviditas vitae,* the wanting to live. It's what makes the wind blow.

Bernard Shaw spoke of it as the inner will of the world. He felt that humans are in the hands of an external, separate force—not God as an anthropomorphic personality, but as a kind of electrical current. At first, Shaw believed the main aim of this energy was as a galvanizer of genetic immortality, as it brings people together to make babies. He later came to see it as an inchoate power trying to self-actualize through us, constantly striving to attain the ability to understand itself. The organ that best performs this function in nature is the human brain, which struggles with its impossible task. Eventually, Shaw felt, the life force will push evolution to replace our bodies with better thinking organisms. Our destiny is to become immortal vortices of pure thought.

The Romantic poets described the life force as a basic energy "that impels all thinking things, all objects of all thought, and rolls through all things." This tremendous internal push is always seeking to transcend itself. Incapable of ever achieving its desired result (variously described as understanding itself or creating immortality), it keeps on blasting its way through inertia, Bergson wrote, "like the fiery path torn by the last rocket of a fireworks display through the black cinders of the spent rockets that are falling dead."

"We can't understand, so stop trying," Gervais continued, going on about the life force. "Bergman too was so intent on trying to comprehend it. But it can't be demystified. That's why we have mass. It's a poetic repetition of Christ giving himself up and we followers share in what he did on the eve."

As he spoke, lying in bed, his hand wavered in the air, trembling, fingers splayed. "The life force is such an interesting way of talking about it. When in the grips of the life force we feel what a gift it is to be alive. A beneficent fluid rains down upon us, whence we draw the resources to labor on. What is the sacrifice that each of us makes in our own life? Sacrifice means doing things for others, but in giving we can ourselves feel love. Our travails give us purpose. Sacrifice leads to the life force; in a mysterious way, they flow one into the other."

"And the life force has something to do with water?" I asked.

"Oh, yes. Why does Jesus bring 'living water'? As he said, 'I am the life.' Baptism uses water, the life symbol. It cleanses you. For the desert people of the Old Testament, the Hebrews and the Palestinians, water really was life. Without it, you'd just dry up and that's it. Finding water meant staying alive. You can imagine their happiness, their *gratitude*. Somehow a benign God became linked to a stream of water. 'I am the Alpha and the Omega, the beginning and the end,' you know, 'I will give to him that thirsts of the fountain of the water of life.'"

Gervais gave examples of how the mystic faculty is often likened to water in its different states. Wordsworth heard infinity speaking to him in watery words. In the Bible, God's voice is heard as the sound of many waters. Plotinus, the Neoplatonist, experienced visions of the fountain from which everything emanates, a spring that has no commencement whose streams overflow into our hearts.

"So there's an everyday sense in which we knew we needed water in order to stay alive," I asked him, "and then a miraculous sense in which the soul connects with the living waters of faith?"

"It's water, it's life, it's love." He was making a slow, wavelike motion with his hand. "It's vague, I know, I'm sorry."

"That vagueness is what I'm trying to understand, and I'm struggling because it's all so intangible."

"It *is* tangible," he replied firmly. "When that priest comes and gives you the wafer and the wine—it's very tangible. When you believe and receive it, you're eating the body of Christ. It's not just a symbol. You wash away your shortcomings and feel a love for others. It's hopeful. And you can help others. Water into wine. It's vague, and it's poetic, but it really works, you know. Sometimes."

"And you find it at mass?"

"Depends, if it's a good day. If your conscience is clean, then yes. I believe very much in confession. It forces you to face up to things you shouldn't be doing. It's a very positive thing. It cleans things out. Lord knows the number of friends and great people—atheists, artists, and that sort of thing—I've met. But, no, I wouldn't change places with any of them. . . . Now how did we get to speaking about this again?"

"We were speaking about the life force."

"Oh, yes, immortality. I think about it often." He looked up at a laminated poster on the wall, of Christ on the crucifix, blood weeping from his wounds. Gervais turned back to me, bewildered not only that I happened to be in his room, but that I was taking notes. "What on earth are you doing? Why are you writing this stuff down?"

"For my book, remember?"

"Really? Oh."

"Who else'll document all of this?"

"Posthumously!" he stated.

"Or at least humorously," I countered.

We both laughed.

"That's right: we were on immortality." He nodded. "Have you heard about scientists trying to make us physically immortal?"

"Yes, of course. How does that strike you?"

"It seems to me that people who want to live forever through science are in effect saying the same thing Catholics are saying. They're asking what we're asking—in a slightly different way. Us believers, we have our beliefs. Limitless, of course, unprovable, but still . . . And scientists also have their beliefs. They're speaking of immortality with the jargon of their particular approach. Somewhere along the line, however, they're arriving at conclusions that science, by its very nature, cannot consider. They're mythologizing."

"But they don't see it that way."

"The difference is that we of faith say our myths are true, whereas they don't even admit that they have myths. But they, too, believe their respective path will lead to the truth, right?"

"I suppose."

"Yet science, as they insist, only deals with proof. Do they have the data to prove that man can be immortal? Scientists have their foundational assumptions, whereas the religious thing, it's a whole different heartbeat. We from the start act in faith. We don't deal in data or proof. Our terrain is immaterial. This a priori of faith is rejected by scientists, who claim to contend only with material things."

"And they're speaking about *material* immortality."

"Whatever that is! They've cleverly taken it for granted that if believers can do it, then so can they. But any terrain in science can only take you so far. The scientist reaches a point where they can't go any further. His system can't do it. In pursuing knowledge, we always come to a realization that we cannot understand. After that comes the leap. Of faith. Can data make that leap? As we say in Swedish, *förlåt mig*. It means 'forgive me.' The scientist can't make that leap, the leap that we make at the very beginning. The need to prove never concerns those of faith."

There came a knock on the door. "Father?" queried a woman's voice.

"I'm not here," he groaned.

"Father, it's time for your dental appointment."

"What?"

The door opened, and the nurse walked in. "Father, remember, I told you this morning: you have to be at the dentist's office by two o'clock. Now, let's get up and go."

"Well," he exhaled, completely puzzled, "I wish I'd known."

The nurse rolled her eyes and took his hat and overcoat from the closet. She said they'd be back in an hour or so, and that I could wait for him if I wanted.

"I'm very sorry about this," he offered. "I really had no idea."

"That's okay." I smiled. "I'll read your book while you're gone."

"You have my book? What? Did you bring it with you? How did you find it?"

"You gave me a copy, remember?"

"I gave *you* a copy of *my* book?"

There it was, on the desk, next to him. He opened the front cover and saw that he'd half-inscribed it to me: *To Adam, surely one of the finest*—he had no recollection of starting to dedicate it, but he picked up another pen and continued—*of my best . . . !?! Come back and I'll try to give you something else.*

In the parking lot, an elderly Jesuit in a wheelchair rolled up and said something in French to Father Gervais, who responded with a Quebecois witticism. The man turned to me, disregarding what he called Gervais's "nimble but ineffective patois." He proffered his hand, introducing himself as an atmospheric scientist. "I specialize in fog," he added.

"He also specializes in hot air," Gervais added, opening up the passenger-side door of a car. "And in giving people the cold shoulder. Speaking of which, I must be off."

The fog specialist and I watched Gervais drive off the property. "So what sort of fog do you specialize in?" I asked.

"The upward-moving fogs," the man answered, slowly gliding a trembling hand toward my head. "And the downward-rolling fogs." The hand wobbled back down.

He looked at me quizzically, as though speaking in a parable about life's peaks and valleys, about revelations greater than the interaction of rising and descending currents, perhaps about death and resurrection or the mysteries of *kathodos* and *anodos*.

"Are there any other types of fog?" I pressed, just to see.

"Those two are the ones I observe the most," the old Jesuit said, quixotically. "But there *are* other fogs, higher in the atmosphere. Have you ever encountered them?"

Thinking of how Gervais had found God while going for walks in the countryside, I set off into the woods behind the infirmary. A diffuse sun dripped through the branches. Blue jays chirped. Rusty leaves dusted with snow blanketed the forest floor.

A path between the trees brought me to a creek. Two supine branches dangling languidly across each other seemed to be saying something about love and death. The water trickled into mud and disappeared. I kept going. A blur of black fur scampered through felled birch trees.

I came upon a clearing, a white glade whose sparseness was broken only by the columnlike trunks of ancient, dark trees, perfect in their stillness. All was calm. It felt as if I'd stumbled upon an abandoned cloister. I could see why they considered this hallowed ground. The treetops, spinning in their silent vortex, seemed to extend beyond thought. I wondered if I could ever be a believer. Could *I* make that leap? If so, what would I even believe in?

When Father Gervais returned from his dental appointment, we reconvened in his room. It was nearly suppertime and neither of us felt up for more theological talk, so we made plans to meet again soon. I asked him who else came for visits. "The occasional young Jesuit, like yourself," he answered. "My brother, my sister. Now and then Sharry comes by."

He pointed at a photograph on the wall, a headshot of the theater actress Sharry Flett. "We got to know one another when I directed her in *Hamlet,*" he recalled. "The critics said she was the best Ophelia they'd ever seen." He'd also been the officiating priest at her wedding. He used to stay at her home in Niagara during the theater season, he added. They would drink wine and have debates long into the night regarding the meaning of the life force. Perhaps she could help clarify it for me?

When I called her to find out, Sharry graciously suggested that the three of us meet up for a life-force symposium.

Sharry Flett and I met outside the infirmary some months later. Petite, she appeared delicate, almost porcelain, an impression immediately dispelled by her resonant, centered voice. Like all great actors, she exuded as much vulnerability as strength. Her pale skin and brushstrokes of shoulder-length, platinum hair were thrown into high relief by an elegant black outfit. She must have been over fifty, but she gave off a versatile agelessness.

We joined Father Gervais in the chapel just as morning mass was about to begin. He couldn't make sense of our being there together. "Sharry!" he cried, delighted. "But how do you know Adam? It is Adam, isn't it? And how do you know Sharry? I know both of you, I think, but I can't figure out how you two know each other? I'm confused."

"That's a good thing." Sharry laughed, hugging him.

The liturgy began. Beneath a stained-glass backdrop of green shoots bursting from a sea of flame, the priest spoke about living forever and ever. We ate Christ's transubstantiated body and drank his blood. When it came time to offer our neighbors a sign of peace, Gervais popped up spryly from his chair, straightened his jacket and tie, and then shook hands with Sharry, me, and five or six priests in our vicinity.

After service, the three of us went to lunch at a waterside restaurant nearby. As soon as we'd ordered, Sharry asked Gervais about the meaning of the word *mystery*.

He looked at both of us. "*Mystery,* eh?" He cleared his throat. "For me it has always been the center." He told his story about being a young boy looking up at the stars at night. In this version, he was lying on his back in a field.

"And was that mystery to you?" Sharry asked. "Even at that young age?"

"Yes." He nodded. "It was also God. The infinite. You know."

"Was it the life force?"

"Perhaps."

"Well?"

"What?" he asked, looking from her to me.

Her posture was resolute. "Can we speak about the life force?"

"I'm not sure," he said, a little slyly. "We can try."

"But what does it mean for you?" she said, trying to get him to explain.

He laughed and shook his head. They both had blue-gray eyes, and despite their age disparity, they could have been siblings. "This little creature here—" He put his hand on her shoulder, turning to me as a means of deflecting the topic.

"Let's stay on track." She held fast.

"I'm going to let you guess what the life force is," he taunted, dancing around the question like a leprechaun. I wondered if he was thinking of Ignatius's dictum to not speak unduly about grace.

She crossed her arms and looked at him insistently.

"Think of it this way," he offered. "When someone dies, it's not just their entire existence that's snuffed out. There is some life force still out there. It goes on. There's a statement of faith there. . . ."

"So the life force is like the soul?" Sharry asked.

"Not exactly. We can't define it. I know, it's confusing."

"You just can't be linear, can you?" Sharry threw her hands up. "Everything is circles with you: circles upon circles."

"Circles are wonderful!" he retorted. "You don't think circles are linear?" He started speaking about the benign cycle of life, a vortex spiraling downward and upward, expanding and constricting, filled with inward vitality. Sharry said he was making her head go around in circles.

"That's a good thing, isn't it?" he asked. "There's clarity in the confusion. As a Jesuit, you have that. There are fathoms upon fathoms."

"*You're* able to think in spirals, but how do the rest of us find clarity?"

"In the confusion. Confusion is how we balance irreconcilable unresolvables. There's no other way. And faith allows us to accept our uncertainty."

"Whenever we talk about this stuff, it always dissolves into an inability to attain any sort of resolution."

"But resolution isn't . . ." He paused, out of sorts. "What are we talking about again?"

"The life force," she said, and we both looked at him expectantly.

"Oh, yes. You two!" He put his head in his hands and laughed silently. "What you're asking for is impossible. There's no final statement capable of affirming the full ambivalence and meaningfulness of all life. Trying to find it is an infinite quest. A relentless search."

Gervais made a sweeping gesture with his arms, knocking over a glass, which he somehow caught before it fell to the ground.

"Instinct!" I gasped.

"Practice," he corrected, sharp as ever. "You know, I was going to say I had this same conversation with Bergman many times."

"What did he think of the life force?"

Gervais characterized Bergman's best films as being stories about the life force. He brought up a scene in *Through a Glass Darkly* where the family loses their daughter to mental illness. The father consoles his son by telling him to focus on the fact that love exists as something real in the world of men. This love, the father says, is actually God. "So love is the proof?" asks the son. The father answers, "We can't know whether love proves God's existence or whether love is itself God. After all, it doesn't make very much difference." God and love are the same thing.

It reminded me of a friend's wedding several days earlier. I recounted how the pastor had described seeing God in the love between the bride and groom.

"He saw God when he looked at the two of them in love?" Sharry asked, starting to cry, softly, unexpectedly. "That's what God is?"

We both looked at Father Gervais. He sat in silence for a moment as she wiped the tears from her eyes with the corner of a napkin.

"So if we could achieve God in love," she pressed on, "or love in God, then there wouldn't be any more war, would there?"

"I don't know about that," Gervais said, kindly. "There've been a lot of crusades fought in the name of love."

There it was, even on the cusp of some breakthrough, that lack of resolution, again, always.

"Is love the same thing as the life force?" I asked.

"Not exactly. You feel the life force in love. Love is a very strong thing, very powerful. You know, when Jesus rose, he spoke a lot about love—about how love is all we have, all there is, all that's there, so love one another and all that. To feel love is to feel something greater than yourself. To have loved is to believe. One understands when one loves."

"And this all plays into the life force?" Sharry asked.

Gervais looked at me, as though he'd done everything he could to adhere to Ignatius's precepts. "I can't tell you," he said, "but I can show you." He leaned in and gave her a hug and a kiss on the cheek.

He turned back to me and, pointing at Sharry, said, "Have you met my aunt?"

We laughed. But even though he was joking, he was also crying.

"Look at the eyes in my tears," he managed, blue eyes limpid and liquid.

"*'Look at the eyes in my tears,'*" Sharry repeated, seeming to understand for a moment. "That's it. Love—tears, fountains, rain."

"The life force," he said.

"That's it." Sharry nodded. "A feeling."

"The feeling remains," he offered.

After lunch, we ended up at a picnic table on the edge of a ravine near the infirmary. The forest spread out before us, standing sentinel in its autumnal robe.

As Sharry and I went to sit next to each other, Father Gervais looked at me in an effort to ascertain what I was doing there. "Where are you stationed?" he asked, confusing me again for a young Jesuit.

"I'm actually a writer. One of your former pupils."

"You mean you're not here doing your novitiate?"

"No, I'm just a heathen," I half-joked.

"You don't mean that!"

"Really, I'm a pagan."

"No! In what way?"

"I love all religions, but I don't practice any of them. Belief systems are great, as long as they aren't hurting anybody. I believe in mythologies without having faith. I'm fascinated with religion, but I'm not attached to any one faith. Isn't that paganism? Maybe it's polytheism, or pantheism. I just like stories that remind us of the power of the elements—"

As I spoke, he went to sit down opposite us, and we all heard a groaning, creaking sound. The wooden bench had rotted, and he started falling through its bowed, worn-out, splintering beams. Before either Sharry or I could move, the final connecting fibers gave way, depositing him on the ground, where he landed in a bed of leaves. We both jumped over to help him as he lay there, looking up at the sky, laughing.

Father Gervais died that winter. His funeral ceremony took place at the Saint Ignatius parish, adjacent to the Loyola campus. A quote from an Ignatian prayer adorned the program: "Give me only your love and your grace. That is enough for me."

When his coffin was brought in by family members, priests in ankle-length white robes and creamy scarves clustered up against it, seemingly excited to be close to someone going where they wanted to go. They huddled together like white petals around the stamen of his body. During the memorial mass, the father described him as a steward of the Lord's mysteries on earth.

The reception afterward was attended by many filmmakers, artists, publishers, and media types. I hung around for a few minutes and then slipped out. I entered the old chapel Gervais used as a theater for screening films. Inside the foyer, I found some of his film posters still tacked to the wall. On the main floor, noontime mass was under way, with nine or ten people in attendance. The priest lifted a chalice to his lips. His words echoed into the sunlight: "For ever and ever."

6

Beneath the Gaze of Eternity

A rabbi would never exaggerate. A rabbi composes. He creates thoughts. He tells stories that may never have happened. But he does not exaggerate.

—Rabbi Krustofsky, *The Simpsons*

I didn't know! I was more Hasidic Jew than I thought.

—Oprah Winfrey, in conversation with Rabbi Motti Seligson

THE NEIGHBORHOOD I live in is primarily Hasidic. My male neighbors wear long, dark robes and imposing circular fur hats. They look majestic in the snow, like eighteenth-century Ukrainians who've landed on Pluto. Seen from the back, they're head-to-toe black, lines redacted from contemporary life. Some denominations demarcate themselves by tucking their pants into knee-high white stockings. Others have wider-brimmed hats or extrashiny coats or tassels that drag along the ground. All the young men are often in a hurry, run-walking wherever they're off to, busy as ants. In line at the bank, they read the Torah while twisting their payos and swaying from side to side. My rare conversations with them always prove insightful. They're smart, funny, wry. They pace the streets at night, lost in contemplation, either calm or wound so tight they seem to be sparking.

In their world, God comes before everything. They live lives centered around family and tradition. For the most part, they don't fraternize with others and can affect a frosty, standoffish demeanor. But so would anyone adhering to such an unmodern mode of being.

In recent years, the local borough council has prevented new synagogues from being built for their expanding population. As a result, covert temples are everywhere—in my back alley, at the end of the block, across the street. The men cluster around unmarked doors leading to subterranean congregations. Sometimes in summer, the windows open and passersby can glimpse Hasidim praying, eyes closed, moving intensely. Their singing is beauteous: ancient, Eastern melodies pouring from the hearts of whiskered men.

Most of the Hasidic businesses in the neighborhood prefer not to cater to outsiders. Their grocery stores, full of kosher salt and kosher yogurt and kosher meats, stock totally different soft drinks (Mayim Chaim Cola), snacks (Lieber's Onion and Garlic Chips), and sweets (Osem's Cookies & Dreams). Hasidic fishmongers and butchers stoically tolerate—rather than welcome—outside customers' presence. Most shops are signless. Some cover their windows with brown paper. Others stack their storefronts high with aluminum containers.

One of the few businesses fully welcoming Orthodox shoppers and gentiles alike is around the corner from my apartment. Cheskie's Bakery bills itself as *heimishe*—a Yiddish expression meaning homemade, hospitable, and friendly. They specialize in Hungarian-style pastries (cheese crowns and babka), as well as New York–style classics such as poppyseed rugelach and black-and-white cookies. The owner is Cheskie Lebowitz, a good-natured, kind-eyed baker who moved here from Brooklyn to live with his wife, a Montrealer.

I'm always happy to bump into him. We commiserate about our Hungarian fathers. When I need a ladder, Cheskie lends me his. He tells me about his passion for auto maintenance. When informed of my carlessness, he arranged for one of his bakers to rent out my parking spot. We're neighbors—good ones—but he sometimes balks at my curiosity.

A few years earlier, I'd read something about the Hasidic belief that God is in all things. The idea resonated, and I'd inquired with Cheskie. Naïvely, I suppose I expected him to talk about God's being in the window display for his pastries, in the pastry cream itself, in the flour-dusted apron he always wears. Instead, he crisped up. "I can't speak on such matters," he answered, looking around nervously. "Listen, I just make dough and go to synagogue." But if I really wanted, he allowed, he could put me in touch with an authority. Having apparently crossed a boundary, I didn't pursue it then—but this time, when I walked in and saw Cheskie taking orders for marble cake and rye bread, I thought of Auntie Tiny and the fountain. As much as I'd learned with Father Gervais and the Sufis, surely Hasidim must have some Old Testament insights?

"How are you, Mr. Cheskie?" I asked.

"Thanks God," he responded, as he always does. Finishing up with the customer, he indicated for me to follow him into the back room. I walked through a doorway to the other side, where employees were braiding tresses of challah bread. A choral wash of male voices chanting nigunim streamed from a CD player. "Nice music," I commented sincerely.

He scurried into his office and emerged with a copy of the CD. Before I could protest, he said, "Take it, please, I have many. It's for those who appreciate."

"Wow. Are you sure?"

"Yes, yes, and also, thank you for your advice."

A couple of months earlier, Cheskie had asked for some guidance on a media request. The biggest francophone daily newspaper in Montreal wanted a journalist to spend a few days working as a bakery employee. They planned on running a feature about the writer's experience. Wary about their motives, Cheskie wondered whether I thought he should accept. He normally avoided publicity, but we both knew the story might help bridge the divide between Quebecois and Hasidim.

"If you allow them to do it," I counseled him at the time, "something positive could possibly come of it. There's always a chance that the report could be negative, but it's unlikely. If you refuse, however, nothing positive—or negative—will come of it." He'd decided to take the risk. The result was in the morning edition, which he handed me. "It's okay, right?" Cheskie asked.

"Seems like it," I said, scanning the text. "Inclusive. Even-handed. All positive."

"All positive." He shrugged.

"Speaking of which," I ventured, handing the paper back, "remember when I asked you about Hasidic beliefs in God, and you said you could refer me to an expert? I'd really appreciate it if you could put me in touch with someone who could tell me about the idea of immortality for Hasidim."

"Imm . . . ?" he responded blankly. "*Imm*—what?"

"Immortality."

"What is that word? What does it mean?"

"Well, that's actually what I want to speak to someone about: the meaning of immortality in the Old Testament. It's a word that has many interpretations—"

"Oh, a morality! I know what morality is."

"No, not morality—*im*mor*tal*ity," I clarified. "Think of mor*tal*ity, which means dying. Immortality means not dying."

"Oh, well, if you mean heaven and hell"—here his face lit up, and he became cheerful—"for sure I believe in it that it's real."

"That's wonderful." I nodded.

"Is that what you mean by *imma*"—he scrunched his face up—"*immortality*?"

"Yes, well, every religion has an approach to what happens when we die. And I'm trying to learn more about that spectrum of ideas. For example, some scientists today say they have the technology to make us not die."

He recoiled and grew serious. "For me, God made everything the way it is. And if God intended for humans to die, then that's how it is. Listen, I'm a Jew who thinks the world is 5,771 years old. In my world, the Torah is everything. For Jews like myself, whatever isn't in the Bible isn't true. So when scientists say that there are dinosaurs that are twenty thousand years old, I don't believe it. For us, dinosaurs are creatures who didn't get onto Noah's ark during the flood. When people say that a guy named Jesus came around two thousand years ago—well, for us it doesn't mean anything. But it's hard to talk about this because it's touchy and becomes an argument so quickly."

"I know it's a sensitive topic, but maybe you know someone I could speak with?"

He twirled his hair and looked at the floor, deep in thought. "Someone who has knowledge, yes . . . someone open-minded."

A few days later, I arrived at a three-story strip mall in the Sri Lankan/Filipino part of town. Walking into a dingy foyer, I double-checked my voice mail; yes, Cheskie definitely provided this precise address for Rabbi Haim Sherrf. One section of the stairwell's beige walls was overlaid with white roller strokes from a long-abandoned repainting attempt. I walked down a corridor of closed office doors on the second floor. One, ajar, had an ornate silver mezuzah on its frame. I peered in.

Two yarmulke-wearing men at the far side of the room looked

up. One waved me in, then lifted an index finger, indicating for me to wait while they finished their business. I'd never been inside a rabbinical chamber before, but whatever I may have expected, this wasn't it. Instead of some burnished-wood, leather-bound, soundproofed enclave of midrashic contemplation, I found myself in the colorful clutter of an artist's atelier—a mess of easels and palettes, stacks of paintings, cardboard boxes filled with T-shirts, Styrofoam coffee cups, and plastic utensils. The bookshelves were lined, haphazardly, with monographs by Dutch masters, as well as Jewish encyclopedias and Israeli travel guides. Tambourines hung on hooks. An acoustic guitar idled in its stand. Some pots demarcated a kitchen area, complete with hot plate and toaster. The place needed a serious dusting, but its warmth and color were immediately inviting.

As the two men spoke, I perused the works on display, from depictions of shtetl fiddlers to scenes of the Red Sea parting. One surrealistic painting caught my eye, of a stone cube hurtling through space, a cosmic rock-box engraved with Hebraic letters. In the background, satellites drifted through the spiraling Milky Way.

"The tablet," intoned one of the two men, walking over. He was big and tall, in his late fifties, strong, bearish. Wizardly. A Samson type. His chest-length, frizzy gray beard hung over a massive, regal, purple shirt.

"'The tablet'?" I parroted.

"The Commandments given to Moses." Under his bushy eyebrows, he had sharp, powder-blue eyes.

I looked back at the intergalactic block. "The famous ones, on Mount Sinai?"

He nodded, subtly arching those thick eyebrows.

"Wasn't that a piece of stone with a rounded top?" I asked.

"It's very clear that the tablets were square," he answered, peacefully yet forcefully. "Moses wasn't holding them; they were holding themselves. They were divine, after all. The text inscribed into the rock passed through and through, on all four sides, meaning no matter which side you looked at, the letters were the same, yet each side's letters miraculously penetrated through the rock to the other side."

I let the awesomeness of his description sink in.

"Welcome, Mr. Adam!" Rabbi Sherrf said, shaking my hand firmly and slowly. He had a calm, powerful presence, a contained intensity. "In this particular painting"—he picked up a paintbrush—"I wanted to show that the teachings are current, not old, that they are rejuvenated every day, so you see the satellite there." I watched him dab a bit of space-blue pigment onto a chipped corner of the painting as he spoke rhythmically about the laws bestowed to Moses.

I made an aside about the miraculousness of letters on the stone.

"You want letters?" He told me how a book of psalms called the *Tehillim* came to King David as a harp next to him played itself. Letters floated up from the strings, arranging themselves into words that David transcribed. The *Tehillim*'s text is so powerful, Sherrf continued, that it can actually lure souls back into bodies in near-death situations. As a soul rises away from a newly deceased body, the letters of the *Tehillim,* if spoken by someone close by, can swirl into the air, attach themselves to the soul, and pull it back down into the body. "That's a true story," he said, leading me across the room. "There are many like it."

We sat near his object-strewn, paper-covered desk, through which he rummaged briefly, emerging with a piece of foolscap decorated with Hebrew letters as well as other, unfamiliar letters. He held it up. "This is the key to the language of angels," he said seriously. People occasionally ask him to intercede between themselves and angels such as Gabriel, Rafael, Machiel, or Nuriel. Sherrf uses this Rosetta stone to help him translate and transmit messages. Before I could look at it for too long, he filed it away. "It's almost dangerous to know this language," he added. "It's like giving a knife to a murderer. You give a knife to a chef and you get a nice meal; you give it to a killer and he kills."

Sherrf continued by outlining the history of Hasidism, started by an eighteenth-century Polish rabbi called Baal Shem Tov. The essence of Tov's teaching, Sherrf explained, is that God resides in all things. He viewed everything in the universe, down to the smallest caraway seed, as a manifestation of the divine, as a form in which God reveals Himself.

I told Sherrf that was what I'd asked Cheskie about a couple of years earlier, and that I'd ended up here after trying to speak with him about Hasidic immortality beliefs.

"Of course, Cheskie will give you a nice piece of cake, but for the rest, for questions of 'belief' . . ." Sherrf shrugged, then fixed me with lupine eyes. "And you, you have Jewish ancestry?"

"Yes, way back, hundreds of years ago."

"On your mother's side?" He leaned forward.

"No, actually she's Irish," I apologized.

The rabbi let out a sigh and then recounted his own backstory. A Sephardic Moroccan born in Israel, he became a Lubavitch Chabadnik after being in the air force and working as an Israeli intelligence officer. "I've heard bullets whistling in my ear, so God could have taken me long ago," he told me. "But he had other plans in store for me." Aged twenty-two, fresh from the military, Sherrf visited Salvador Dalí, who encouraged him to paint. He'd achieved success with his art. "Even Madonna owns one of my paintings," he said proudly. He also made mezuzahs, designed clothes, and balanced creative pursuits with rabbinical work, doing chaplaincy in jails and servicing the Jewish Eldercare Center. "I'm a loving husband, and my wife and I have eight beautiful kids," he added. "My name—Haim—means 'life,' in Hebrew. Definitely that's me: I embrace life in every way. Mind you, I'm not afraid of death at all. I've done good and I know the process."

As we spoke, two ladies walked in, looking to buy a painting. Sherrf went to help them, and I struck up a conversation with the man Sherrf had been speaking to when I arrived. He introduced himself as Albert Hakim, owner of a gallery in the space next door. *The art that will touch your soul,* read his card. He gave me a tour of his gallery, explaining that he sells primarily to an Orthodox clientele.

"Interesting topic for your research," he mentioned, as we walked back over to Sherrf's side. "If you don't mind me asking, what do you think about death?"

I stammered a few syllables before he followed up with another, more pointed question: "Do you think you just come here to earth and then die and that's it?"

"I don't know."

"Okay, that's understandable. But I'll tell you what Judaism thinks about death. This life is a hallway before getting into the big room. This is a passage for improving yourself. We're all souls that were here before as other people. And we're all here to repair our soul. But there are so many temptations, it's easy to derail. I used to live a very open life, *très flyée*. I now know the truth: we come here to repair things. Man has to work on himself to become better and better—or worse and worse."

He drew me a little picture about how we're on a straight and obvious line to heaven, but that somehow we take detours and that our work is to get back to that main road by rectifying the wrongs we've committed. The whole point of the Torah is to help orient people toward the broad highway to heaven. When I asked him which specific passages in the Torah describe this road, Hakim shook his head and drew a distinction between reading and learning. "If you read the Torah as a book, you won't understand any of it," he assured me. "But if you learn it, then you can know God."

Sherrf, having concluded the sale, came over and joined us. "Where are we at?"

"We're on the path," answered Hakim.

"Beliefs," I added.

"Beliefs?" Sherrf sounded dubious. "You can believe in anything, but then there's what's true—facts. For example, we know the world is 5,771 years old. Fact."

"Just last night I saw someone on TV saying that mountains are millions of years old," Hakim recounted, incredulously. "How do they know?"

Sherrf shook his head. "How can you calculate time? Who are they? They must be out of their mind."

"To be more specific," I sidestepped, "we were talking about the afterlife."

"We definitely believe in life after death, if that's what you mean," Hakim said.

"Yes, Judaism clearly states that righteous people start living after they're dead," Haim chimed in. "'The righteous in death are called alive.'"

"So you believe in the soul?" I asked.

"More than 'believe,'" Sherrf clarified. *"Know."*

"You know?"

"Absolutely. The body is confined, limited. If you are well connected to God, you have the ability to go outside of your body to God, even when living. At the end of life, you are no longer limited by the body. You don't simply continue; you can become a full servant of God."

The Lubavitch movement places more of an emphasis on the afterlife than other Jewish denominations. Because the Torah does not clearly discuss specific details of postmortem states, Judaism incorporates a variety of attitudes toward the soul's fate. Many mainstream Jews feel that what happens after death isn't important. Some believe in Gan Eden, a post–Last Judgment paradise; others don't. At the End of Days, when the Messiah appears, will all purified souls end up resurrected in a perfected realm? Depends on your perspective, or not. One of my Jewish acquaintances keeps a packed suitcase under his bed for the moment it'll happen. But bringing a carry-on into the beyond isn't an agreed-upon convention among modern Jewry.

In Sherrf's view, which is in keeping with much Orthodox Judaic belief, a soul's duty on earth is to purify itself from sins and temptations. He calls this "the process." (Hakim termed it "the path.") The basic point is that each soul assigned to a body is on a mission to perfect itself. When elements are lacking in a soul's perfection, it has work to do on earth before it can return to its source in God. A soul can make mistakes—or it can attain beatitude. "It depends on our struggles," Sherrf explained. "If we overcome them, we fulfill what we were meant to achieve. If a person fails in their temptations and falls into a place where they don't believe in God, then the soul starts to drift to places it shouldn't be. It needs to come back."

If it doesn't get back to the right track before death, Sherrf noted, it ends up in Gehenom—"a washing machine that puts the soul through a spin cycle, reconditions it, leaves in the good parts, and then sends it back into another body. A soul has three chances. Three times it can come back into a human body on earth in order to perfect itself. This is our version of reincarnation."

"Is this what's called Gilgul?" I asked.

"Exactly. The soul is given tools to perfect itself because it has a challenge to complete. Whether you're rich or poor, God ensures that each soul has the correct situation in which to go through its requirements. What happens next depends on what happened here on earth."

Every soul leaving a corpse ends up in a purgatorial waiting zone where it goes through a process of judgment and ordeals, contending with what Sherrf described as "armies of angels, millions of angels, some of whom can be very harmful." Those souls deemed good and perfected move on to the infinite light; those flunking angelic cross-examination are sent to Gehenom for a refurbishing. Gehenom isn't a bad place, Sherrf reiterated. It's a place of cleansing, of purification. "There are far worse places to be," he added, solemnly, which is where souls that strike out three times end up.

Surviving family members can help a loved one's soul as it deals with celestial border guards and immigration-agent angels. Reciting Psalms of Ascent assist with *aliyah,* the soul's upward trajectory. "The process of a soul detaching itself from a body is very painful," Sherrf avowed. "To alleviate its tribulations, mourners say the kaddish. The newly departed soul is hungry, but cannot eat. The only thing that can satisfy and soothe it is the words of the Torah. Every kaddish said ensures one and a half hours of safety for the soul. Saying sixteen kaddishes protects the soul for twenty-four hours. If you keep going, and do it properly, you can protect a soul for the full period of return."

As the soul, or the God-Breath, journeys from this life space toward the One Source, it gets inhaled back into the Breath of the Divine Breather. But this process is fraught. Souls crossing the Sea of Finality can get stuck in raging waters, pulled under the dark waves into chaos. Meddlesome angels can wreak havoc, which is why the kaddish must be recited in Aramaic. "Angels do not speak Aramaic," Sherrf confided, "so they can't interfere with Aramaic prayers on their way to heaven. In Aramaic, there's no filter between the praying person and God."

Sherrf then walked me across the room to pick up a T-shirt from one of the cardboard boxes. He'd designed a fashion line, called Celestial Chariots, based on kabbalistic imagery such as the ten heavenly realms surrounding this world. The shirts looked exactly like Ed Hardy's clothes, which I mentioned.

"Yes, but I think everything Ed Hardy does is morbid, violent, and negative," Sherrf said. "And the kids think it's so cool. One of his shirts says 'Love kills'—can you imagine? There are always guns and knives and bleeding skulls in the designs. So I created my own line of Ed Hardy–inspired clothes, but filled with very positive messages. It's all divine and kabbalah. The clothes are an extension of my rabbinical work."

I looked them over and complimented the intricacy of his designs.

"And now that we've covered all this terrain"—the rabbi stood tall—"I'm wondering if you can help me as well?"

"I would be glad to," I responded, curious how a goy like me might possibly be of assistance.

"At Celestial Chariots, we've got design and imagery covered. But we need a little help with the writing."

"Writing what?"

"To be more precise, I would appreciate it if you could help me write a letter to the actress Angelina Jolie. She works for good, and I think she would connect with our clothes. I'd like to make her want to be a part of our project—in a very polite, gentle way, of course."

"Well, I could try." I laughed tentatively. "What exactly would you want to say?"

"You're the writer. . . ."

"Yes, but what do you—"

"Something about how we hope that she'll wear it. 'We are contacting you because we know that you are very into positivity'—you know, like that."

I told him I'd be happy to do it.

"Perfect. I'd very much appreciate it. And once it's done, I'd like to invite you and whoever else you want to Shabbat dinner at my house."

That week, an old friend named Himo Martin returned from a three-month trip to Morocco and came to crash in my spare room while apartment hunting. Himo, a Quebecois jewelry designer, had been interested to hear about Rabbi Sherrf's silver mezuzahs. He'd be the perfect dinner companion.

On the bus ride over, Himo told me about an experience of immortality he'd had on an Indian reserve, about five years earlier. "Some friends and I went out to an island. You have to take a canoe to get there. We stayed in a cabin next to a waterfall. That evening, after the moon rose, I walked out alone and lay down at the tip of the island. You couldn't tell where the water ended and the sky began. I closed my eyes, listened to the loons and the fireflies in the silence, and then it happened."

A dreamy look came over his face.

"What happened?" I asked.

"It's hard to describe." He shook his head. "It felt like there was a haze over everything. The sounds, the smells, the wind—everything had been united into this one thing. Although my eyes were closed, I began envisioning the stars. I started seeing sparks of light inside my eyes. They were in motion. And I was in motion, too. I was part of the movement. And all of a sudden, I fell."

"You fell down?"

"Well, not physically, not in reality, but I felt the sensation of falling through the lights into water. It was like plunging into a pool, and being surrounded by a glow of liquid. Every piece of my body felt embraced by water, completely enveloped. All sounds were very far away, the way they are when you're underwater."

"Could you see anything?"

"Yes, all the specks of light I'd been seeing formed themselves into a pattern. The spaces between the lights moved apart, and within each of the spaces were more specks of light. At first it seemed as though they were mirroring themselves, but then I realized they were duplicating, or replicating. Every heartbeart brought exponentially more shining points of light. And I was a coordinate on that grid, a radiant dot that kept falling through dimensions. Things were getting smaller and smaller and bigger and bigger, infinitely so. It was infinity."

"Wow. You saw the mesh of reality. Were you on drugs?"

"No. Nothing. When I opened my eyes, I saw a candle, lit, on the table back at the cabin." He paused again. "Vibrations of heat emanated off the candle. I could see heat waves interacting with the space around it. Like fractions of light. Everything became pixelated. In those pixels, I saw how everything is connected to everything else. I realized that I am but a pixel, that this body is a vessel. Every body, every thing, contains an inner pixel. Whenever something dies—a tree, an animal, us—its form is taken up by something else. Transformation is the law of nature. I saw energy become matter."

"How were you feeling at this point?"

"I was almost hyperventilating, but I tried to just keep breathing. I felt I was dying, but I wasn't scared of dying. The experience was so beautiful. Whatever happened, I knew it would be okay. Even if my body died, its vessel would be consumed by nature. Its energy would be released and new forms would arise. I knew that we have a larger interaction with the unknown beyond life and death. I all of a sudden understood. Death is an interaction with the unknown. The unknown is the beauty of the interaction."

Twenty minutes before the 7:00 p.m. start time, Himo and I walked into the Quality Hotel around the corner from Rabbi Sherrf's home in Côte Saint-Luc. My girlfriend had helped me begin the letter to Angelina Jolie, but it remained unfinished. I figured I'd be able to complete it and print it out in the hotel's business center. Luckily, the bored front-desk receptionist gave me permission to use a computer—with a printer attached—in the lobby.

"Dear Angelina Jolie," I started typing. "I hope this letter finds you well. As a rabbi who strives for positivity in all facets of life . . ."

A minute later, Rabbi Sherrf strolled into the hotel lobby. I leapt up, amazed to see him. Sherrf smiled calmly. He, too, was accompanied by a friend, a deeply tanned, fortysomething Moroccan guy with a kippah and a bright, toothy grin. "What a coincidence!" I laughed. "I'm just printing the letter to Angelina now."

"Don't worry about that, it's no problem," Sherrf said, a mildly pained expression on his face.

"It's no worry at all, it'll just take another moment," I reassured him.

"No, really, there's no need to do that now, so don't bother," he insisted. "Let me just find out about the guests we're meeting here."

He walked over to the front desk, leaving us with the tanned Moroccan man. While Sherrf wore dark, baggy clothes, his friend had on boot-cut designer jeans with long, white, square-tipped dress shoes. His shiny shirt, buttoned low, exposed a gold chain nestled in thickets of chest hair.

"So," Himo attempted, genially, "how do you two know each other?"

"Me and Haim? We're brothers from another mother, man!"

Himo chuckled tentatively, unsure about the expression's meaning.

The man noticed and clarified it. "Meaning we go *way* back. And you?"

I explained about the interview I'd done with Sherrf, and about Himo's recent trip to Morocco, and about the letter to Angelina, which I just needed to print up.

"Hey, bro," he spoke up, nigh antagonistically, as I sat back down to the keyboard. "It's Shabbes, stop working on the computer."

"I'm not working," I clarified, taken aback by the tone.

"Yeah, but it's not done. Not now."

"Oh, I see—*you're* not supposed to work on Shabbat. But I'm not Jewish, actually. And I'm just printing up a document Haim asked me to write for him."

"Well, you shouldn't. You're disrespecting Haim by doing that."

"Wait; I don't mean any disrespect here."

"Then get off the machine."

Apologizing, I moved away from the computer terminal. "No offense intended."

Across the lobby, Haim stood greeting a pretty young woman. Although she lived in Montreal, she'd booked a room because she didn't want to drive home after the dinner. It seemed as though she'd come to meet the tanned man, as they behaved effervescently around each other. Another friend of hers was still getting her makeup on in the room above, so we stood around making small talk for several minutes. The topic turned to a Chabad conference taking place the next day. When Sherrf disclosed that they have great food at the conference buffet, I blurted out a joke about buffets and shrimp cocktails, not realizing that shellfish are *trayf*.

The woman frowned. "Ummmm, we don't eat shrimp."

"No, no, there won't be any shrimp," the Moroccan man said, soothingly.

"Oh, of course not," I muttered, mortified. First the computer insult, now a crustacean-joke misfire.

"Actually, they do make excellent mock-shrimp," Sherrf, unflappable, added. "Which I'm sure is what our friend Adam here is referring to."

"Yes, of course!" I managed.

"Aha," sighed the man and the woman, looking at each other, relieved.

"Good!" Sherrf clapped his hands together. "But now it's getting late, so why don't Adam, Himo, and I walk back to my place together to get things under way, and you two can wait for her friend."

"As soon as she's ready, we'll walk over, too," agreed the woman.

The three of us crossed a parking lot, where Himo attempted friendly banter. "So I gather you two are brothers from another father?"

"'Brothers from another father,'" the rabbi repeated, considering it. "No. Nope, that isn't possible."

"But that's what your friend said inside," Himo protested.

"Well, I'm not sure what he meant, but for him to be my brother—from another father—would mean that my mother was impregnated by another man, which I assure you wasn't the case."

"Wait, I think what Himo meant," I leapt in, trying to salvage the situation, "is 'brother from another mother.' It's an expression meaning very tight-knit. Because Himo is Francophone, he just misunderstood the meaning, right?"

"Mmm," Sherrf mused. "Brother from another *mother*. Yes, that makes sense."

Inside Sherrf's home, three dining tables had been pushed together in a long row connecting two separate rooms. Around sixteen Orthodox kids were running around, half of them the rabbi's own offspring. Another dozen adults were clustered throughout the living room and the kitchen. Even though Himo and I were outsiders, obviously clueless about their ways, we felt welcomed.

Sherrf's pretty, young-looking wife oversaw the food preparation, pausing only when her husband began the proceedings with blessings and a recital of *kiddush*. They clearly loved each other very much. We all washed our hands by filling a cup and pouring the water over our hands in a prescribed manner. Then Haim cut into some loaves of challah bread, his wife and children brought out platters of Middle Eastern food, and everyone feasted.

A young Sherrf boy seated next to me asked me why I'd come, and I told him about my research into the idea of immortality. As I spoke, he poured me a small glass of purple grape cola, which he then diluted with seltzer until it was a pale lilac-lavender. "There! That's the color of immortality," he announced triumphantly.

I looked at him in astonishment.

"It's true." He smiled, genuinely.

"So you're writing about immortality?" the man opposite me said. He had a receding hairline and a yarmulke. "Want to know what I think? All religions are crap."

"Honey!" cried his wife, looking around. Fortunately, Rabbi Sherrf was deep in conversation with the two women who'd just arrived from the hotel.

"No, I mean it," persisted the man, a lawyer in his forties. "Religions are set up to keep us further away from God rather than bring us closer. Rabbis preach about our connectivity with the beyond, but what they really do is put a buffer between us and God—and they're the buffer."

"Could you please keep your voice down?" his wife pleaded.

But there was no stopping him now. "God is within us, he's in everything, not just in religious leaders. If you want to feel God, all you need to do is lift your arms to the sky and you'll feel it."

"But not everybody is like you," the wife added.

"Have you felt God in that way?" I checked.

"Yes, for sure. If you align yourself with what's above, you allow God's energy to flow in. When I feel anxious or upset, I do it. It's like a medication of sorts. And I experience its benefits."

"That's very interesting," the boy next to me said, "but if you think all religions are 'crap,' then why are you here tonight?"

"I'm here tonight because I'm strongly Jewish," the lawyer retorted, smiling. "I grew up with a Sephardic Jewish education. I want my kids to have the same upbringing I did. And we all need to belong to something, so this is my community. I know it sounds strange, but I'm anti-organized-religions while also being religious."

"Beliefs are often paradoxical," I suggested.

"The sages and rabbis have always told me 'do and then you'll hear,' *naaseh v'nishma*. It means 'follow the precepts, and then you'll understand.'"

"Or 'believe, and then you'll know'?" I asked.

"Precisely. I guess it's like a math problem. You really need to get into the crux of things and apply the formulas. But I've never been ready to make that leap of faith. Why is Moses's death written about in the tablets that 'God' gave to Moses? It makes no sense. We're not wired to have all the answers. I don't think we'll ever have the answer to why we're here until we die—and even then, I'm not sure. We're so limited; God is *illimited*. How could we ever understand?"

"When we don't understand, we believe," I said.

"I suppose so. I mean, I don't know what I'm talking about. I don't know if I'm right or wrong or if I'm stupid or going to hell. I'm not an authority, but I don't think anyone is. I don't even know what I believe. Nobody knows the Truth. Whoever professes to know is lying."

"Okay, so in the context of what I'm working on—different attitudes toward immortality—may I ask how your family spoke of the afterlife when you were a child?"

"In my family, there wasn't much talk of death," he answered, testily. "We were much more about life. My father, a religious man, believed that there's nothing after death, and that's how he raised us. 'When you die, that's it,' he told me. That's where it stops. You disintegrate back into the earth; from dust to dust. Death is just the end of your span, like what happens when a flower dies."

"So your family, devoutly Jewish, didn't believe in an afterlife."

"Not at all. But we were spiritual. I've always felt that, if you want to see God, you just need to look at a tree."

"That's for sure," added the boy sitting beside me.

"Even better"—here the lawyer picked up an orange—"peel an orange. Quarter it, and then pull off the pith and the white skin, and gaze at that flesh, at its extraordinary color, at the intricacy of all its compartments."

"It's interesting that you bring up trees, fruits, and flowers when speaking about God and death," I said. "You mentioned our end being akin to that of a dead flower. But when a pollinated flower dies, it becomes a fruit."

"Really?" He looked at his wife, doing her best to ignore us. "I had no idea."

"Yes, petals wilt and fall and die, but the flower's pollinated carpel grows into a fruit. And all trees come from dead fruits whose seeds fall into the ground. I really think that our species' ideas about the after-life came from observing those cycles in nature. If a seed goes into the earth and grows into a tree that gives flowers and fruits, we must have thought, then what happens when we go into the earth?"

"The same thing happens with water," chimed in the young boy next to me. "It evaporates up into the air, fills the clouds, and then falls down as rain, over and over again. Just look at water: it never really dies."

7

Technical Interlude: Writ in Water

The sea is back there, back in the reservoir of memory. The sea is a myth. There never was a sea. But there *was* a sea!

—John Fante, *Ask the Dust*

So, something you can only speak by saying you can't speak it. And when you did try to speak it . . . you found yourself talking water—it was the obvious metaphor—abundant, flowing, crashing water, ultimate antidote to thought.

—Tim Parks, *Teach Us to Sit Still*

ALL RELIGIONS are based on a relationship with immortality. A religion is a consolidated system of thought regarding mysteries such as the world's origins and the individual's death. Eschatology is the branch of a religion that concerns itself with the soul's final state. No religion is eschatologically unified, as different denominations posit their own interpretations of the end point, but all religions seek to illuminate the end. And they often use water when doing so.

From a Taoist perspective, each of us came from Tao, is a manifestation of Tao, and returns to Tao upon death. The way of the Tao is *li,* the path of liquidity, the watercourse way. The Sikh soul is like a water droplet that merges back into the ocean of God. The Celts spoke of Tír na nÓg, an island of eternal life beyond the western waters. Souls in the Islamic afterlife of al-Jannah, the Garden, are anointed with the water of

life; those in al-Nar, or Jahannam, face an eternit
promises an eternity in heaven or hell, with bapti
the pearly gates in rapture, or lying in state until Arma
ing the Savior's return, at which point the Kingdom occurs
earth. Judaism is primarily focused on terrestrial existence, bu
tells of resurrection, of life continuing after death, whether in pu
torial Sheol, or Gan Eden, a paradise for purified souls, or Gehenna, as Haim explained. Other Kabbalistic traditions tell of souls transmigrating through different vessels, a phenomenon called Gilgul.

Mormon weddings aren't just a simple, temporary exchange of vows: they are called sealings because each soul agrees to be fused with the other for eternity. In Jainism, the soul freed from karma savors the four infinities. Metempsychosis, a soul's cyclical journey from life to death to incarnation in a new body, is taken for granted in Hinduism, but also appears in gnosticism, druidism, Manichaeism, Orphism, and also Buddhism, notably Tibetan. (Each Dalai Lama, or "ocean teacher," is the reincarnation of a bodhisattva called Chenrezig. When a Dalai Lama dies, the High Lamas search for a new embodiment of Chenrezig's mindstream based on oracular visions at a lake called Lhamo La-tso.) In Japanese Shintoism, everything contains a spiritual essence called the *kami,* with the human *kami* sometimes compared to a fireball that becomes visible following death as it blazes its way toward eternity. Some Shintoists consider water to be capable of bringing us into direct contact with gods.

In many religions, the beginning is also aquatic. Water flows through creation narratives, from Iroquois to Yoruba, Mongolian to Incan. The sea is there first; then, a God emerges. When Eurynome, the Goddess of All Things, shimmered out of chaos, she danced lonely upon the waves. The epics of Sumeria and Babylon, whence sprang the Judeo-Christian Bible, describe the two primordial gods as entities made of water, Abzu being fresh, Tiamat salty. Blue water was the basic ingredient of creation, explained Moses. Allah created man and animals from water. Brahma, the Hindu progenitor, was self-born in dreamwater. The universe came into being when a Bindu drop coalesced into the Word.

Science, too, incorporates water into its creation myths. The Big Bang may have been the beginning of it all, but what about life on Earth? Some theories suggest that life originated in volcanic vents at the bottom of the ocean floor. But how did oceans get here? Were they always here, or were they delivered from outer space upon comets and asteroids? The world's oceans contain enough deuterium for us to hypothesize that they are made up of at least some melted comet ice. The ratios of drogen isotopes in lunar rocks are also an indication that cometary ater was battered into the Moon. The Moon itself used to be part of Earth (it is thought) until a shattering impact with a protoplanet called Theia broke it off into orbit. And despite its dusty appearance, the Moon is covered with so much hydroxyl (one hydrogen atom bound to one oxygen atom) that astronomers speculate it might be possible to extract water simply by heating the Moon's soil.

Western philosophy begins with Thales saying everything starts with water. The oldest sources of Hellenistic thought traced the generation of all life to Okeanos, the body of water encircling the globe. This river was also a god: "he from whom all gods arose." In classical antiquity, to be alive was to be in a state of watery wetness. Young people were described as "abounding in liquid." As they aged, the moisture dried up. Death was waterlessness. The expression *during one's water* referred to one's allotted life span. Richard B. Onians's analysis of ancient Greek attitudes and behavior, *The Origins of European Thought,* argues how "in this thought, that life is liquid, and the dead are dry, we have found the reason for the widespread conception of a 'water of life.'" This conception began before Athens, though, and before philosophy, even before history. It is hardwired into our hearts, as is the hunt for its this-worldly equivalent, the fountain of youth.

The hunch that some real liquid might prevent death has always been part of humanity's lore. The earliest Mesopotamian cosmogonies suggested that the fountain that cures any ailment was located at the source of all rivers. The Aleutian islanders who lived between Alaska and Siberia believed that early men were immortal: when they grew old, they only had to dive off a mountain peak into waters that renewed their youth. The Hindu idea of *soma*—a drink granting godlike immortality—is at least three thousand years old. Sikhism tells of *amrita* (*a* means "not," and *mrita* means "mortal"), a creamy liquid of deathlessness skimmed from the milky ocean of life everlasting.

Bronze Age Greeks called dead bodies "the thirsty." Parched souls ended up in a place called "the dry country." To protect themselves from ending up in a liquidless afterlife, Mycenaeans and Minoans worshipped Dionysus—a suffering god who dies, disappears beneath the waves of the sea, and then returns to life. They believed that the god's fate was their fate, that they, too, would find another life in death, that each of them was Dionysus: the one who dies yet does not die. His presence saturated early theater, with its contrasting masks representing the tragedy of death and dismemberment and its attenuating comedy of rebirth and renewal.

Ancient Greek thinkers distinguished between two forms of existence: *bios* and *zoe*. *Bios* is the life of the flesh, as in *biography,* the story of a life, or *biology,* the study of what we can know about life. *Bios* is finite. *Zoe,* on the other hand, is endless, indestructible, ongoing. *Zoe* is "non-death," untouched by mortality. We all die; a part of our soul life lives on. The *zoe* continues after the *bios* ends. "*Zoe* is the thread upon which every individual *bios* is strung like a bead," as one description has it. Even the Bible uses the term *aionios zoe* to refer to eternal life. In symbolic terms, *zoe* has wet qualities, to contrast it with the aridity of death in *bios.* The god of *zoe* was Dionysus. Worshipped as a sea deity, he was the "lord of all moist nature." Today he is better known as the god of wine and intoxication, but at the outset, he was the mystery of liquidity itself.

Actual initiation into the Greek Eleusinian mysteries required—before the tasting of death and the divine encounter—a purifying bath. The *mystai* (initiates) came to see the end as just another beginning. The inevitable could then be embraced with *amor fati,* the "love of fate." Christianity was in direct competition with pagan cults for the first few centuries, and their sacraments also revolved around aquatic immersion. Those who'd been baptized were fearless in the face of death. As Elaine Pagels has shown, the peacefulness of Christians when thrown to the lions became a major impetus for the spread of monotheism.

Just as holy water connects the faithful today to their God, and as born-again Christians submerge themselves in purifying baptismal water, in classical times water was a connection to the sacred depths. Greco-Roman sanctuaries invariably boasted flowing fountains or mineral springs. When Jesus performed a miraculous healing at the pool of Bethesda, it was still an *asklepieion,* a sort of hospital-temple devoted to the ancient Greek god of medicine, Asklepios. Sick patients seeking cures checked in and soaked in the medicinal baths. Long before Christianity, water was turned to wine in the mystery cults. In Dionysus's presence, streams would miraculously gush forth from rocks, as they also do in the Bible. In tearing animals apart and devouring them, the bacchants were doing what is done today at mass: drinking the blood of Christ and eating his body, a symbolic participation in God's pain, demise, and rebirth. Their actions had—and have—healing powers.

For much the same reasons, Indians still drink water from the Ganges. Having one's ashes scattered into that river helps secure an auspicious afterlife. The whole essence of the Hindu religion is to reassure followers that there is no reason to worry about dying—as long as the correct oblations are performed. As a result, the subcontinent is rife with rites meant to allay or alleviate believers' fears of death. Even the most devout may find themselves doubting the efficacy of their offerings, which is where mythology comes in. Stories bring an extra layer of protection. The Brahmanas are, according to Sanskritist Wendy Doniger, "myths that attempt to tame the ritual that attempts to tame the fears."

In Hinduism, death is itself merely a pit stop. In the words of Lord Vishnu, man "is not born, nor does he die. Having lived, he does not cease to exist. He is not slain when the body is slain but is unborn, eternal and ancient." The Hindu soul is an imperishable selfhood temporarily occupying a perishable body as it shuttles through successive cycles of death and rebirth (*samsara* means "to flow on") until it attains soul liberation (*moksha*) in a reunion with Supreme Reality. To arrive there is to conquer recurring death, to experience freedom from the trammels of existence.

The deathright of immortality helps worshippers come to terms with mortality. Hindu karma theory states that every person alive has lived previous lives. We are all reincarnations. Rebirth is a forgetting. What we can remember of past lives (if anything) are traces of prior experience that sometimes surface into our conscious minds. Each of these traces is called a *vasana,* meaning "perfume," an almost imperceptible reminder that we've possibly been here before.

The suggestion that there's something more to come after death is built into the fossil record. Tens of thousands of years ago, Neanderthals dug premeditated gravesites. Dead kinsmen were intentionally arranged in the fetal position, indicating some hunch about posthumous rebirth. Food was placed next to entombed bodies, so they'd have something to snack on when they awoke. And the same Paleolithic humans who painted on cave walls reverentially buried their kin alongside objects to be used in the afterlife.

In evolution, human verbalization began with grunts or signals and became a way of transmitting information, of conveying wishes and thoughts, of saying that something not presently here had at one point been here, of saying that tomorrow will happen. Talking about the past entails a shared assumption: even though it is not in front of us, in speaking of it we conjure it and believe that it occurred, that it was and therefore is real. It's a natural leap to speaking about the future, about what will happen. We began naming intangibles, using words to explain the inexplicable: *life, time, eternity*. Having coevolved, communication and credulity became codependent.

We can't survive without convictions that allow the brain to navigate time. Any relationship with the future is a belief. There are no guarantees that things will occur again the way they happened today, but we make assumptions about continuities in order to survive. Predicting is believing.

We've grown so accustomed to the sun rising each morning that we take it for granted. Early humans, who worshipped nature, considered this occurrence deeply miraculous. We can imagine them gazing out at the empurpled horizon, trying to make sense of it. If the deadening sun, extinguished each night by the sea, bursts back to life the following morning—why can't we? In 1887, E. B. Tylor, who founded the scientific discipline of anthropology, pointed out that every early civilization that contemplated the sunset near a body of water came up with their iteration of a fountain-of-youth tale. To see a similarity between ourselves and our environment is the foundation of what Tylor called "associative thinking," a way of projecting our magical fantasies onto real phenomena. We still do it. Genuine aging discoveries are being made; that doesn't mean we're about to cure the disease of aging once and for all. We have no idea *why* we age, let alone *why* we're here, or *why* our minds evolved consciousness. Such is life. And part of it is dying.

The reality of death is something we all have to face or story away, whether through science or faith. We'll never understand.

Water is a symbol of all we don't know. The ocean, with its impenetrable fathoms, contains unknowable mysteries. But it's also a source of revelation. Water represents the subconscious, the simultaneity of knowing and unknowing. "As every one knows, meditation and water are wedded for ever," wrote Melville. "We ourselves see in all rivers and oceans. It is the image of the ungraspable phantom of life; and this is the key to it all."

Water's basic attributes are already astounding: it vanishes into steam, freezes into crystals, melts into gas, and fluffs into clouds. That ice floats is an utter abnormality. Perpetually changing, water's very mutability suggests indestructibility. To our ancestors, it must have seemed capable of anything. We deemed it the *mysterium tremendum,* the substance from which all else is formed, the understandably incomprehensible essence of creation, the primal secret, the protean escape from finality.

Whatever we think of the ocean is what we think of death. Is it formless terror, the chaos of endless tumult, the violence of flux; or is it calming, entrancing to gaze upon, perfect in its limitlessness? The choice is ours. The sea is not quite a blank slate, more like an undulating surface rippling with potentiality. Gems of light, dissolving glass. We can be terrified by bodies of water, or we can revere them, but it is impossible to remain indifferent. Unending waves are reminders of our insignificance.

The motion of the ocean—it rolls uncontrollably, with a rhythm of its own, pulsating to melodies beyond our comprehension. Are we at peace riding the oceanic precipices and abysses, or does its senseless churning give us motion sickness? Whitecaps are fingers reaching out for our souls, trying to pull us under. Can we feel a sense of acceptance about those watery hands summoning us, about the inevitability, the inconsequentiality? We can't control a sea. We can only follow its movements, tremble before its swells, admire its mute constellations of silvery foam.

Nobody knows exactly when the idea of immortality-inducing waters first trickled into consciousness. Perhaps it started with the realization that there is so much water it will never run out. The illimitability suggested immortality. Gazing out at the ocean, at that unarticulated space, we must have suspected that it has always been there, that it is endless and eternal. Water grants a perception of something indestructible that reflects back the perceiving subject's own indestructibility. On the open sea, we still feel the timeless circumference of infinity around us.

Early humans couldn't help noticing that water is in everything that lives. Water allows things to grow, to thrive, to survive. It replenishes and refreshes, renews and rescues. It brings fecundity. We feel like ourselves again after drinking it, or when immersed in it. These prosaic powers of regeneration are mirrored by their poetic double: rebirth. Water keeps us alive, our ancestors realized, maybe it can keep us from dying?

Water has long been what religious anthropologists call a hierophany, an element of the physical world that reveals something of the divine world. As rain, it passes from the heavens to the earth, a nourishing messenger of the gods. Evaporating, it transmits our prayers up on high. Its transparent depths help us contemplate the inconceivable. Something about water transposes infinity into the finite.

Water is hydrogen and oxygen, but it is also a third thing that completes it, as D. H. Lawrence wrote, "and nobody knows what that thing is." That three-in-one conjunction of two H's and an O is its own unity within a trinity. Without those molecules, there is no existence. All biochemical processes occur in an aqueous medium.

The ocean boils with life, and just as terra firma bursts out of it, so did we. Humankind evolved from sea creatures. The tide remains in our blood. Babies in the womb don't breathe oxygen, they breathe fluid. The sound in the placenta is the sound of the ocean. We are all, like the world, primarily water. Water enters us, courses its way through our system, enriching and being enriched. It is then passed on to the earth. The entire biomass operates through water's circulating. It is constantly moving through the soil and the sky. It uses flora and fauna as filtration devices, as means of traveling from sea to stream, as temporary containers, rest stops on a never-ending journey. The earth is a planet-size water-transformation system. Flowing, changing, cycling through clouds to flesh and vaporizing back again, eternally dying into new life.

In Western civilization, medicine begins as a water-delivered gift of the gods. At his mineral-water temples, Asklepios would appear to the sick in dreams, offering prescriptions or banishing ailments with his magical staff. The Rod of Asklepios, still used as an insignia for the medical profession today, consisted of a snake coiled around a stick. It was a symbol of transformation, representing the way believers "undergo a process similar to the serpent in that they, as it were, grow young again after illnesses and slough off old age." Asklepios was a manifestation of Apollo, the god as sunrise. He emblematized the notion that we, too, can rise from darkness, from illness, rejuvenated by the waters. And just as we can recuperate after getting sick, Asklepios suggested the possibility of dying's being a new beginning. Healing waters and serpents shedding their skin were linked to the mystery of rebirth, itself indicative of the promise of life after death.

That premise still resonates in homeopathy. Tinctures dilute active medication to infinitesimal levels, yet the memory of water is enough to cure some patients. It makes no sense scientifically, yet here we are. We've wanted to comprehend water's powers forever. In the fifteenth century, seeking to understand the medicinal attributes of hot springs, Italian physicians called balneologists started systematically investigating mineral baths, performing experiments in hopes of explaining the waters' curative powers. (The term "spa" goes back to the Roman acronym for *Salus Per Aquam,* meaning health by water.)

Dondi, Savonarolo, and Ugolini of Montecatini each tried to develop a way of testing different springs' properties. Their efforts are what we today speak of as the scientific method—a notion not yet extant then. Their treatises, often written in the first person, are early examples of nonfiction books. But they couldn't find any satisfactory explanation for what caused the mineral baths to have healing attributes. The subject, Savonarolo concluded, "is not conducive to demonstration." It wasn't—and still isn't—something amenable to scientific grasp.* They could note observables, but these didn't explain why one spring had therapeutic effects on a certain ill patient and another didn't. Their empirical inquiries, he noted, were never fully reliable. All they could do was "give knowledge approaching the truth."

We're still in that boat today. In countless ways we can't achieve certainty; but we *can* live with uncertainty. This faculty of accepting our not-knowingness, of not needing resolution, of simply allowing reality to be all it is without attempting to reduce it to human comprehension, is what's known as *negative capability*. Keats employed the term to describe that headspace wherein one is "capable of being in uncertainties, Mysteries, doubts without any irritable reaching after fact & reason." To be in uncertainty is to be in a transitional state, inhabiting a question without answering it, embracing paradox and realizing that any response is destined to be incomplete.

Symbols allow us to navigate that uncertainty. Water is just water but its constant flow is also suggestive of souls passing from the temporal world to some eternal place. In nature, symbols of continuity are everywhere. Snakes are reborn after shedding their skin. The moon, full orbed or tending to decline, wanes and dies and comes back. The sun is

---

*In Italy, "taking the waters" is not an anachronism; it's medicine. Trips to medicinal hot springs today are covered by health care. Many Italians spend two weeks each year at thermal establishments.

vanquished by darkness, but each night perishes into daylight. The tide rises and washes out. What are seasons if not cyclical?

There are as many hierophanies as there are stars. Stars! Shining up there, so far away, stars suggest realities beyond the here and now. Whatever put the stars up in the sky, it occurred to us eons ago, can only have done so before we were around. This realization led to a sense of instability, of having been deposited into an incomprehensible world not of our making, and so the thought arose that perhaps someone, an efficient presence, a God or the gods, put them there. Seeking to regain our footing, we concluded that whoever did so must also be watching over us. By making ourselves important—the center of it all—we restored the illusion of purpose. If the stars are here for a reason, that means we are here for a reason, and perhaps we can understand that reason. We matter, we assured each other, because we matter to whatever put the stars there.

We still try to convince ourselves that the world is meaningful, but we can't, logically, unless we accept this very unknowability as a gateway to the sacred. Believing not only helps us cope with our inability to comprehend, it elevates and ennobles that seeming inaptitude. In belief, stars become holes in the firmament with paradise shining through. Stars can be heavy-lidded eyes glowing in darkness, seeing through the veil of endless night, blinking promises of daylight beyond. Stars might be what we turn into when we die. Stars are probably friends awaiting us.

Indeterminacy is a beautiful word. All our interpretations of the noninterpretable are just that: opinions. We cannot put the Truth into words or images. That hasn't deterred us from trying everything from the water of life to the starry skies. Symbols, mythologies, and religions all help us come to an acceptance of the unknowable. As the Sufis say, "A person who seeks God through logical proof is like someone who searches for the sun with a lamp."

8

The Magical Fountain

Of all the world's wonders, which is the most wonderful? That no man, though he sees others dying all around him, believes that he himself will die.

—*The Mahabharata*

You cannot go with your pitcher to this fountain and fill it, and bring it away.

—Herman Melville, *Moby-Dick*

SHORTLY BEFORE 9:00 a.m. on a blustery spring morning, the fountain of youth's gatekeeper walked into a café near Gramercy Park in Manhattan. Tilting her Armani shades up into a graying-blond bob, she scanned the room. I was reading the newspaper at a table near the entrance. A Brazilian priest attached to a thousand helium-filled party balloons had gone missing somewhere over the Atlantic Ocean.

"You must be Adam," she said, extending her hand. A graceful, slight woman in her early sixties, Martha Morano owned a public relations firm specializing in ultraexclusive luxury-travel destinations. I'd contacted her to discuss one of the stranger gems in her portfolio: a seven-hundred-acre, eleven-island archipelago in the Bahamas belonging to the magician David Copperfield.

"It's truly a fairyland," said Martha. She handed me a press kit and a rattan-wrapped stack of postcards bursting with palm fronds, white sandbars, and aquamarine expanses. "May I ask how you heard about Musha Cay?"

The answer was in my briefcase. I pulled out the moldering newspaper clipping, a two-year-old wire story about Copperfield's discovery of the "real" fountain of youth. "We found this liquid that in its simple stages can actually do miraculous things," Copperfield told the Reuters writer. "You can take dead leaves, they come into contact with the water, they become full of life again. Bugs or insects that are near death come in contact with the water, they fly away. It's an amazing thing, very exciting."

According to the report, Copperfield had hired biologists and geologists to examine the fountain's potential effect on humans. Until the tests were carried out, the magician said he was refusing anyone else access to the water. Its location—a spot where "everything is more vibrant, ageless, and full of life"—was being kept secret. Getting there wasn't simply a matter of chartering a catamaran or attaching myself to helium-filled party balloons. Renting the property cost $32,250 per day (approximately my net annual income). The only way in was right in front of me.

"I understand that David wants to protect the fountain while completing the research phase," I acknowledged. "Do you think enough time has passed that he'd be willing to speak to me about his findings?"

Martha sat there, arms folded, eyes narrowing. She'd agreed to meet me because a friend of hers was an editor of mine at *Gourmet* magazine. That didn't alter the fact that it was a touchy time in Copperfield's camp. Media relations had been strained ever since his engagement to supermodel Claudia Schiffer ended in 1999. They'd worsened in 2007, when a twenty-one-year-old beauty-pageant runner-up accused him of raping her on the island. The subsequent FBI investigation, including a raid on his Las Vegas warehouse, was widely publicized. He'd canceled shows, an entire tour. At the time, the case remained pending. (As I will discuss later, the investigation was ultimately dropped.) Those tabloid reports hadn't escaped my notice, but my mission was the fountain.

"I don't know if David will go for it," Martha said, adjusting her fringed cashmere stole. "When he went out there with that fountain thing, he was scorned by the media."

"I don't intend on ridiculing him. But there's clearly a story here."

"David would agree with that—the island is all he thinks about."

David reminded me of Prospero, the deposed ruler and sorcerer exiled on his tropical island. Only he wasn't a Shakespeare creation: he was an aging prestidigitator embroiled in scandal who'd issued a press release about finding a miraculous body of water on his private island. He hadn't yet hung up his flying shoes, but his days of walking through the Great Wall of China were over. The crown duke of late twentieth-century magic had been replaced by edgier pretenders to the throne like Criss Angel and David Blaine. It had been years and years since he'd made any national monuments disappear on a CBS special.

As kids, my brothers and I would gape at videos of David Copperfield dangling upside down, ten stories above a pit of burning spikes, freeing himself from a straitjacket, from the confines of insanity. He looked longingly into the camera as though into a lover's eyes. Somehow, his expensively coiffed mane always seemed to be billowing in the breeze. Even though the lothario shtick was cringingly corny (rival magicians griped about his act's being to entertainment what Velveeta is to cheese), we'd still get all goosey when he came on the TV. Whatever anybody thought of him, during those filet years, it was impossible not to watch his latest stunt. Copperfield went beyond good or bad: he was something we grew up with, like microwaved hot dogs or Bazooka Joe comics. As dubious as that Reuters report may have been, the mental image of brown leaves turning green in the fountain had a hold on me. I wanted to believe that he'd found something important.

"Did he really hire scientists to study the liquid?" I pressed Martha. "It says here that it 'rejuvenates simple organisms.' Is he serious?"

"He's very serious about this," she said. "It's about a belief that there's something other than science."

Copperfield's conviction that some aspects of life are impervious to scientific inquiry was a sentiment I—and everyone else I'd spoken to for this book—shared. The accomplishments of science over the past few centuries have led to the perception that believing is a primitive means of thinking, an anachronism. But rationality cannot access the unseen. We want to believe that no aspects of life are resistant to proof—that we simply haven't yet devised the proper means of testing and then understanding them. Some physicists still speak of formulating an all-encompassing theory of everything, as though the entire universe could be reduced to a single equation. There's a utopian fantasy that, with enough rigorous scrutiny, everything everywhere can be explained. Such scientism scoffs at belief without realizing its own faith-blindered shortcomings.

Civilization has indisputably accumulated vast databases of findings, but never before have we been so perplexed about what it all means. Despite our fetishizing the assumption that everything is ultimately knowable, countless aspects of the human experience will forever remain under the jurisdiction of belief. Time and consciousness are indefinable, and nothing in that ever-widening body of knowledge can tell us what happens, if anything, when we die.

To be human is to be preoccupied with hopes of comprehending why we're here and who we are, but the deeper we delve into any avenue of inquiry, the more fathomless it all becomes. At a certain point, we have to confront the realization of how little we actually know; we have to take that leap—to the fountain.

"To be perfectly honest with you, I've never seen the fountain," Martha admitted. Her body language intimated that it couldn't be real, but she didn't know for sure. I wavered. Yes, Copperfield had amassed a fortune by making people believe that the impossible just might be real. But this seemed different. "It isn't one of his magic tricks, I can assure you of that," Martha added.

We're all programmed with a need to believe, which is why magic can still enchant us, even when we're aware it's an illusion. By suspending our disbelief, an entertainer permits us to switch from our logical, empirical, analytical framework into a sense of awe, a sense that is also at the heart of religious worship. The make-believe can make us believe. To counter this susceptibility to suggestion, we fortify ourselves with cynicism. Yet beneath our skeptic's armor, what we all seek, whether in books, movies, love, drugs, dreams, consumer goods—or God—is a glimpse of the beyond, another world, something greater than ourselves.

Martha and I started speaking about the idea of belief in a secular age, about the ways knowledge and belief seem to be opposites, about how we only believe in things we can't know.

"And what's the difference between faith and belief?" Martha asked sincerely.

"I think of beliefs as personal truths that cannot be verified or unverified," I said, trying to sound believable. "And I suppose faith is a ritualistic, repetitive practice that deepens those beliefs."

"But you can have faith in someone," she interjected, uncertainly.

"You're right." I nodded, treading semantic water. "Like having faith in me and this story and letting me visit the fountain?"

"Well, you do have a different way of seeing this." Martha got up to order a coffee. "But David's the one who decides. You'll have to write him a proposal. All I can do is make sure it gets into his hands."

As Martha stood in line for our coffees, I opened the glossy folder full of promotional materials she'd brought along. Musha Cay was billed as "the most private private island experience in the world." Copperfield claimed he'd found it by drawing a line from Stonehenge to the statues of Easter Island and another line between the Great Pyramid of Giza and the Pyramid of the Sun in Teotihuacán. The lines intersected at the exact latitude and longitude of his Caribbean hideaway.

In aerial photographs, the main island resembled a bat with its wings outstretched. The archipelago's other outposts had names such as Isle of Wonder, Alchemy Bay, and Forbidden Island. The information kit noted that each of the five guesthouses has its own private beach. Forty beaches are scattered throughout the eleven islands. The main house, where Copperfield spends much of his free time, is a ten-thousand-square-foot mansion perched like a black eagle on Musha's summit. Having purchased the cluster for $50 million in 2005, he started renting it out soon thereafter. Google owner Sergey Brin married Anne Wojcicki, cofounder of a personal genome sequencing service called 23andMe.com, on one of the island's sandbars. Other high-profile visitors included Oprah Winfrey, Bill Gates, and various socialites able to spend six figures on a weekend getaway. The rental fee, soon to be increased to $37,750 per night, entailed a four-night minimum stay and allowed up to twelve guests. There were enough beds to sleep twenty-four; bringing an additional twelve people raised the total daily rate to $50,250 (plus a 5 percent surcharge for Bahamian room tax). Not included in the booking fee: international phone calls, fireworks, steel-drum concerts, or treasure hunts. The brochure recommended traveling there by private jet or yacht. If necessary, one could always take a commercial airline to George Town. From Exuma International, it's an hour-long speedboat jaunt or fifteen minutes on a Super Twin Otter amphibious aircraft.

The pamphlets referred to Copperfield as "the master." Because he lived there when not on tour or in Vegas, the master was available for meet and greets. "Guests if they wish can ask for some time with him, just to chat or, if they're lucky, to experience a slight [*sic*] of his hand. Should the guests prefer, the master will be as invisible as he becomes in his international shows."

"I'd love to know the real story behind that fountain," Martha said, returning with espressos. Getting caffeinated, she started steering the discussion into questions of mythology, history, and culture. We spoke about Kublai Khan, whose stately pleasure dome in Xanadu cast its shadow over a mighty, sacred fountain. That segued into talk of Michael Jackson's Neverland, about the lumpen fascination with unfettered wealth and fame, how the scrutiny must be so warping. "David just escaped," she said in a hushed voice. "And Musha is where he escaped to. I really see him as an other-dimensional character because he doesn't live in this world. He occasionally steps into this realm in order to manipulate it somehow, but he left this life a long time ago and has been living in another one. He's just not here."

Finishing her coffee, Martha mentioned that she'd seen *Armida,* the Rossini opera in which the sorceress turns a forest into a pleasure palace.

"And that reminded you of Copperfield's island?" I asked.

"Yes, although there's an evil component there, in the opera, so you probably don't want to go there . . ."

I raised my eyebrows.

". . . or do you?"

"I'd just like to describe Musha and the fountain, as they really are, without diluting anything."

It was clear that she'd help. But the master still needed to be sold. Everything hung on a letter.

"It really is so bizarre down there," she said, leaning in. "And *he's* bizarre beyond belief. He's such a bizarre person. He always has a bevy of . . . strange characters . . . with him. You know, David *is* the fountain of youth."

9

Letters upon Letters: Dividing the Invisible

What is the unknown force which lies within your mysterious steeds?

—Nikolai Gogol, *Dead Souls*

Come, ye wheeling cumuli, ye clammy condensations, come!

—Aristophanes, *The Clouds*

IN 1165 CE, a letter arrived in the hands of the Byzantine emperor Manuel I of Constantinople. A grump who favored jewel-bedecked robes of imperial purple, the emperor likely read the note with the same distrustful frown he sports in gold-leaf portraits. The letter contained world-changing information. Copies had also been sent to Pope Alexander III, Holy Roman Emperor Frederick I, and various other European monarchs, all of whom were equally startled by the epistle's contents.

The sender claimed to rule the Orient. His empire, centered in India, extended westward to the Tower of Babel in deserted Babylon and all the way to the wastes beyond the sunrise. His realm was called the Magnificence. A direct descendant of the magi from the Gospels, he boasted of surpassing everyone else in virtue, riches, and power. He even owned the fountain of youth.

His name was Prester John, and he was offering to help.

The letter arrived shortly after the failed Second Crusade, which ended with defeat in the orchards of Damascus. Christendom's plenipotentiaries were desperate. Even though Prester John dared to suggest that his Magnificence was closer to God than the Vatican, news from a Christian chief reigning somewhere on the far side of the Holy Land sounded heaven-sent.

Europe's leaders had been alerted to the correspondent's existence about twenty years earlier, when a Syrian bishop arrived bearing tales of an emerald-scepter-wielding Eastern despot. Prester John had already taken Persia, the bishop enthused, and he wanted to assist the Crusaders in their efforts to control Jerusalem. Unfortunately, his army hadn't yet found a way to ford the Tigris.

The arrival of the letter reinforced what everybody at the time believed: that paradise was a place on earth, somewhere around India. And as Prester John's detailed inventory of his resources attested, something heavenly was indeed bubbling out of those sacred lands: "From hour to hour, and day by day, the taste of this fountain varies; and its source is hardly three days' journey from Paradise, from which Adam was expelled. If any man drinks thrice of this spring, he will from that day feel no infirmity, and he will, as long as he lives, appear of the age of thirty."

Prester John's kingdom also contained other marvels, including cyclopes, horned men, and people with eyes on both sides of their heads. Rivers brimmed with gemstones. There was a sandy sea with no water. "The sand moves and swells into waves and is never still," he wrote. "It is not possible to navigate this sea by any means." He told of citizens who weren't afraid of death. When one died, their compatriots would eat the corpse. At the top of a spiral staircase built out of lavender crystal was a speculum that reflected all the happenings in the provinces under his purview.

The letter created a furor. On September 27, 1177, the pontiff Alexander III addressed a papal response to Prester John and sent his physician into the unknown to deliver it. Dr. Philip was never heard from again. King John II dispatched men fluent in Oriental languages through Abyssinia and Egypt to find the Magnificence. Vasco da Gama set out carrying letters of introduction addressed to Prester John. In an attempt to establish contact, Bartolomeu Dias ended up discovering the Cape of Good Hope. A big reason the Fifth Crusade flopped is that Cardinal Pelagius and other field marshals gambled that Prester John or one of his descendants would bail them out.

Even Marco Polo was convinced of his existence. *The Travels,* undertaken more than a century after the letter made its rounds, described the regions conquered by Prester John and told of his descendants still reigning in Tartary, somewhere between Siberia and Manchuria. While visiting the court of Kublai Khan, Polo learned of battles ostensibly waged between Prester John and Genghis Khan (Kublai's grandfather). But the Venetian explorer never actually saw Prester John's wonders. Awed by the sight of Kublai hunting game with a pet leopard, he was less into fact-checking than relaxing in Xanadu, its golden palace overlooking grounds dripping with opulent fountains.

Another thirteenth-century visitor to the Mongols, however, cast doubt on Prester John's veracity. William of Rubruck suggested that the whole yarn may have been spun by Nestorians, a sect of Christians exiled from Europe since the Council of Ephesus in 431 CE. Condemned as heretics for purportedly believing that Jesus Christ was actually two men—one fully human and the other fully divine*—the Nestorians spread their teachings outside the Roman Empire, throughout the Middle East, into India, China, and Mongolia. Rubruck's encounter with them in the Far East left him convinced the Nestorians had fabricated Prester John and the fountain of youth: "For this is the way of the Nestorians who come from these parts: out of nothing they will make a great story."

My letter to David Copperfield didn't mention Prester John's letter. Instead, it recapped the conversation Martha and I had had about knowledge and belief, emphasizing her point about the limitations of science. I described a journey that would explore the fountain's symbolic value, and the nature of belief itself. "Stories, now more than ever, grant access to deeper truths, as Mr. Copperfield has shown us time and time again," I wrote, suitably sycophantically. To tell the story of the master's Caribbean discovery, I proposed spending a few days on Musha Cay with him. He could teach me about the fountain, and I would document what it meant to him. The story of its discovery was begging to be told. "If possible," I added, "I would like to speak with some of the biologists and geologists investigating the islands' waters."

---

* Nestorius was charged with "dividing the invisible."

While trying to understand the links between Prester John's fountain of youth and the waters of eternal youth alluded to in the Qu'ran, I read about the origins of Islamic mysticism in an ancient text called *Drops from the Fount of Life*. I also perused the writings of Farīd al-Dīn Attar, an esteemed Sufi poet, chemist, and perfumer. "If you can drown in a drop of water," Attar wrote, about our inability to understand certain things, "how will you go from the depths of the sea to the heavenly heights? This is not a simple perfume."

Attar's most famous work, *The Conference of the Birds,* tells of a mysterious prophet called Al-Khidr. "When you enter into the way of understanding, Khidr will bring you the water of life," he wrote. Sufis consider Al-Khidr, also known as the Green Man, to be an immortal being. He drank from the fountain of youth five thousand years ago, and he still walks on earth. He's the teacher of all prophets and messengers through the centuries. He has the power of multiplicity: he may appear in different shape, with a different face, in many different places at the same time.

In hagiography, Al-Khidr, the Hidden Guide, is deemed the patron saint of the Sufis. An emissary from the unknown, he appears, transmits a divine message, and then vanishes from cognition. Those praying fervently without receiving any outward response might encounter Khidr in dreams. He emerges from a thick tangle of foliage and explains that one's longing *is* the reply from God.

When he alights upon barren lands, they turn verdant, hence his moniker: the Green Man. He is a kind of spiritual fertilizer. "He is at once the guardian and genius of vegetation and of the Water of Life," explains the Indologist Ananda Coomaraswamy. Khidr moves with greenness, bringing rain and growth to arid regions. In Carl Jung's essay "Concerning Rebirth," he considers Khidr emblematic of the individuation process's goal of psychic transformation and self-realization.

In the Qur'an, a Khidr figure teaches Moses difficult lessons. But Khidr is best known from mythology as the old man who accompanies Alexander the Great into the Land of Darkness in search of the Water of Life. Khidr drinks from the spring, but Alexander doesn't. Instead, the king learns that no matter how insatiable his ambitions may be, he is destined to die. "You will find satisfaction only through the earth, when it covers you," Khidr tells him.

Nizami's *Iskandar N ma* decodes this story as being an allegory about grace. Alexander has misinterpreted the meaning of the fountain, thinking it is an actual thing when it is but a symbol. Its metaphorical waters arrive unsought, bringing with them a sense of rebirth and renewal. It's what Father Gervais called the life force.

In the Sufic system, there's a name for grace: *baraka.* It is a blessedness that descends, deservedly or not. It arrives from on high, a something given, a messenger of sympathies. Lent only for uncertain moments, it's not perpetually available, yet it is imperishable. It is an example of what Sufis call *tajalli,* a manifestation or "shining through" of the sacred into this dimension. It is an infusion that whips through our world, uniting the divine and the profane. And within each of us, distinct from the self, is a symbolic splash of bubbling water that connects to its source through the medium of *baraka,* dispensed by Khidr.

Grace, like the fountain of youth, is what scholars call a mythical signifier, something immaterial and full of meaning whose form can be envisioned as a way of approaching the even more meaningful Truth it refers to. Stories show this rather than telling it. In *The Conference of the Birds,* the parrot, dressed in his brilliant coat of green feathers, speaks of himself as a Khidr among birds. But even he cannot get to the fountainhead. "I should like to go to the source of this water," admits the parrot, "but the spring of Khidr is enough for me." Perhaps acknowledging grace when it alights in our lives is also enough.

As Father Gervais hinted at, the water of life spoken about in the New Testament, and alluded to on Auntie Tiny's funeral invitation, is itself a metaphor for grace, something vague yet tangible: "Whoever drinks of the water that I shall give him will never thirst; the water that I shall give him will become in him a spring of water welling up to eternal life."

Alongside being a prophet in Islam, Khidr has been linked to the Green Knight of Arthurian literature, to the Wandering Jew in medieval lore, and to dragon-slaying Saint George. He is also venerated in Hinduism, where he is known as Khizr, or Jind Pir, the Living Saint. Khizr's main shrine is located on the Indus River, and he has long been the divinity of the Bengal boatmen. Some worshippers consider him an incarnation of the river, a protean water-god who assumes human form, among other guises. Leaving his watery abode, he can also appear as a ball of colored light, mist, a presentiment, leaves shimmering on branches. For Hindus, as for Muslims, Khizr's role is to lead seekers deeper, to demonstrate the Marvelous to those of little faith, to thrust upon the needy silken bolts of grace. He initiates solitaries. He brings a new mode of knowing. He embodies the flowing reality of something metaphysical. He is the waves.

Beyond Khizr and the Indus waterway, several other South Asian rivers have Hindu deities attached to them, as well as various apotropaic attributes. The sacred Ganges plays a part in helping release souls from the cycles of rebirth, and the nearby Yamuna River can even, it is thought, exempt some believers from the necessity of dying. In Brahmanic tradition, the semilegendary Sarasvati River* is the site of a particular fountain, called *saisava,* or the Place of Youth, which restores bathers to whatever age they desire.

Perhaps the earliest example of a fountain of youth in history, the *saisava*'s origins predate the historical record. Sacred waters figure in the earliest Vedic hymns, and Brahamanic tales about the Place of Youth were already in circulation twenty-nine hundred years ago. Scholars who've traced the fleeing fountain back through time agree that the written Place of Youth stories were certainly preceded by an oral tradition in India.** Even the word *fountain* seems to come from there: its winding etymology seeps from a delta of possibilities, but one of its primeval tributaries is indisputably *dhanvati,* the Sanskrit word for "flow." If that weren't complicated enough, bathing in the *saisava*'s flow merely has a rejuvenating effect; living forever requires drinking another beverage, called soma, celebrated in the *Rigveda* circa 1500 BCE.***

These are, of course, mythologies. If we believe in them and practice their attendant rites, they accrue meaning. Hindu scholars have a name for such stories: *terum*. These are "stories about something that could not or should not be true." All immortality stories fit into this category. We know that we have to die; *terum* tell us that there may be ways to get around the inevitable. "When what happens in the myth is not physically

---

*Which may have overlapping connections with the Indus.

**In a 1905 essay for *The Journal of the American Oriental Society,* Yale professor E. W. Hopkins concludes that the idea of a Fountain of Youth derives from India, but also notes (regretfully) that his sixty-eight-page attempt to pinpoint the source of the fountain, "though longer than at first intended, is yet still too short to be definitive."

***A beverage of immortality was also spoken of by Zarathustra, the Persian prophet who lived around the same time the *Rigveda* was composed. His teachings formed the core religion of Iran until Islam arrived in the seventh century. Their elixir is *haoma,* best drunk at the end of time near the junction of the great gathering place of all the waters.

possible in this world," writes the Indologist Wendy Doniger, it alleviates our fear of the unknown. "It enlarges our sense of what might be possible. Only a story can do this."

History, too, is always a kind of story. In the history of Western history, miraculous Eastern waters have been there from the start. Herodotus, in the first ever book of occidental history, told of his journey to the distant orient, where he found a violet-scented spring that kept those who bathed in it alive to an average of 120 years.

In the fourth century BCE, another Greek historian, Ctesias of Cnidus, wrote *Indica,* a compendium about the wonders of India. A physician for the Persian court, Ctesias claimed to have reliable information on India, even though he'd never been there. He distinguished between his own meticulous, honest reporting and that of his predecessor Herodotus, whom he characterized as a peddler of deceptions. *Indica* states that the average Indian lived to be 120. A longer life, he said, was somewhere between 130 and 150, with the really elderly making it to 200.

Ctesias's writing overflows with miraculous fluids. His India had rivers of honey, lakes covered in cooking oil, and a stream whose water, when drawn off, fermented into wine. Another fountain's water curdled into an aquatic cheese that could be used as a truth serum; one wonders what would've happened if Ctesias's successor Theopompus had consumed it and was then interrogated about his fabulations concerning Meropis, a land he claimed lay beyond India. It was crisscrossed by two rivers, he wrote: one of pain, which made you die in anguish and misery; the other of pleasure, which killed you in ecstasy. People entering its waters would forget everything and grow younger and younger until they became a baby and were then "quite used up." Then they'd die.

In a linguistic coincidence that suggests a familiarity with ancient Indian texts, Ctesias mentioned a river that could cure leprosy, called the Hyparchus, which meant "bearing all good things." India's Sarasvati River, mythical site of the fountain of youth, was said to "provide all good things." In Vedic lore, it, too, could cure leprosy and other ailments.

Living alongside the banks of this Hyparchus, continued Ctesias, was a race of men with dog's heads who spoke in barks. There were also pygmy shepherds who tended miniature flocks, people with ears dangling down to their feet that they used as blankets, and griffins that lived for a thousand years. Past them, in the very north, dwelt an ageless people called Hyperboreans (*hyper* means "beyond," *boreas* "the north wind").

These foreigners captured the early Western imagination. Hesiod, Homer, and other poets seemed to understand that such places weren't actually real. As Pindar wrote, "neither by ship nor on foot would you find / the marvelous road to the assembly of the Hyperboreans." But others considered them a genuine people. Greek ethnographers disputed the precise whereabouts of a mythic continent called Uttara Kuru ("beyond the north") whose sap-drinking populace lived for ten thousand and ten hundred years. The *Puranas* said they existed and they didn't exist. Euhemeros wrote of a healing spring called "the water of the sun." It was located near India, on an island where gods like Zeus, Hermes, and Apollo had at one point lived as actual human beings. From him comes the term *euhemerism,* which means interpreting myths as accounts of historical persons and events.

History started out messy, and as time went on and more historians joined the fray, the line between the factual and the mythical got increasingly blurred.

Once Alexander the Great actually made it to India in 326 BC, that land's wonders ought to have been relegated to the Sci-Fi/Fantasy shelves. Instead, stories about the waters of immortality became even more widespread. The king of Macedonia traveled with chroniclers ostensibly charged with documenting his advances. Their accounts of India were as outlandish as ever. "All those who wrote about Alexander preferred the marvelous to the true," admonished Strabo, centuries later.

Alexander's official historian, Callisthenes, claimed that the sea prostrated itself before them, allowing safe passage. That book, like every other contemporaneous text, is now lost. What did survive, however, is a fictional and fantastical historical narrative known as *The Alexander Romance*.

In the oldest extant versions of the *Romance* (the earliest of which appears to date from the third century CE), Alexander's foray into India doesn't have anything to do with the Waters of Life. As of the fifth or sixth century CE, recensions of the tale depict Alexander searching for the fountain of youth in the Land of Darkness beyond the northern limits of the world (with Khidr as his guide). These stories weren't yet known in Europe, and they are what inspired Prester John's letter. It's not quite clear how the fountain trope entered the picture. From the start, however, *The Alexander Romance* does explore the theme of immortality in India.*

At one point in the original story, Alexander bumps into some Indian Brahmans. "Ask me for whatever you like," he tells the cave-dwelling gymnosophists.

"Give us immortality, that we might not die," they respond.

Alexander explains that he cannot grant them eternal life, as he himself is merely mortal. Shortly thereafter, he encounters a god named Serapis who informs him that his name will live on forever. This "dying and yet not dying" is the only immortality available to him. Alexander, wanting some here-and-now answers, demands to know when he'll actually pass on. The oracle refuses to indulge him, explaining that it's better not to know: "to be ignorant of it brings the secret forgetfulness of not remembering that one is ever going to die."

In Europe, *The Alexander Romance* became one of the most popular stories from the tenth century until the fourteenth century (only the Gospels were translated more times). Its success spawned countless fairy tales employing the Indian motif of rejuvenating waters. Variations on the *fons juventutis* appeared in numerous pre-Renaissance fables, from *Swan Knight* to *Le Bestiaire* of Philippe de Thaun. It ended up in the land of Cockaigne as well as *The Arabian Nights*. Ballads about people throwing themselves into magic springs and coming out transformed were on heavy rotation for traveling minstrels.

Just as heads of state did when confronted with Prester John's letter, many people continued confusing a divine Hindu construct with literal reality. Others stepped in to exploit that credulity. The story of the fountain of youth—always a story about belief—became a story about our willingness to be deceived, and of the lengths some go to deceive others.

In the fourteenth century, Sir John Mandeville published his *Voiage and Travails,* a travelogue in which he claimed to have actually drunk

---

* *The Alexander Romance* depicts India as a place boasting fleas as big as frogs and ship-size lobsters that could swat fifty-four soldiers off a deck with one flick of the claw. The human inhabitants were taken straight from Ctesias's imagination, including a race of headless people with eyes and mouths on their chests. Some men were lion-faced, others had a half dozen feet, and there were giants whose hands ended in serrated knives.

from the fountain of youth on the Malabar coast of India. A smirksome jester in pointy shoes, Mandeville was thoroughly unreliable, claiming, for example, that cotton came from tiny sheep that grew on the end of plants and that watering diamonds makes them grow bigger. His Welle of Youthe smelled like every spice imaginable, and its taste changed every hour. "Whoso drynkethe three tymes fasting of that Watre of that Welle, he is hool of alle maner sykenesse, that he hathe. And thei that duellen there and drynken often of that Welle, thei never han sykenesse, and thei semen alle weys yong."

Fountain-mania mounted when the Americas were discovered by Europeans looking for a shortcut to India. If we can discover a new world, the thinking went, why, surely we can find the fountain of youth as well? Conquistadors were instructed to keep their eyes peeled for any "sprynge of runnynge water [that] maketh owld men younge ageyne." Medieval cartographers inscribed the words *Here nobody dies* next to Insula Jovis, the Island of the Young, located somewhere in the Atlantic. Christopher Columbus thought he'd found the site of paradise when he spotted the Orinoco delta in Venezuela. "I believe that this water may originate from there [Eden]," he wrote. "I believe and I still believe what so many saints and holy theologians believed and still believe: that there in that region is the Terrestrial Paradise."

In the Age of Exploration, the West Indies were considered part of South Asia, which intensified the hope of stumbling upon paradise. Most Europeans were convinced the fountain was somewhere in the Bahamas. "And here I must make protestation to your Holiness not to think this to be said lightly, or rashly," Peter Martyr d'Anghiera wrote to Pope Leo X, "for they have so spread this rumor for a truth throughout all the court, that not only all the people, but also many of them whom wisdom or fortune hath divided from the common sort think it to be true." D'Anghiera, who coined the term the New World, said people thought the fountain was somewhere on the island of Boynca—not far from David Copperfield's place.

Long after the discovery of the western hemisphere, rumors about the fountain's possible location persisted. There were rumors of revivifying waters known to the Maasai people in the steppes east of Kilimanjaro. Others went looking for it around the mouth of the Nile. Perhaps it was in the South Pacific? In 1831, a missionary's collection of *Polynesian Research* told of the *wai ora roa*: a life-giving fountain in the Pacific that healed any internal malady or external ailment. Its salutary waters also granted immortality. The mythologies of the region told of idyllic spirit-lands in the underwater ocean. Samoan chieftains looked forward to a place called Pulotu, where the "water of life" bestowed eternal vigor. *The Sexual Life of Savages in North-Western Melanesia* describes an island called Tuma, where the inhabitants bathe in a spring called *sopiwina* (washing water) whose brackish liquor morphs them into kids. This can be done repeatedly, allowing for a never-ending return to childhood. "When they find themselves old, they slough off the loose, wrinkled skin, and emerge with a smooth body, dark locks, sound teeth, and full of vigor. Thus life with them is an eternal recapitulation of youth with its accompaniment of love and pleasure."

Nowhere had as much potential as America, though. The *Nineteenth Annual Report of the Bureau of American Ethnology,* published in 1900, recounted the Cherokee tale of an invisible lake near North Carolina and Tennessee that could cure any wound. Unfortunately, the report concluded, "no man can see it." Five years later, two springs in Florida alone were claiming to be the veritable Fountain of Youth sought by Ponce de León. One of them was around Ocala; the other was Green Cove spring. When I started looking for it, the fountain that got the most Google hits was located in the town of St. Augustine, Florida. While awaiting Copperfield's response, I decided to find out whether it was worth a sip.

10

Almost Real

"You see," he went on, "it's very much like your trying to reach Infinity. You know that it's there, but you just don't know where—but just because you can never reach it doesn't mean that it's not worth looking for."

—Norton Juster, *The Phantom Tollbooth*

The history of America begins, like that of the Ancient World, with legends in which it is not easy to recognize the exact proportion of reality and imagination.

—Leonardo Olschki, "Ponce de León's Fountain of Youth"

THE VOICE on the other end of the line sounded pleasantly corporate: "Fountain of Youth, this is Michelle Reyna speaking."

"Hi, Michelle," I said. "Could I please speak to the fountain's media department?"

"Why, that would be me!" she said, her excitement suggesting a dearth of such requests. "I'm the creative marketing director and special events coordinator for the Fountain of Youth. How can I help you?"

I told her I was a writer considering a visit.

"If you come to St. Augustine, I can promise you'll learn everything you ever wanted to know about the Fountain of Youth," Michelle explained. Clearly, this fountain would be a lot less challenging to see than David Copperfield's.

Michelle embarked upon an account of the fountain's origins. "It was opened over a hundred years ago by Diamond Lil, based on evidence she uncovered that Ponce de León—who came looking for an eternal spring—found it right over here."

"So Ponce de León actually found the fountain in St. Augustine?"

"Absolutely. Although some people claim he may not have made it this far north. We agree to disagree with those people. In either case, because this is known as *the* Fountain of Youth, and it has been here for over one hundred years, we're going to say that he was here, because no one else is taking the lead on it. If no one else can produce any evidence that he landed elsewhere, then why not?"

Her argument had a kind of simplistic complexity that momentarily suspended my critical faculties. I just wanted to keep listening to her speak. "So why was Ponce looking for the Fountain of Youth?" I asked.

"As far as I know, he heard about it from some Indians, and then King Ferdinand sent him to find it. You know, Ferdinand was being fed arsenic—their version of Viagra at the time—trying to get healthy enough to make a male baby—and so Ponce set sail. We have some historians here who'll be able to tell you more about that."

She recommended a handful of authorities I'd be able to interview, including the fountain's owner, somebody from the city's tourism department, and an amateur archaeologist. The number one historian she wanted me to meet was named David Nolan. "Some institutional historians look down their nose at him because he doesn't have a PhD, but he's extremely charming," explained Michelle. "He's more of a storyteller and less of an academic. He's very popular. And I'll also give you a personal tour of the property. Did I mention I do occasional work as a costumed storyteller? I *become* Diamond Lil."

"Diamond Lil—you mentioned her name earlier. Who was she?"

"Doctor Lil was quite the little lady," continued Michelle coquettishly. "She came here from Alaska with a diamond in her teeth. She had lots of jewelry and ermines, so you can imagine what it was like when she blew into town. She kind of put on airs. Lil would tell people that she was related to Napoléon."

"Was she?"

"She said she was. But we never found any genealogical evidence. She did own one of Joséphine's evening gowns, made of spun silver and mother-of-pearl, which was a gift from Napoléon."

"Just so I'm clear on this: Diamond Lil found the Fountain of Youth?"

"*Ponce* found the fountain," Michelle said, her benevolent rebuke making me feel like a kid. "Lil *re*discovered it and opened it to the public. She found the cross and the parchment proving that Ponce had been here. Then she started selling the water for twenty-five cents a glass."

Michelle and I agreed to meet in a few days. In the meantime, I started researching. A government website devoted to important Floridians noted that Diamond Lil "fabricated stories to amuse and appall St. Augustine residents." Her 1906 memoir is unrelentingly melodramatic. Brawls, murders, hangings, fits of spleen, apoplectic rages, rascality—*Tragedy of the Klondike* has it all, including a notarized statement testifying that none of it is false or exaggerated.

She had traveled to the Klondike during the gold rush, wearing sealskin coats and fox-fur robes. Her companion—a massive Saint Bernard named Prince Napoleon—saved her from drowning in icy waters but was then swept into a whirlpool. When panning in the bullion fields didn't pan out, she became a doctor, a hotel owner, and a treasurer overseeing hillocks of gold dust. She had affairs with men who'd found walnut-size nuggets and married a hustler named Edward "Easy Money" McConnell.

The money may have been easy, but life in the Arctic was hard. She was accused of libel and of robbing sluice boxes. Once she found steady work, she said, she had to endure frequent poisonings. On one occasion, she was stabbed with a veterinarian's horse hypodermic full of cyanide. On another, she was fed raw eggs laced with arsenic. Her stomach rapidly rejected them: "When they came up, the whites of those eggs were really cooked."

The story of a woman with a diamond in her teeth rediscovering Ponce de León's Fountain of Youth and selling it for twenty-five cents a glass seemed to make even more sense once I came across a passage in her book about gold rushers paying twenty-five cents a bucket from a waterhouse in Dawson City. Diamond Lil had actually sued the Canadian government to remove the well on the grounds that it was a public nuisance. Although her lawsuit flopped, she seemed to have learned something up there about selling water for a quarter.

Almost every other document I came across regarding Lil was peppered with contradictions. Nothing about her was certain—not even her name. Official papers referred to her as Dr. Louella McCollum until somewhere shortly before her death in a car crash, by which point her name had been changed to Luella Murat Day (to reflect her direct link to the Napoleonic Murats). When she died, two copies of her will surfaced: one leaving everything to her ex-husband Edward McConnell, the other leaving everything to her ex-husband's blind brother. The officiating judge claimed that he had never before been associated with a case with so many multiplying complications.

The only thing that seemed certain was her eccentricity. She placed classifieds offering a liberal reward for any information leading to the capture of her forty-five-year-old gardener (he disappeared, the ads explained, while sailing off to get seed potatoes). She was arrested for shooting at a policeman she believed had tried to feed her tainted apples. Writer Theodore Dreiser's Florida diary from 1925 recounts the time she insistently told him how the new Messiah was going to land on the property—presumably the same way Ponce had. She told Dreiser that people were trying to poison her with watermelons, gas, and Coca-Cola. Once she was dead, she said, they'd wrest the property from her hands.

Legal documents characterized her as a meddlesome paranoiac with a complex of self-pity. Her occasional public-speaking engagements were described as "the maunderings of a disordered brain." In a 1909 speech at the St. Augustine Tourist Club, she claimed to have proof that local families were being killed for their land. In the talk, she told of being unafraid of death, and of being able to read the shadows reflected from seditions coming from the four quarters of the globe. The best part was her opening rhetorical salvo: "There are three things, my Christian friends, which I have not got. I have not got Insanity. I have not got Hysteria. And I have not got any Ladylike complaints. But what I have got is indisputable evidence which I brought with me from the Rock of Gibraltar in 1908 . . ."

As soon as I arrived in St. Augustine, I was lost. The map I'd printed brought me to a defunct beauty salon called the Fountain of Youth Spa. One sign said TAKE YEARS OFF YOUR FACE—WE HAVE THE ANSWER! Next to it was another: OUT OF BUSINESS.

After passing signs for the oldest schoolhouse, the oldest jail, the oldest pharmacy, and a host of other "shoppes," I stopped to ask for directions. A bearded man decked out in galleon attire handed me a flyer for some *ye olde tyme* butter-churning events and gave me a complicated set of instructions to go the wrong way down several one-way streets. "It is confusing," he said. "This is St. Augustine."

After driving along a canopied avenue, I pulled through the arches of the Fountain of Youth National Archaeological Park. At the end of the parking lot, beneath elms draped with Spanish moss, two male peacocks were going at it. Formidable plumage fanned out, Mohawked heads bobbing, they circled each other and then charged, butting chests in midair, swirling into a metallic-purple cloud of beaks and talons. One ran out of the dust, shedding feathers and cawing *oww oww owwwwww!* The victor strutted daintily along toward a gaggle of peahens down the lane.

A stocky, redheaded woman in a gold silk top waved at me from across the lot. She was standing next to a concession stand offering ICE-COLD PICKLES ON A STICK in dripping, ice-blue letters. I walked over and she finished up a call on her Bluetooth, waving aside the $7.50 cover charge.

"Welcome!" Michelle Reyna said, shaking my hand with fingers covered in oversize rings. She wore so much jewelry she seemed to be shining. Her emerald bracelets matched her emerald heels. She was in her fifties, with a solid build, and smelled bracingly of perfume. "Did you find your way easily?" she asked, taking off the headpiece and adjusting her big crystal earrings.

As we turned toward the garden, another peacock leapt off the roof. "That's a male indigo," she told me, raising a painted-on red eyebrow in its direction. "They're pretty feisty. They'll even attack their reflections in shiny cars. We also have white peacocks. People think they're albino, but they're not. When they're together, they're a *party* of peacocks. I'm really precise about that."

Several honked in approval.

"Also, to be precise again, the Spanish moss on these trees is not Spanish, and it is not moss," she said in her genial, no-nonsense manner. "It's called Spanish moss because it looks similar to an old, gnarly Spaniard's beard. It's actually an air plant related to pineapples."

As we walked down a stone path into the park, I hoped we'd be getting more precise about the Fountain of Youth. "The grounds are bigger than most people imagine—fifteen acres full of fountains and Florida flora," Michelle alliterated, as I craned around at all the water features, wondering which, if any, was the actual Fountain of Youth.

Descended from Spanish Minorcans who'd settled here in the eighteenth century, Michelle had been raised Catholic. As a child she was groomed to be the next Singing Nun. Recounting the story, she burst into the chorus of "Dominique," the unlikely hit song from the sixties about the saint who founded the Dominican order. But then, in her teenage years, she'd discovered boys.

We came to a little shack overgrown with vines. "You're not allergic to sulfur, are ya?" Michelle asked, pausing in the doorway.

"I don't think so."

The springhouse was lit low to convey gravitas. A large cross lay on the floor. A diorama showed Spanish ships rolling across the ocean. Pelicans soared motionlessly through the wave troughs. On land, an Indian in a loincloth was holding an ax that lifted and dropped mechanically.

All through the shed, not a fountain could be spotted. In the middle of the room, a hole in the ground had a rickety little gate around it. Plastic mouthwash cups full of water were laid out next to the cranny. A few silver jugs were there to give the impression of old Mother Spain. A whirring fan rotated back and forth.

"Think of it as a small, underground river," said Michelle earnestly.

I looked at her. I looked back at the fountain. It was actually a hole in the ground.

She launched into a convoluted story about the fountain's having gone beneath sea level and how they'd put in a pump similar to one you'd find in a goldfish pond. "People have this image in their mind of a fountain, so they used to leave unhappy," she said, handing me a small plastic cup. "Since we put the pump in, everybody is satisfied. Anyways, take a sip! It has forty-two minerals in it."

It tasted like bathwater.

"People say it tastes like old well-water from their grandparents' backyards." Michelle smiled. "How was it?"

"Pretty good," I lied. "You can really taste the sulfur."

She beamed.

"I haven't turned into an eight-year-old yet, though."

"You just took ten minutes off your life with one sip." She laughed. "People say, 'Is this going to make me younger?' 'Yes,' we say, 'but only in ten-minute increments.'"

She then brought me down to the waterfront, where we came to a statue of Ponce de León wearing a silver hat and puffed-out pumpkin pants. The paint was peeling near his scabbard. His skin was a shiny gold color.

"It's not real gold," said Michelle. "Probably plaster or something. We just paint it gold. But it looks cool and people like it, so . . ."

I tried to focus on certainties. "Do we know whether Ponce de León really did land here?"

"That's what we can't prove," she answered, puppy-eyed. "But the cross shows that he did."

"You mean the cross that Diamond Lil dug up?" I asked skeptically.

"Yes." Michelle held her arms open and shrugged.

Just across the lawn was a church with an enormous cross standing high into the air. "That's the cross of Our Lady de la Leche," Michelle said. "It's the second-largest freestanding cross in the world, after Rio. It's two hundred and eight feet tall. It has nothing to do with Ponce's cross, but it's another historically important cross. When JFK came out here, he said this is the most sacred acre in the United States."

"Why did he say that?"

"Because this is where Christianity began in the US."

"Okay . . . who was Our Lady de la Leche?" I asked.

"Our Lady of the Milk is the patron saint of breast-feeding. The church was started by a friar whose wife was having trouble breast-feeding, so he named it after her. It may have roots in a fourth-century grotto in Bethlehem. When Menéndez arrived, he laid a cross right here and proclaimed the land in the name of Spain. That marked the beginning of America's colonization by Europeans."

"I'm confused," I admitted honestly. "I'm trying to figure out who landed here, who was first, what was first. Are you saying that Ponce landed right here and Menéndez arrived here, too? Both at the same place?"

"We know Menéndez landed here," said Michelle. "There's no dispute about that. Well, not much dispute about that. As far as Ponce goes, nobody else has any evidence that he landed anywhere else in the state either, so the Frasers' attitude is that until someone else comes up with convincing evidence that he was *not* here, we're going to continue believing that he was."

"The Frasers own this place?"

"That's right."

"And when did Menéndez arrive?"

"In 1565. Ponce was here in 1513."

"And they both landed exactly here?"

"Somewhere here—this property is fifteen acres," she said.

I took a closer look at the bronze sign under the statue: "On Easter Sunday, March 27th, Ponce de León sighted this land and named it 'Florida.' On April 3rd, 1513, he entered the harbor and landed in this vicinity."

"That's true, you know," said Michelle. "Well, it's quite possibly true. It's corroborated by an old priest's diary."

"What would historians say about this sign?"

"Depends on the historian. See, history is evolving, and as we find more documents, we keep learning more. I'd hate to say *definitely* because next week it could change and I'd have egg on my face. I'd like to say we do a pretty convincing job here, with our evidence such as it is."

Empathizing with her situation, I wanted to stop being an accountant of facts, but couldn't help persisting. "It says that he was in St. Augustine without any doubts."

"He was here, and that's what the Frasers want to believe, so that's what I say, too. So say I."

Michelle's cell phone rang. She put on her Bluetooth. "Were you able to get an Indian?" she asked, before quickly ending the call. "Sorry about that—I'm looking for an actor to do a reenactment this afternoon. Everybody's calling in sick today. Where were we? Ponce. Yes, we do get into some arguments with historians who are convinced he did not land here. You could go to the historical archives of St. Augustine and see what you turn up. But the information on these signs is right on."

We looked at another plaque together. It quoted fragments from some nonattributed historical document and used ellipses for a heightened aesthetic authenticity. "Juan Ponce de León went to discover . . . He went in search of that fabulous fountain . . . That the Indians said turned old men young."

Michelle then told me a long story about how some Indians in Cuba told Ponce about the fountain in order to get rid of him. "They said, 'Hey, little guy, go get that water, and you'll even be able to bring back some tall slaves in your big ships.' And of course Ponce believed it because he *wanted to believe it,*" she said, emphasizing the last words as though that were the moral of my visit.

You just need to want to believe, she was saying, to have faith.

Michelle had to go to a meeting, but not before signing me up for the guided tour beginning in half an hour. In the meantime, I decide to research the fountain's history in the Spanish Gift Shop.

A bored attendant named Jill* was flipping through a book about

* I've changed her name so she won't lose her job.

dream catchers. She directed me to a photocopied booklet called *The First Landing Place of Juan Ponce de León*. The contents were written in overly footnoted, academic-sounding prose intended to lend an aura of credibility. The narrative described how, in 1900, a few years after buying the property, Dr. Louella Day McConnell made her "scientific discovery."

First Diamond Lil found the cross. Subsequent excavations yielded a small silver saltcellar containing a parchment. This document, which Lil had translated by a local Spanish-speaking public-school teacher, supposedly told of Ponce's 1513 discovery of a "fountain good and sweet to the taste." There was no way of verifying it: the photocopy of the parchment in the book was utterly illegible. It looked like a Rorschach blot photocopied onto crumpled black carbon paper.

"From internal evidence the document could have been written only between the years 1506 and 1516," concluded the book, citing Dr. McConnell's professional training and her extensive investigations in the General Archives of the Indies.

I walked over to the shopkeeper and asked her where they kept the parchment and the silver saltcellar.

"We have a pewter replica of the saltshaker on display, but nobody knows where the original is anymore," Jill said. "Maybe the owners have it somewhere? I think Diamond Lil lost it."

"What about the parchment?"

"The parchment is in a safety-deposit box. The Frasers are the only people who've seen it. Apparently it's written on sheepskin."

"So neither of them are available for viewing?"

"Nah, sorry. I keep telling 'em they should at least have a photo of the darned parchment up on the wall."

Mounds of trinkets were for sale near the checkout counter, including key chains, place mats, and mugs that said DRINK FROM THE FAMOUS FOUNTAIN IMMORTALIZED BY PONCE DE LEÓN. That seemed hazy enough, but then I came across a postcard claiming that this very spot "is where Ponce came ashore to landmark and record for all time, the first moment of our nation's history, the discovery of North America."

"Is there any proof that Ponce actually made it here?" I asked Jill, still trying to piece together what had actually happened.

"We think he may have. We believe he did." She handed me another brochure that described this land as *the very property on which US history began*. "Whether you believe it or not, it's interesting stuff, right?"

I picked up a small glass bottle full of Fountain of Youth Water™. Water from the Fountain of Youth, the label said, *is considerably different from that of the wells*. Jill told me that the Frasers owned the patent for Fountain of Youth Water. "You can't bottle Fountain of Youth Water or the Frasers' lawyer will come after you."

On cue, a managerial type strolled into the shop. He introduced himself as Brian Fraser, grandson of Walter B. Fraser, the entrepreneur who had taken over the Fountain of Youth after Diamond Lil died in a car crash in 1929.

"There are twenty-five thousand patents attached to the phrase *fountain of youth* so we copyrighted 'Ponce de León Fountain of Youth Water,'" said Brian, handing me a business card. He was the operations manager of the Fountain of Youth Archaeological Park and a sales associate for Colonial Saint Augustine Travel.

"Give him your other business card," gibed Jill.

He rolled his eyes and shook his head indignantly.

"Do you have several business cards?" I asked.

"I only have one Jill," he answered, glaring at her. "She's the highlight of my day, every day."

And that was that. He walked out, leaving the business-card issue as unresolved as my understanding of Florida's Fountain of Youth.

"It all began out there over 495 years ago," said the twentysomething guide in a baseball cap. I was back in the springhouse, alongside seven or eight other visitors to the fountain. "The year: 1513. The first steps taken in North America happened right here on April second. Spanish discoverers docked out there on the First Coast and called this land Florida—*the flower*."

Aspects of what he was saying were true. The oldest continually inhabited European settlement in the continental United States is indubitably in St. Augustine. But the first European steps in North America were those of the Norse Vikings a thousand years ago. And John Cabot made his discoveries in 1497.

The more the guide spoke, the more ridiculous it all seemed. I'd been dubious from the outset, but now I was just along for the ride. "Good old Juan came here to find the fountain of youth—this *very* fountain. After he drank from the spring, he doubled his expected life span and lived to be sixty-one. We can measure Ponce—he's buried in the cathedral on San Juan, so we know he's four foot eleven inches, and that's confirmed. It's not an exaggeration. It's true." The crowd looked on transfixed. A bald man's eyebrows seemed to be climbing up over his head like two millipedes.

Among the clippings on the wall was a pixelated shot of Helen Keller being driven around the property. A newspaper story related how, in December 1995, Gregory Peck pulled up in a limousine, got out, took a sip, and left immediately afterward.

"How do we know this is the right spot?" the guide asked. "This cross right over here! This property's former owner Diamond Lil found it buried under three feet of dirt. She had been digging up a dead palm when she found the stone cross."

"But that could have been done whenever," a child piped up.

The tour guide ignored him. "But anyways, there are thirteen stones in a row, and laid across them are another fifteen stones, just like the year 1513. They're laid perfectly from north to south and from east to west. *La Cruz,* the Cross, is how they referred to this area. But that's not all: Diamond Lil had more proof that she had found that fountain of youth. She dug up an egg-shaped, silver saltcellar that contained a parchment. It was a sworn affidavit signed by the king's witness offering hard evidence that this is the fountain Ponce found."

"Wow," whistled a curly-haired woman with a pink backpack, without irony, it seemed to me.

"The salt holder vanished mysteriously," continued the guide, "but you can see a replica over there. More importantly, we're all standing in what would have been a pond . . . the *fountain*. Everyone expects a bubbling spring, but the water levels dropped when people dug wells into the aquifer. It used to bubble out of the ground, spurting out to create a chest-deep pond for thousands of years. By 1875, the boil stopped bubbling and the pond disappeared. Now we have a pump—so we can turn the Fountain of Youth on or off whenever we want. Come on down, people, grab a cool glass of the Fountain of Youth, and check it out while you're here. Don't worry 'bout falling in. And if people drop their sunglasses in, I can fish 'em out with a swimming-pool net."

The group shuffled over to try their dentist's-cup shots of water. A boy grimaced and washed it down with a gulp of Sprite. Not needing another taste, I headed outside.

Inside a domed building nearby, an attendant was just getting ready to start the Celestial Planetarium show. He came out to say a few words.

"Good morning, everybody, and welcome to the only manually operated planetarium in the world. It's a sometimes-working antique, so if there's smoke and sparks coming out of it, don't panic—that's normal at our planetarium."

The lights went down and the crazy old machine sputtered to life. It started rotating and groaning, the cosmic apparatuses blossoming out of a central steel pot like mechanical flowers. "A 1957 Chevrolet's ball bearings and a Singer sewing machine are what turn this thing," said the attendant, speaking through a microphone. Hexagonal lamps covered in perforated sheets projected lights on the ceiling resembling constellations—sort of. "These were the stars that guided Ponce de León as he searched for the fountain of youth," explained the host. "You are now under the exact night sky that Ponce saw when he made North America's discovery." I loved that he called it the *exact* night sky and loved it even more when he started the strobe show, which consisted of his turning a handheld flashlight on and off. Then the machine broke. He kept the flashlight on, punched some buttons, and soon the contraption started up again, casting its shadowy, strobe-spiced spectacle onto the circular ceiling.

A small theater next door was called the Explorer's Discovery Globe. A few rows of seats were set up in front of a big, two-story-high globe in the center of the room. Black lights came on and the orb started glowing in the dark. A sonorous baritone voice intoned facts about Spain's discovery of the New World. The recording told of how, in the beginning, the earth was without form and void: "There was a multicolored planet radiating its own subdued brilliance as it rotated through the cosmos. Then the western hemisphere arose." When America was discovered, civilization really kicked into high gear. "Florida is synonymous with the Fountain of Youth," the voice stated. "And the name Florida was later changed to the United States of America."

As I walked out, a lanky, somewhat melancholy man approached and introduced himself as Harry Metz, the amateur archaeologist Michelle Reyna had told me about.

"Did you like the show?" he asked. "They're going to be putting a computer in the planetarium soon to have it programmed by astrologers—I mean, astronomers," he said. Given what I'd seen so far, the distinction's porousness felt perfectly site-appropriate.

Metz walked me through the small museum next door. The star attraction was a small boat billed as a replica of Columbus's famed ship.

"That's not really a replica of *La Niña*, is it?" I asked.

"Not technically. It's actually a small shrimping skipper."

On the wall was a display about the cadavers of prehistoric dogs that had been found on the site. Radiocarbon dating showed them to be eight hundred years old. When they were disinterred, their stomachs contained the remains of deer hooves and catfish. "I want to put a dog skeleton on display, one of the dog skeletons we dug up," Metz said. "I'd like to flesh the skull so that visitors would be able to see what a real Indian dog looked like."

The next room was devoted to the Native Americans who'd inhabited this land prior to the arrival of Europeans. There were paintings of scalps hanging as trophies, dripping blood. The room's centerpiece was a large photo of skeletons, eye sockets glowing. "This is where they found the skeletal remains of Indians who had been given Christian burials," said Metz. The lurid, oversize photo of bones piled alongside skulls grinning hideously looked like a Misfits LP cover.

A framed letter on the wall was from M. W. Stirling, director of the Smithsonian Institution, Bureau of American Ethnology, on June 1, 1951. It stated that if people "undertake further excavations of the site, they will find it to be of more than ordinary significance." Metz told me that he'd done numerous excavations himself, but that legitimate archaeological digs were overseen by Dr. Kathleen Deagan from the University of Florida. As bogus as most of the displays had been thus far, Deagan had made numerous non-fountain-of-youth-related finds on these grounds. "It's always been interesting and ironic that the site is, in fact, one of the most important historical sites in Florida," she told *National Geographic*. Her digs revealed, with certainty, that this area was at one point a Timucua Indian village called Seloy. Metz volunteered on some of the official digs and told me he'd been allowed to keep some of his discoveries. "Would you like to see some of the artifacts I've found?" he asked. "They're in my lab, or my cave, as I like to call it."

He brought me behind the planetarium to a small storage area. The shelves were lined with his finds: pottery shards he had washed in his dishwasher, crumpled pop cans, whelk shells, horseshoes, a number of pipes, a doll's head, and musket balls. He hoped to one day bulldoze under the planetarium. "I'm convinced there are things in the ground from when Ponce landed here, and I say *here* because I think it could have happened."

"So you believe Ponce de León really was here?" I asked.

"I don't want to tick anybody off, but it looks to me like he very well could have been here." He started speaking about the conjugate graticule, the geographic-coordinate system that shows that Ponce's ships arrived at latitude thirty. Each latitude being sixty nautical miles, he reasoned, the landing probably happened somewhere within thirty nautical miles. "Ponce will always remain controversial. Naysayers think he couldn't have landed here, but it all could be—y'know? Look at the Indian burials—they're not phony," he argued, as though that proved the validity of everything else by proxy.

He then told me a convoluted story about Ponce de León's possible reasons for coming. It had something to do with the lunatic daughter of Ferdinand's, named Joanna, who kept her dead husband's corpse, dressed it, spoke to it, and ate dinner with it. "Isabella and Ferdinand were hoping for an heir to prolong the bloodline," said Metz. "Physicians were feeding him potions to keep him alive, but they were making him sick. They thought maybe the fountain would help him out."

St. Augustine's historical archives are housed on the second floor of a building near the center of town. Its filing cabinets are filled with disputes relating to Ponce and the fountain, going back to where he actually landed. There are numerous passionate arguments on behalf of various latitudinal and longitudinal positions, but no way of knowing where it happened with unassailable certainty. "Once you've got that one figured out, you can sift through the trial transcripts about whether the Fountain of Youth is real or not," the librarian told me, thumping a half dozen crates onto the table.

People have always claimed that its water is plain old well water. A decades-old report from *Harper's* magazine called it "horrible tepid water drawn from . . . a very dubious source." Tepid, yes, but a certificate from a bacterial analysis in Gainesville found the water to contain absolutely no traces of fecal streptococci. In 1978, a St. Augustine centenarian,* Allen Stowball, who attributed his longevity to meat, coffee, and pineapple juice, told journalists that he'd been to the Fountain of Youth

*I looked into whether St. Augustine boasts a higher proportion of longer-lived residents than elsewhere, but couldn't find it on any of the top-twenty lists of American cities where people live the longest. However, a purportedly secret society billed as "protectors of the Fountain of Youth" claims to have been granted superlongevity resulting from their work.

twice—only to make deliveries. "No, I didn't drink any of that water," he said. "That water's all a bluff."

The St. Augustine Historical Society stated it explicitly in 1929: "There is no fountain of youth as we all know, and it is silly and quacky to carry on the inventions of a woman who was not in her right mind." Document after document expounded on the fountain's complete lack of historical standing among scholars. A National Historical Committee in the 1930s judged that none of the spring's claims were factual.

Charles B. Reynolds, of the Audubon Society, was one of the most tireless debunkers of the Fountain of Youth. "It being a well, to speak or to write of it as a spring is to violate the truth and to deceive," he wrote. His pamphlet "Give Back the Lost Dignity to Historical St. Augustine" deemed the cross dug up by Louella Day—the single piece of evidence that the Frasers could provide—a hoax and a "flim-flam proposition."

In the bowels of one box I came across a sworn statement testifying that a handyman named Benjy Pacetti had laid the cross at the Fountain of Youth sometime before the First World War. Then, a few boxes later, I found a notarized deposition from 1928 by the then seventy-six-year-old Pacetti. "It has been stated promiscously [*sic*] that I am the one who built the stone cross," he declared. "I positively swear that I had nothing whatever to do with the building of this cross."

There were folders full of information pertaining to the dozens of lawsuits surrounding the fountain, and its various owners over the years. In 1952, Walter Fraser filed a $750,000 libel suit against the *Saturday Evening Post,* which had published an article titled "St. Augustine: Its chief industry is still the preservation—and fabrication—of historical landmarks." Fraser felt that his reputation had been smeared by the story, in which he purportedly showed the *Post* reporter how to make new things look old by mixing mortars and paints to imbue surfaces with the appearance of antiquity. Fraser was quoted as saying that history had to be "presented in a dramatic way to attract more people to St. Augustine." The journalist questioned the need to create showy fake antiques when the city was full of genuinely old, if a little boring, monuments.

The trial centered around the authenticity of the oldest wooden schoolhouse. A number of respected historians testified that there was no evidence it had ever been a schoolhouse before the Civil War. An official of the St. Augustine Historical Society and Institute of Science ("dedicated to the preservation and accurate interpretation of St. Augustine's rich historical heritage") accused Fraser of continually misrepresenting history. Fraser denied doctoring anything, but did admit that he'd published documents about the schoolhouse that contained factual errors.

In the end, Fraser won. Although he was granted only 10 percent of what he had asked for, it was still a huge sum: $75,000. The more I read about Fraser, who'd been the former mayor, the more his political prowess became apparent. In one interview, when asked whether Ponce felt he had found the mythical spring, Fraser answered that "he was satisfied he had found something."

As the lawsuits increased, the Historical Society switched tactics and went on the defensive. I found an internal memo ordering all employees "not to knock" the Fountain of Youth or anyone in the business of selling history. Employees were warned that saying anything derogatory about the fountain would lead to their being summarily discharged.

"Luella Day's fabrication of stuff she found in the ground does not count as evidence that Ponce de León landed here," said the librarian. "But we tiptoe around the word *fabrication* 'cause we like the Fraser family and we don't want their land sold to developers."

We spoke about the way reality and fantasy is so fuzzy here, how the real archaeological excavations and the fact of Ponce's discovery of Florida blended with the weirdness of a diamond-toothed con woman opening an attraction selling well water as an elixir of eternal youth. "Everybody in town knows the fountain is bunk, but nobody wants to be quoted," the librarian said, laughing. "The Frasers had a real litigious grandpa. But you should visit the original fountain of youth. I've never slogged around in it, but it's near Spring Street. Apparently late-night drug smugglers come in on boats over there."

"Wait"—I paused—"there's more than one fountain of youth in St. Augustine?"

"There's at least three over here, and a bunch all over the state. There's one in St. Petersburg, another up near Alabama, one around Daytona Beach. Venice, Florida, claims they recently uncovered the real mineral springs sought by Ponce de León."

Outside the archive, the sun was starting to set on streets teeming with tourists, all as glazed over as Cinnabon cinnamon rolls. The roads and walls were made from old clam and mussel shells mixed with mortar, just like the cross Diamond Lil had dug up. The seashells poking out of the ground created the impression of walking on water—although perhaps that was just a reaction to the dizzying amounts of disinformation I'd been subjecting myself to since my arrival.

On one nondescript modern building, putatively the erstwhile residence of the royal Spanish treasurer, were two signs: one claimed the home was from the early 1700s; the other stated circa 1750. To my eyes, it could only have been built in the decades following World War II. Nearby, a gaggle of tourists huddled around a monument. It was a sundial, "undoubtedly," the sign said, left by Ponce de León.

Heading into a corner store to buy a bottle of water, I felt a surge of happiness about being in such a real-yet-artificial place. I picked up a cigar labeled hand-rolled; an asterisk lead to a tiny footnote: MACHINE MADE.

Behind the store, a group of young men were skateboarding, and I asked them about the Fountain of Youth water. "It's just . . . *blech,*" said a guy in cutoff jean shorts. "Once you try that Fountain of Youth water you'll decide to age gracefully."

"Everybody who's ever lived here and who's ever come here gets that cup of water," explained his friend. "Most of them throw half of it away. I mean, it tastes like rotten-egg sulfur water. But this is where America *began,* you gotta taste the history."

"I drank it at ninety-one," said the first one.

"You were ninety-one years old?" I said, laughing.

"Nah—*in* '91," he said, "1991."

They told me that every fourth-grader in the state comes here to learn about the history. "Don't they teach you about St. Augustine where you're from?" he asked.

I told them how, in Canada, we don't learn anything about St. Augustine, but that most schoolkids consider Canadian history to be totally boring. "Maybe if we jazzed it up a bit, like you do in St. Augustine, students would be more interested?"

"History," he answered, "is definitely better that way."

I still had half an hour to kill before meeting the historian David Nolan, so I sat down on a park bench to review my findings. Within seconds, a shoeless, fat, balding black woman approached. "Can I use your phone to make a call?" she asked, placing a fist on one hip. Her huge breasts were barely contained by a stained T-shirt. I said yes—on the condition that she let me interview her about the Fountain of Youth.

"I love the fountain: it's excitement, it's money, it's pretty, it's *life,*" she emphasized. "Don't write that down. Gimme your phone and I'll tell you what you want to hear. You want to know what I really think? I'll tell you 'cause you my friend. I ain't gonna go there. When I was a six-year-old, my mama told me, 'Don't ever drink that water; it will give you disease. Keep your hands in your pockets and don't touch the water.' What? You crazy."

She said "you crazy" because I was writing it all down. I handed her my phone; she dialed a 904 number. "Don't write down that number," she said, waiting for it to start ringing. A moment later, someone picked up and coughed on the other end.

"I heard a cough," she said, her bloodshot eyes beaming. She coughed back as though communicating in code. She waited for an answer. None came. She coughed some more. A minute went by and then she gave me the phone back. "I ain't gonna tell you no more," she said, hiking up her jogging pants. "Can I get five dollars? I'll tell you things that'll make the angels come down in light."

"I don't have five dollars," I said. "I'm just writing about the Fountain of Youth."

"All right then—everything I told you was a lie." She strutted away. "I was lying."

Heading back to the Fountain of Youth, I passed a Disney-like castle: Ripley's Believe It or Not! Museum. I took a quick stroll through its chambers, filled with oversize gallstones and distorting mirrors, and mused upon our attraction to things we know can't be true, whether it be stories, myths, or people with tusks. The earliest museums were cabinets of curiosity; only in the nineteenth century did collections start being divided into art museums or science museums or technology museums. At the beginning, all museums were dedicated to the basic sense of wonder, a passion dampened by the Enlightenment. "What we commonly call being astonished," wrote Descartes, who got people out of their hearts and into their heads, "is an excess of wonder which can never be otherwise than bad." As of around 1700, the sense of wonder became linked to foolishness. People wanted understanding, to brush up against omniscience, to feel that everything was ultimately within our grasp.

Leaving Ripley's, finding myself lost again, I popped into a bustling leather-goods shop to ask for directions. The woman behind the counter asked why I would possibly want to go to the Fountain of Youth. "I've never been, but I drink that water every day," she said.

"What do you mean?"

"It's the same water everybody drinks in St. Augustine, except the water in the Fountain of Youth is not treated. You should have smelled the showers at my university dorm: sulfur water, just like what they give to visitors over there."

"You know, it really did taste like bathwater."

"It *is* bathwater!" She said that I could quote her in the story—on the condition I didn't use her real name: "Call me Sabrina, 'K?"

For me, her eagerness to be somebody else embodied all of St. Augustine's ambiguities. Why be truthful when you could pretend? Being something you aren't is fun, she seemed to be saying, especially in a town where everything is almost real.

There's little evidence that Ponce de León ever sailed off in pursuit of the fountain of youth. The patent or charter signed by King Ferdinand authorizing the exploration makes no mention of it. Instead, it guarantees Ponce de León will be named *adelantado,* or governor, of any lands he conquered—power being a more plausible reason for his journey. In the logbook of the seven-month-long trip, there's not a peep about any rejuvenating waters, let alone any talk of asking Indians for directions to the fountain.

Twenty years after the voyage was completed, Oviedo's *Historia General y Natural de las Indias* made the first recorded mention of Ponce de León's searching for any magical waters. Ponce was, Oviedo claimed, looking for the fountain of youth as a remedy for *el enflaquecimiento del sexo*—for his sexual impotence. Oviedo was an author of chivalric romances whose historical work was denounced as containing as many lies as pages. Still, the gossip stuck.

It became ingrained as fact nearly a century later, when another court-appointed chronicler published his findings. Herrera had access to all of Ponce's logbooks and, despite finding nary a murmur about the fountain, still chose to build on Oviedo's rumor, saying, "There was not a river, or brook, nor scarce a lagune or puddle in all Florida but where they bathed themselves in." Peter Martyr added fuel to the *enflaquecimiento* rumors. After washing in the fountain, he heard, even old men were able to "practise all manly exercises."

Historians started accusing Ponce of covetousness, of taking a chimerical cruise. It scarcely mattered that he'd discovered a land frilled with flowers; his reputation, once soiled, was sealed. The memoirs of Hernando D'Escalante Fontaneda, a Spaniard marooned on Florida for seventeen years, characterized Ponce's search as ridiculous—"cause for merriment." As merry as that may be, Fontaneda also submerged himself in a bunch of puddles hoping to find the one.

Until the Enlightenment, a rigorous presentation of archivally verifiable research was considered less important to historical writing than confecting a riveting narrative. Readers wanted stories about the wonders of the world, whether fully imaginary or only semireal (as with Marco Polo, who may never actually have gone on any of his travels, and the pedantic Thucydides, who inserted speeches by public figures rewritten in his own words). "History was a literary genre in which truth took second place to rhetorical effectiveness," writes Oxford's John Burrow in *A History of Histories*. Strictly fact-based history was considered "uncouth, useless, pedantic . . . dirty and ungentlemanly."

A "scientific" approach to history only became entrenched in the past few centuries. Before then, historians believed their art entailed storytelling—including making up quotes, fudging details, skewing facts. "History used to mean stories even more than fiction," explains Jill Lepore, chair of History and Literature at Harvard. "In the eighteenth century, novelists called their books 'histories,' smack on the title page." The Spanish word *historia* still means both history *and* story.

Once writers impose narratives on history as a means of interpreting reality, rather than merely recording the happenings of an era, murkiness ensues. History began as a way of documenting wars, and from the very start the relationship with factuality was strained. Herodotus conducted first-person research in Egypt and returned with tales of people who hibernated for half the year in a land full of winged serpents, cows that could only walk backward because their horns were so huge, and fox-size ants that dug gold out of the ground. "My business is to record what people say," he wrote, "but I am by no means bound to believe it." His successors, those spicing up their accounts of the Indies' discovery, were telling a historical tale, not writing empirical history. As they knew, *enflaquecimientos*—and all other things *sexo*—have always made for good stories.

In the time before photography, excursions in search of new geographical lands weren't too different from expeditions into the spiritual quadrants. Only those who'd been there could know what was true and what wasn't. Even today, with no way to document the validity of the spiritual, we can only heed the accounts of voyagers—at least until we go there ourselves.

Back in the Fountain of Youth's parking lot, a sign had been posted over the TODAY'S SCHEDULE: LIVE CANNON FIRING! sign saying SORRY, NO CANNON FIRING TODAY. David Nolan let out a deep belly laugh when he noticed it. Nolan was a pleasant, boyish man whose smooth face and sharp eyes had a gentle, otterlike quality. An advocate of real history, he seemed to find the wackiness of the fake history more than merely humorous.

He'd been part of a commission that had dated all the city's buildings. "Most of St. Augustine was built in my lifetime, but that's not what they're telling the tourists," he said. "Nobody took down their signs saying BUILT IN 1586 when we told them that it was actually built in 1964. People still put bronze plates up with any old date."

As we stood in the parking lot, Nolan told me about the rumors that Diamond Lil had been a drug addict who ran a brothel here at the fountain. "There are lots of tacky places here with tawdry tales. There are several 'oldest houses' here in town. Not just one. The 'authentic old drugstore' was actually part of an old jail."

"It seems like there are many Fountain of Youths in Florida as well," I added.

"Basically every body of water here was a Fountain of Youth." Nolan laughed. "Real estate promoters have always been able to generate buzz by suggesting they had 'the' fountain, whether in Bal Harbour or Sarasota. In fact, I want to take you across town to the original Fountain of Youth."

As we drove, he told me about growing up in St. Augustine. When he was a kid, a friend of his used to be paid five cents a bucket to bring water to the Fountain of Youth when they'd run out or when they weren't able to pump any more out. Nearing Ponce de León Boulevard, we stopped as a mother hen and her two chicks ran across the road. As though sensing the dumb joke forming in my mind, Nolan grew serious. "You know, I have nightmares about them selling the Fountain of Youth."

"What do you mean?"

"Waterfront condos. The fountain is so linked to our identity. Every culture manages to tie this age-old symbol into their own needs and appropriate the idea. Florida has always been about getting tourists and selling land. There's something so essential to our history wrapped up in alligator farms and all these other tacky tourist attractions. They're how Florida developed itself."

He started telling me about the state's oldest businesses, how early settlers had invented glass-bottomed boats so that tourists could watch women dressed as mermaids floating around in the bottom of springs. Sixteenth-century accounts already told of Florida's mermaids and crocodiles with sweetly perfumed breath. Nolan's intimations of a dignity within the kitsch connected the city to an obsolete form of historiography, one where fictitiousness and inventiveness weren't simply tolerated—they were valued.

"The idea of turning something into an attraction that you wouldn't normally think to be an attraction is so crucial in the history of cracker ingenuity in Florida," Nolan noted. He explained that he considered himself a cracker, in the positive sense of a wise, humorous person who is close to the land. "Our town's motto is 'See things differently.'"

He directed me to pull my car up off the road onto a grassy embankment. A sign said Spring Street. We got out and walked down a path festooned with vines, stopping when we came to an overgrown swamp. "Not exactly a spring of life," I remarked.

"When you talk about this area, it's one scoundrel after another. They all promoted the idea that economic salvation lay in promoting our history. What sort of history? That's another issue."

Something about the fictional dimension of history in the Age of Discovery almost legitimizes St. Augustine's entire tourism industry: it's simply basing itself on a premodern approach to historiography. There's little difference between Fraser's booklet on the fountain and Herrera's conclusion that Ponce sought the fountain of youth—both share a disregard for objective evidence.

At the time of Ponce's discovery, *objective* actually meant the opposite of what it has come to mean. Descartes, in the seventeenth century, spoke of "objective reality" as that which exists in our mind, rather than out there in "formal reality." Before the Enlightenment, the word *subjective* was barely used, but when it was, it referred to things (subjects) in and of themselves. According to historians of science Lorraine Daston and Peter Galison, the modern connotation of the word *objectivity* only took hold in 1817, when Coleridge—a Romantic poet, of all things—used it to describe things in nature that exist independent of human observation. That same year, Coleridge also coined the term *willing suspension of disbelief,* the notion (perfected by David Copperfield and other magician-storytellers) of allowing ourselves to believe in something while also knowing that it can't be true. Since then, our faith in objectivity has rivalled our faith in God, yet both remain equally elusive.

Our pursuit of objective reality is precisely that—a pursuit. Reason is a horizon, as Kant explained. We can't ever get there. It's something we chase but never attain. In attempting to take it all in, we experience the pleasure and agony of the sublime, the sense that things are bigger and more awe-inspiring than we can ever begin to imagine. The sublime is what happens when the dream of trying to comprehend reality, of getting tantalizingly closer to some ever-receding understanding, leaves us dazed on some dusty Florida side street of the mind.

11

Let's Run into the Waves and Spring Back to Life

I don't think that all the Buddhas and Bodhisattvas of the three times will criticize me for giving you a little secret, that there is no need to go somewhere else to find the wonders of the Pure Land.

—Thich Nhat Hanh, *A Guide to Walking Meditation*

I felt as the dead feel. . . . I suspected myself to be in possession of the reticent or absent meaning of the inconceivable word *eternity*.

—Jorge Luis Borges, "History of Eternity"

On a summer day several months after I first met with Martha Morano, my request to interview David Copperfield on Musha Cay was officially declined. A representative from his public relations agency, Polaris, sent the following note: "Martha forwarded us the information on your forthcoming book, and your request to interview David Copperfield. He's been in the research phase since acquiring the island, and until that research is completed, he's not ready to discuss his findings with the media. We do appreciate your interest, however, in talking to David for the book, and wish you the best with the project."

The morning that e-mail arrived, I sat there reading it over and over. *His findings. The research phase.* What had he found? Checking Google news for any developments, I came across an interview in which he jokingly claimed to be eighty-two years old. "I want to be the first 150-year-old magician," he was quoted as saying.

While I was exploring the possibility of hiring a Caribbean privateer to help me find his fountain, an e-mail arrived from Martha: "I would not give up on this. It was necessary to include his personal PR agency, but I can work on David from my end. There is time and I will be with him in the next few months and will talk to him about it. He really should do this."

While she did that, I kept hunting for the fountain. I tried Saratoga Springs, the sulfur springs of Saint Lucia, the mineral baths at Clifton Springs in the Finger Lakes region. My sessions in these baths were relaxing, but all I really found was a laminated poster of Night trailing her train of stars, *shhh*-ing with a finger to her lips, as though saying, "Keep the mystery sacred." Giving it one last shot, I visited California's fountain of youth. It is found on the grounds of the Esalen Institute, a New Agey retreat center in Big Sur affectionately known as the Court of Last Resort. "At the heart of this temple of Nature, the springs pour forth warmth from the womb of the Earth," as one description has it. "The baths are literally fountains of youth."

The Institute offers such courses as "Vision Seeker," "Wild Serenity," and "The Spiritual Ecology of Business." With its creativity-unleashing seminars and other participatory workshops, Esalen is a mecca for avant-education that, according to its website, offers no assurances of change. The lack of guarantees hasn't stemmed the hajjes for transformation. Each year, around ten thousand pilgrims make their way to Esalen's university of the waves. Disciples visit for various reasons: new beginnings, sunrise meditations, high-octane contemplation, to get certified as massage therapists—or simply to soak in the hot springs. (The mineral pools are open to the public from 1:00 a.m. to 3:00 a.m. nightly.)

Those baths, like the rest of the area, are portrayed as "dramatically charged" in Jeffrey J. Kripal's 2007 academic survey *Esalen: America and the Religion of No Religion*. "Both the personal risks and the promises of adventure are quite real here," he writes, "and the powerful currents that flow just under the surface of things, like the explosive hot springs, should never be underestimated."

Esalen is considered a world navel, or an energy center, as any place where people pay for change usually is. Change is, after all, painful. The charge for a seven-day workshop and a private, premium room is $3,515. (Sleeping on a bunk bed in a dorm room brings the price down to $1,360. Offering to do grunt work on the grounds knocks off another $100.)

Esalen was founded by Americans who'd spent time studying under the yogi Sri Aurobindo, who said that coupling spirituality with eroticism leads to "a supernal fulfillment and an infinite satisfaction in the all-possessing bliss of the Infinite." His teachings suggested the possibility of achieving physical immortality through sexual-spiritual transformations. Some devotees interpreted this as meaning they would literally never die. These diehards were astonished each time someone in the ashram died. "I firmly believed that death was impossible here," explained one of them, when another died. At funerals, sadhaks would turn to each other and say, "I have the feeling that this will not happen to me." When Aurobindo died, in 1950, a segment of the movement remained convinced he hadn't actually passed on.

As a central gathering place for the Human Potential Movement, Esalen still attracts those willing to explore what onetime lecturer Aldous Huxley called the "human potential," the belief that within each of us lies the latent possibility of reaching the divine source of all existence. Fulfilling this potential is challenging, but not impossible, argued the psychologist Abraham Maslow, another early Esalen figurehead. For Maslow, our greatest potential is self-actualization, the ultimate state in his hierarchy of needs. One can only ascend to the pyramid's apex, he wrote, through the intensity of peak experiences, with their "feelings of limitless horizons opening up."

A prime place for such a peak experience is Esalen's baths. "The operative word is *hot,*" as Maslow noted. "This place is hot."

My drive up the vertiginous Pacific Coast Highway to Big Sur took place during a rockslide thunderstorm. Clumps of unfiltered mountain dribbled from the hillsides onto the dark, wet road. Afternoon became night. I went slowly to avoid all the stones in the way. It took me nine hours to get from Silver Lake to Esalen, double the normal time.

If the storm hadn't been so intimidating, I might have been better able to appreciate its beauty. Electrical currents pulsed through the cloud base over the ocean. Green, white, and orange Valkyries galloped across the sky. At one point, my car felt as if it had been hit by lightning. I started imagining the obituary headline: "Canadian Nonfiction Writer Dies Researching Immortality."

Trying to stay focused, I listened to Daft Punk's *Alive*. It started with two computerized voices, one saying "human," the other saying "robot." As the tempo increased, the words overlapped: humanrobot, humobot, hrubot, rhoman, until man and machine merged into cyborgs. I'd been driving so slowly for so long I, too, was turning bionic.

Shortly after 1:00 a.m., a demure brown sign popped up from the foliage: ESALEN: BY APPOINTMENT ONLY. At the bottom of the sloping driveway, an attendant in a candlelit wooden shack checked me in and gave me the key to my room. I parked next to a sea-foam Jaguar with a SEEKR license plate. Outside the car, the ocean was screaming. Swaths of senses were missing. All I could smell was wet salt. My hands were still shaking from the drive. I groped through the rain-strewn obscurity, threw my belongings into the appointed cabin, and headed out to find the fountain.

It was still pouring, but the clouds were breaking slightly, allowing the moon to peek through. At the bottom of the stairs, near a picnic table, a young man stood alone, smoking a joint. I asked for directions to the springs. A large, brown moth sat perched on his shoulder. The man pointed to a path along the cliffside. As I thanked him, he gestured at my shorts. "By the way, you won't be needing those."

Nudity without shame is the sort of goal Esalen helps its visitors attain. I knew the springs were clothing optional, but it hadn't quite dawned on me that everybody would be *in naturabilis*. As I picked a towel from the stack in the open dressing room, a Rubenesque woman strolled past wearing nothing but an oversize hat that flopped around her neck like a big fried egg. Outside, around twenty wet, naked bodies steamed in the rain.

I hung up my jacket and was unbuttoning my shirt when an East European couple walked in from the "silent" baths area toward the "quiet" baths area. They stopped and said hello. He was tall and had a broad smile; she was petite and had squinty eyes. They were both naked. Despite the faint light, I could tell they were tripping.

"How's the water?" I asked.

"The hot springs will make you tall—she doubled in size!" he said, putting his hand on her head.

She looked up, toweling off. "Soon I'll be so big I won't fit anymore!" she managed, starting to laugh uncontrollably.

I hung my pants up, folded everything else, showered quickly, and made my way toward the shiveringly cold baths area. My hands kept instinctively covering my loins. A few silhouettes lay bobbing in the first pool, so I self-consciously climbed into a larger one at the end of the path.

The liquid was unlike anything else I'd ever been in. The oily heat felt muscle-melting. The minerals, viscous at first, dried on my skin within seconds. Submerging myself, they'd fade back into primordiality. It was like being in a womb, slathered in placental gel. The vitality was tangible. The water couldn't, obviously, morph anybody into a twenty-three-year-old, but I could see why people considered it a fountain of youth. Something about its fluidity was volatilizing. Sitting in the slippery, sulfuric waters while the Pacific Ocean unfurled into a sooty horizon, I felt the life-giving destructiveness of it all, civilization behind me, gnashing waves in front, howling rains above. Water on all sides: beneath, around, within. The darkened ocean seemed to be teeming with galleons. Two black clouds were the sky's unblinking eyes.

The springs had an undeniable effect. Seeping naturally from the craggy cliffside, they have medicinal attributes that natives knew about thousands of years ago. This was considered a power spot by indigenous North American tribes such as the Salinan, Rumsen, and Esselen—after whom Esalen was named. The place had almost been called Tokitok Lodge to commemorate the Esselen word for "the god in the springs."

Two of my pool's other denizens had become quite talky-talky. Middle-aged and gay, they spoke in petulant lilts about recent parties at the institute. "I was watching you the other night," said one of them, "and I was wondering how you felt having all three of those guys on you at the same time?"

His friend, chin perched on the tub's rocky side, considered the question at length before answering. "I felt . . . ," he said, finally, "*grounded.* So grounded."

"Grounded how—like an electrical socket?" asked his companion.

I stepped out of the tub. It was around 2:00 a.m. My spume-covered glasses gave everything an aquarium-like quality. It had been frigid getting in; with those minerals caked into my skin, it didn't feel cold anymore. I took a shower and then went for another soak in a nearby empty pool.

A few minutes later, an olive-skinned lady with pale-gray wolf eyes and frizzy hair joined me. We started talking. The topic turned to immortality. She said that every LSD trip she'd ever done had granted her an experience of eternity. "This much is certain: immortality is something you feel when you are in the creative womb. And coming to Esalen means entering the creative womb."

She herself had come to learn shamanism 101. She told me that gathering the events of one's life into a scrapbook or writing a memoir can be a means of preparation for the afterlife. Such work isn't done for earthly posterity, she said, but rather to facilitate admission into the concrete realm of infinity. I asked her how one goes about becoming a shaman in this day and age. She said that the first step is being initiated.

"In some sort of ritual?" I asked.

"There aren't too many authentic shamanic initiation rituals left," she explained. "The initiation usually takes place in your dreams. Even in the past, the process often happened asleep. In dreams, the shaman would be transported to a place beyond living and dying. Then they'd meet some god or goddess who would initiate them. And then they'd wake up back on earth, beginning a new chapter of their lives."

That night, I dreamed of a woman with long, white hair. We were in a rainy garden. She started telling me about a book full of wisdom, knowledge, and secret insights. "Where can I find this book?" I asked.

"Right here." She pointed at a dot in the air. The spot flashed open, becoming a rectangular portal. Her other hand moved invitingly toward it. Light poured into view. As I stepped through the hole she had opened in the sky, I woke up.

A French ethnographer named Arnold van Gennep coined the term *rite of passage* in 1908. He considered portals to be symbolic of the entire shift from one stage of life to another, as we can only move from the old reality into a new one by crossing a threshold. "Whoever passes from one to the other finds himself physically and magico-religiously in a special situation for a certain length of time: he wavers between two worlds," van Gennep wrote. These areas of transition, he added, are found in all ceremonies accompanying the passage from one social position to another.

The roots of the word *initiation* are "to begin." It ushers in a new phase. In many religions, initiatory rites have to do with immortality. The ritual of baptism, to pick one well-known example, prepares

Christian souls for rebirth in the afterlife. And at ancient Greek mystery schools, aspirants were initiated into the basics of eternal life. Although we don't know exactly what happened at those rites (divulging the mysteries was punishable by death), it is clear that worshippers died, symbolically, and were then reborn with assurances about immortality to come. "Initiation and death correspond word for word and thing for thing," explained Plutarch.

In many cases, those being hazed would attain a state of ecstatic madness in which they became one with their God. This apotheosis brought a foretaste of spiritual immortality, an intimation of a birth in death, proof of the indestructibility of the soul. The profound realization that dying begets new life didn't come painlessly. There was often, Nietzsche wrote, an "eruptive character" to these initiations. They could involve mutilation and scarification, being burned by torches, cut with fangs, or stung by ants. These ordeals were designed to be sufficiently scary for participants to feel sure they had died. The deprivation and torture endured was meant to cause "a regressive disorganization of the personality." Once broken down, they could be reassembled into a newly formed grown-up no longer afraid of dying.

An extraordinary example of such a coming-of-age ritual was documented in the 1920s among Australian aboriginals, specifically the Pitjantjatjara people of the western deserts. The first stage entailed being covered in the blood of elders for a year or so. The old men cut open their veins with sharpened wallaby bones and drenched the young initiate. They made extra sure the running blood covered his penis. For the next lunar cycle, the youth could eat nothing but coagulated blood.

Some initiates would then cut an incision into the underside of their penis, so that it resembled a menstruating vagina. Géza Róheim's account of a urethral-cutting ceremony in *The Eternal Ones of the Dream* notes that this "penis womb" permitted the initiate to give birth to himself as a grown man. "We are not afraid of the bleeding vagina—we have it ourselves," cried the initiates. "It does not threaten the penis, it is the penis."

Undergoing all of these blood communions allowed initiates to become something greater than themselves. They got so gone—matted down with bloody feathers, producing quartz crystals from their mouths, their subincisions rendering them fully indifferent toward death—that they started to feel an at-one-ment with their immortal ancestors. The old men lifted the young initiates' fear of death by bringing them as close as possible to their nightmares.

Simply reading about such harrowing trials may also affect us, even though we don't undergo the ritual ourselves. Consuming a tragedy (as a viewer or a reader) can be as powerful as any initiation. The ancient Greeks called this vicarious purging *katharsis*. A story's cathartic moment has the ability to thrust us into a new phase of life. Transfigurative images can arise with equal strength from within or without, whether in actual rituals or in dreams or in stories.

In the past, initiates the world over were forced to come to terms with death as a way of growing up. Our society today no longer has such clear-cut rites of passage. Certainly sports hazings, as well as bands and gangs, offer contemporary adolescents an opportunity to undergo the ancient magic of disentangling themselves from childhood. But such experiences rarely, if ever, implicate society as a whole, and the youths going through it often have no one guiding them. Nor is anything waiting for them on the other side of the rite. As a result, the innate urge to boggle the senses (through drugs, piercings, or tattoos) becomes devoid of any context or transformative effect.

Lacking proper initiation rites, our civilization lacks adults. This may be why our society is so fixated on youth, despite the plethora of studies showing how youths are, on the whole, unhappier than adults. Most grown-ups never learned how to sever the psychic umbilical cord chaining them to infancy. We long for perpetual adolescence rather than maturing into subsequent phases of life. We want the fountain of youth, not reality.

Being initiated means bearing what we can't change: dying. But there will always be something within the collapse, a fountain in the rocks. This geysering energy is renewal, transfiguration, Eros: the life force.

With my dream of the portal in mind, I signed up for an Esalen seminar led by the depth psychologist Bill Plotkin. His bio said he specialized in nonordinary states of consciousness and nature-based individuation programs. He'd guided many people through initiatory rites in the wilderness. Our workshop would be a form of initiation.

It took place across campus, inside a former family home, in a big living room with ocean-view bay windows. A black swath of silken fabric spangled with golden stars and crescent moons had been spread over the middle of the floor. On it lay an assortment of objects: pinecones, seashells, some pebbles. There were about fifteen students in total.

Bill Plotkin sat meditating with a rigid intensity until everyone settled in. We positioned ourselves on jumbo, saffron-colored pillows. Plotkin, whose chiseled, gaunt face was topped by an almost glam-rock shock of white hair, began by telling us how he had once met death in an alcove. I, alongside a few others, sat taking notes. As he spoke, he shifted his gaze around the room. He didn't seem to be looking *at* us, but rather through us, or past us, the way cats peer into other dimensions, seeing shadow selves.

He explained that he would be our vision-quest guide, our underworld escort, but we'd need to do the actual transformative work alone. He warned that such experiences are often powerfully life-changing. Some of us, he said, might collapse into something called "caterpillar soup." Here he licked his lips. He's the sort of person, he explained, who hears the word *dismemberment* and thinks, *Oh, yummy!*

Plotkin told us how, in August 1980, he'd trekked into the Colorado wilderness looking for a magical healing stick—"an instrument of great power, capable of transforming others by the simple wave of my wizardly arm." Instead of a wand, after four days of fasting, he found himself. It happened when a large yellow butterfly flew toward him, fluttered right up into his face, brushed his cheek with its wing, and whispered "Cocoon Weaver."

As he shared the story, I thought he might be joking. But then he said the name had saddened him—until he came to understand his calling: "My soul wants me to weave cocoons of transformation for others. People have to decide to go into the cocoon. I wanted a magic wand to change people, but they need to do it themselves. I can only point the way. Once I saw the power of it, I was terrified. And I still am."

The waves were rioting outside. When a gale of wind blew the shutters open, a fellow workshopper exclaimed, "Whoa, that's a lot of power." Moments later, the lights in the compound went out.

"Yes, we are here on the edge," intoned Cocoon Weaver, his voice rising. "The edge of the continent, the edge of consciousness. Allow yourself to feel dizzy." He wanted us to get discombobulated, to revel in the vertigo. When the lights came back on, my head was tilted upward. I opened my eyes and noticed the word *bliss* etched into the paint on a grate in the ceiling. Getting into the spirit of things, I imagined removing the panel and crawling on in.

Weaver read a poem by Robinson Jeffers, a poet associated with these parts. It was about how the first living cell had echoes of the future in it, how the first bit of life knew its direction: forward.

Then, the time came to start the initiation. "What's the price of admission in the new life?" Weaver asked. "The price is the life you've been living so far." He took out a box of clay and told us to grab a handful. "I want each of you to sculpt the shape of your life. Don't think about it too much, just see what comes out, but be mindful that you are giving a form to your life up until this point. While you make it, think of all the people you have loved, do love, will love, and want to love."

We each took our allotment of clay and molded it. I made a water drop: a smooth, imperfect droplet the size of a small pear.

When we were done, Weaver told us that the rest of the ceremony needed to be completed alone, each of us by ourselves, outside in nature. "You are going to take the shape of your life that you have made and you are going to give it to the waters of life. You are going to sacrifice it—and your present-day self—to the mystery."

We all shuffled out into the afternoon rain. I made my way to a bridge over a rushing creek whose waters poured into the ocean. Raindrops dripped down on the clay droplet of my life. Closing my eyes, I released the figurine into the rapids, where it was swallowed by froth and vanished into the sea.

That afternoon, I stopped for lunch at a place in Big Sur called Deetjen's. Adjacent to the parking lot stood a little shack with a sign saying LIBRARY. After eating, I walked in to check it out. There was a bookshelf, a desk, a telephone, some knickknacks. I noticed a softcover by the mystic Rudolf Steiner, about something called the "spiritual science." If we just concentrate hard enough, he argued, our soul can detach itself from the body and experience immortality. The way to do it is by focusing on symbolic images unlike anything in the real world (such as a portal filled with light) while simultaneously ignoring all one's senses. With persistent practice, ordinary thinking patterns grind to a halt. A new, altered, sleeplike state of consciousness is attained in which the freed soul can commune with the eternal unknowable.

Putting Steiner's book down, I noticed a piece of paper underneath the alabaster bust on the desk. "Realize you may die at any moment," said the note. "Learn how to keep it in mind. He who has freed himself from the dream of tomorrow has attained what he came here for."

On the short drive back to Esalen, I kept thinking about my car plunging off the cliff to the rocks below. The gray ocean was covered in blankets of fog. Mist tentacles wafted landward. Being freed from the dream of tomorrow; experiencing eternity in the present moment. As the ancient aphorism goes, *Aeternitas est merum hodie, est immediata et lucida fruitio rerum infinitarum*: "Eternity is merely today; it is the immediate and lucid enjoyment of the things of infinity."

How can we lucidly enjoy infinity? The Romantic poets claimed that it was through our imagination. Blake considered the world of imagination to be "what eternally exists, really and unchangeably," that "which liveth for ever." Only imagination permits us to see heaven in a flower. Wordsworth defined the imaginative as the part of us that is conversant with or turns upon infinity. He saw it as a visionary power, a mental glimpse of our indestructible selves.

The word *infinitude* is crucial to Wordsworth's understanding of imagination. We come from infinitude, he felt, and there we end up: "Our destiny, our being's heart and home, / Is with infinitude, and only there." His *Ode on the Intimations of Immortality* calls birth a "forgetting." Our soul forgets that it entered this world trailing clouds of glory from some immortal place. We can only rekindle this connection through our powers of imagination: "Though inland far we be, / Our Souls have sight of that immortal sea / Which brought us hither, / Can in a moment travel thither." Wordsworth often used water symbolism when discussing infinitude. According to Harold Bloom, "He constantly associates the sudden onset of the Imagination with the sound of rushing waters."

Walt Whitman found it in lakes and at the seaside, and elsewhere, too. Merely contemplating a blade of grass gave him an unshakable conviction in eternal life. For what else is grass than life bursting from decay? "The smallest sprout," he wrote, "shows there is really no death." His *Song of Myself* is actually a song of ourselves living forever, "of people, all just as immortal and fathomless as myself. (They do not know how immortal, but I know.)" Whitman is deathless, and the reader is deathless, too. In his world, everything is deathless: "I swear I think there is nothing but immortality!"

Any insignificant object is a connection to this truth. Whitman tells us we can find God in the tiniest, most ordinary places. Grass, some barns, any old gnat, the breeze; all revelations. A little mouse "is miracle enough to stagger sextillions of infidels." Unbelievers, he cries, all you have to do is look around to be convinced. Trees, seaweed, sparrows—all are immortal: "I swear I see now that every thing has an eternal soul!"

This truth is omnigenous—meaning it is in all things. We can find it anywhere, in the hollows at the bottom of the sea, in rocky riverbeds, in dust. If we ever lose sight of it, we can find it simply, Whitman assures us, by looking under our boots. Just look. The message is all around us. "Failing to fetch me at first keep encouraged, missing me one place search another, I stop somewhere waiting for you." Even if we can't see it right away, we have the ability to see it. Finding it doesn't mean we will understand it. "I hear and behold God in every object," Whitman admits, "yet understand God not in the least."*

For him, beholding God is akin to entering the Now. There "will never be any more perfection than there is now," he writes, "Nor any more heaven or hell than there is now." We can access the now anytime. Each moment—this moment—is the source of all creation. And when else shall we live if not now? Walt Whitman's immortality isn't in the afterlife. It's not far off, later, or prolonged—it's now, in every moment.

This same eternal now is the protagonist in *The Story of My Heart,* a cult-classic, 1883 memoir of immortality experiences by the British naturalist Richard Jefferies. It describes him sinking into timelessness while dipping his fingers into a brook, gazing upon blue pebbles on a beach, or sitting on a hillside tumulus. "Listening to the sighing of the grass I felt immortality as I felt the beauty of the summer morning," he wrote. "It is eternity now. I am in the midst of it. It is about me in the sunshine; I am in it."

Modern neuroscience has shed light upon the notion of tapping into the now. In a series of experiments, the physiologist Benjamin Libet (1916–2007), author of *Mind Time,* demonstrated that our brain is out of step with reality. There's a barely detectable delay between something's happening and our awareness of its happening. According to his findings, it takes at least ten milliseconds for neuroprocessors to register sensory information. "We are not conscious of the actual moment of the present," noted Libet. "We are always a little late."

---

*One of my favorite descriptions of God comes from my three-and-a-half-year-old friend Johanna Seligman, who says that God is from the future and wears a sparklesuit 'cause he's a lady.

This micropause is not unlike what computer programmers call latency. When we record music into software, for example, it is never exactly in sync with the performance: there is an infinitesimal delay as the microchips convert the information codes into a binary approximation. We, too, are always a micro-blink off. Our consciousness only rarely plunges into real time, into what mystics call the stream of now. The censoring mind usually creates a schism between itself and outer existence. It is as though our brain is NBC, and reality is Janet Jackson's exposed breast at a Super Bowl halftime show. In mystical experiences, the veil over our conscious mind experiences a wardrobe malfunction that thrusts us into the now, where we glimpse the bare nipple of existence.

During my last soak in the mineral pools at Esalen, I ended up in a conversation with a white-haired gentleman named Cornelius.

"Is this your first time at Esalen?" he asked.

"Yes," I answered. "How about you?"

Cornelius had been here countless times. He used to edit journals such as *World Affairs* and *New Realities,* he said, and had published works by Esalen legends such as Stanislav Grof. Cornelius was here deepening an integrative practice of connecting body and mind, he said, adding, "It's all about consciousness."

"What *is* consciousness?" I blurted out, immediately regretting that I'd asked something so impossible.

"Consciousness means being in the here and now," he answered, perfectly comfortable with the question. "Consciousness is not accessible through analytical reasoning, which is why scientists don't like it."

There it was, simple as a blade of grass. "That's a lot less heavy than what some of the people in my workshop want it to be. They're trying to turn into caterpillar soup."

"You know, it doesn't need to be a bad trip to be a powerful trip," Cornelius said. "In my years here, I've noticed that a lot of people who come to these sorts of spiritual retreats are negatively daemonaic."

"What does that mean?"

"That they are negative daemons. They come here seeking light, but they bring their darkness with them wherever they go."

"What about you?" I asked. "How come you seem so easygoing?"

"I'm an old spiritual eclectic." He smiled. "And I'm here doing what are essentially stretching classes. I'm not undergoing some sort of radical transformative experience—which is what some of your classmates are here for."

And what of my own journey? Had I transformed? I definitely hadn't found the fountain I'd dreamed about. But then it occurred to me, as I took a timeless moment to jot things down, that the precise act of trying to stop time—scribbling naked into a tattered, mineral-water-splattered fig leaf of a notebook—was what I had come here to do. Whether or not I'd gone through an initiation rite, writing is itself an ongoing initiatory process. Crafting narrative is ritualistic, like prayer. It means sitting down day after day, sculpting structure, living the story trying to live its way through us.

The cultural anthropologist Victor Turner made a distinction between ritual liminars who move through the rite and those he called *marginals*: people with "no cultural assurance of a final stable resolution of their ambiguity." For some doctors, monks, psychologists, artists, and others (including writers), initiation can mean learning to constantly waver between worlds. The Greek term *palingenesia* refers to a recurrence of birth. The necessary corollary to continual rebirths is having to die repeatedly. Aldous Huxley called this "a perpetual perishing." He considered it a necessary precursor to realizing one's human potential—that foundational Esalen aim—that "transience that was yet eternal life."

Having explored the outer limits of the belief in spiritual immortality, having followed Auntie Tiny's funeral invitation, having entered the portal, I was now ready for the next part of the journey.

Part 2

Magic

"I must know the real truth, the truth beyond magic."
"There is no truth beyond magic," said the king.

—John Fowles, *The Magus*

12

Mystifier

My job is deceiving people, but I start out by telling my audiences that I am a professional deceiver, an illusionist, and that they shouldn't believe anything they see, but that I'll make them believe anyway. It's a fair game. I don't tell them, "Trust me."

—David Copperfield

O brave new world
That has such people in't.

—Shakespeare, *The Tempest*

AROUND THE time I'd finally given up on ever getting to David Copperfield's fountain of youth, Martha Morano contacted me with a sleight of hand in mind. Could I write an article about Musha Cay for a magazine? My book's description hadn't hooked Copperfield, but if I helped Martha place a piece about his $32K-per-night private-island getaway, she'd help me get the story of the fountain.

A writer friend suggested I contact the editors at *BestLife,* a luxury fitness magazine run by the team at *Men's Health*. They were actively looking for life-extension stories, he said, so Copperfield's fountain of youth could be a perfect fit. They bit: a senior editor responded to my query letter immediately. Over the phone, he explained their demographic split in the following way: "*Men's Health* helps eighteen- to thirty-five-year-olds get the abs that get them the girls. *BestLife* is what they graduate to. It's for thirty-five- to sixty-year-old men who've got the abs and the perfect girl and who want to hold on to both of them while making 150 to 350K and having the healthiest, wealthiest *best life* possible."

Both magazines were published by Rodale, founded by J. I. Rodale, a passionate life-extensionist. At the age of seventy-two, he declared on national television that he would live to one hundred. Moments later, he dropped dead of a heart attack.

"Musha Cay and the fountain of youth sounds like it could be a great story for *BestLife*," wrote the editor in his commissioning letter. "We're really interested in all things longevity, and I can see a bunch of ways that we might be able to turn this into a story. I think the best plan of action would be for you to go down and explore the island and then for us to regroup when you return."

"This would work, I think," Martha replied. "As soon as he meets you, I know he will not be so wary about the book interview."

A month later, Martha Morano got in touch with an urgent message. David Copperfield wanted to meet in person. That very night. After his show at Montreal's symphony orchestra's concert hall. I hadn't even realized he was performing in town, but apparently he'd taken his "Grand Illusion" tour on the road. She suggested I not mention the book and focus instead on the *BestLife* article. He was a huge fan of the magazine. There'd be plenty of opportunities to fill him in on my own project later. For now, I had two tickets for the eight thirty show at Place des Arts.

To prepare, I checked out a library book called *David Copperfield's Tales of the Impossible*. The dust jacket featured a photo of his face partially in the shadows, hinting at a darkness within, while simultaneously recalling a Fabio romance. His mane of black hair had been sprayed into a puffed-up rocker bouffant. The short-story anthology included contributions by authors such as Ray Bradbury and Joyce Carol Oates. One blurb described the book as "more than a book."

An ideological thread ran through the collection: to believe something is real makes it so. Every mystery ought to remain unexplained, Copperfield wrote. Dean Koontz's preface characterized the magician's act as a religious experience, a means of going beyond set limits: "He convinces me, on some deep level of the heart, that I can fly too, if not physically, then in spirit."

In his introductions to each piece, Copperfield shared psychological insights. Emotional survival, he wrote, "depends on the mind's capacity to transform what it sees into what it wants to see, to create from our imaginations an illusion more comforting than reality." In another intro, he discussed how, at the end of life, "the delusion of having been a 'good person' must be faced head on."

Copperfield had also written the first story in the collection. It told of a boy named Adam whose grandfather is dying. Trying to keep him alive, Adam tears up bits of paper and throws them in the air to create a snowlike effect. Alas, the following morning, Grandpa dies. Grieving, Adam heads to the beach, wishing he could bring his grandpa back to life. The story ends with a sheet of paper floating toward him on the breeze. He tears it up and hurls it into the sky; real snow wafts down.

What a coincidence, I thought: a story about using magic to ward off mortality.

The show that night was sold-out. Arriving early and giddy with anticipation, my date and I picked up our tickets and spent a few minutes scoping out the rest of the audience. It consisted mainly of families, children out with their grandparents, as well as a few couples dressed up for a romantic night.

The usher led us to a pair of plush blue velvet chairs set apart from the rest of the theater, a row in front of the first row, somewhere between the anonymity of the crowd and the glare of the stage. I immediately felt exposed. Whispers went up behind us, a hushed mixture of envy and curiosity, speculations about the type of people who'd procure such ostentatious seats.

As we waited for the show to begin, a DVD loop projected lists of Copperfield's accomplishments against the backdrop. "Most tickets sold around the world"; "Only living magician to have a star on the Hollywood Walk of Fame"; "Largest sum of money ever amassed by a magician"; "Most Emmy Awards won by a magician"; "Magician of the Millennium." The slide show was on repeat. It kept trumpeting his wealth, and how he'd sold more tickets on Broadway than *Cats*. The excessiveness of it all felt like an illusion, one I couldn't help seeing as intended to make the other headlines about his private life disappear.

The boasts were as revealing as they were routine. Writing about Prospero, W. H. Auden spoke of magic as a way to blot "the gross insult of being a mere one among many." Copperfield's idol Houdini also excelled at self-promoting. They both had East European–immigrant parents and came from a lower-class, *shmatte* background. Houdini billed himself as the Greatest Magician the World Has Ever Seen. But the boasting also led to his downfall—his claim that he could withstand any punch caused a young man to deal him a fatal blow.*

Copperfield made his entrance on a motorcycle shrouded in smoke. Disembarking, he trotted right up to the lip of the stage. He came so close that I could make out the granularity of his makeup's powder. His long, delicate fingers didn't shake in the least, but he looked tired, baggy-eyed, like a wax figure of himself. His hair had an almost metallic quality. His black dress shirt was crisply pressed, and it billowed off his clothes-hanger shoulders, emphasizing a skeletal slenderness. He wore surprisingly scuffed-up, black leather Prada high-top shoes.

The show opened with another on-screen montage, this time of his appearances on TV shows, from *Seinfeld* to Conan O'Brien to *The Simpsons*. Again: displaying importance. When the video ended, a few minutes later, the actual entertainment began. His demeanor became increasingly businesslike. His gift for speaking quickly and clearly herded the show along at a brisk clip. As he spoke, his eyes roved through the crowd, peering intently at everyone in the first few dozen rows, zeroing in on certain individuals. He recited memorized passages while spending long seconds eyeing audience members. When our turn came, I tried to look unquestioningly enchanted, but not overtly so.

Alongside an icy assuredness, his ability to segregate his thoughts from the words he was speaking gave the performance a mechanical quality. Throughout his TV years, he'd been the opposite of ironic, so vanilla that his manager compared him to white paint on a white wall. Now, still G-rated, his delivery had taken on a sarcastic edge.

As detached as he was, his tricks were fun to watch, complexly orchestrated and mainly inexplicable. Unfortunately, when he made an origami rose float, I was sad to see a string attached, easy to detect from our vantage point even against the intentionally distracting shimmery-silver background. After a few bits, he again played some video footage, this time of a large-scale Houdini-inspired illusion from the glory days. His TV tricks—like the one in which he chained himself to a flame-engulfed dinghy, plunged over the frothing edge of Niagara Falls, and somehow emerged dangling from the base of a helicopter—were significantly more impressive than the small-scale legerdemains he enacted before our eyes.

The show heated up when he picked two women from the crowd,

* The organ-puncturing hook happened backstage here in Montreal, after a performance eight decades ago.

including my date, to come onstage. He handed her a pair of large black rubber gloves—to protect her, he explained, from an African scorpion she'd be handling. She struggled to open the first glove, which was deliberately stuck.

"Come on, you can't get it on?" he joked.

At the end of the routine, in which the real scorpion picked a card, any card, Copperfield pointed to his cheek and asked for an innocent kiss. The audience clapped as my date left the stage.

"How was that?" I asked in a hushed voice, smiling in case he was watching.

"That glove was rigged," she whispered, back in her chair. "He manipulated my stage presence. I was robbed of my grace."

The show continued with a number of solid prestiges. Copperfield moved through them briskly, getting the show done. For his grand finale, he made thirteen audience members disappear. Each of them climbed aboard a wooden ark. One elderly lady took a little long to get into place and he started grumbling about getting paid overtime. Once she was settled in, he covered the boat in fabric. The sound of triumphant horns and strings unfurled from the PA. A moment later, Copperfield pointed into the distance. At the other end of the auditorium, the load of teleported people stood smiling and waving. Indisputably magical.

The curtains fell. A short, impassive assistant named Robyn immediately swooped in and whisked my date and me backstage. We arrived at a set of thick metal doors. "Wait here," Robyn told us. As she went inside, we caught a glimpse of the boat people signing paperwork.

A few minutes later, we were ushered through the doors into a cramped vestibule. Copperfield stood waiting with a towel around his neck and a stack of Musha Cay postcards in his hand. I told him how great the show had been. He nodded graciously, considerately asked my date about herself, and then handed over some glossy postcards. They were the same ones Martha had given me. He started describing his resort, "the most magical vacation destination in the world," in tones as well honed as in his onstage performance. While his charisma was undeniable, other qualities were also on display: wealthy salesman, embattled celebrity, master seducer. An agitation lurked behind his calm façade, a worldview strained through a colander of stress. But as he spoke, it seemed to me as though precious crystals were pouring from his mind. An entire vista unfolded before our eyes.

I glanced down at the postcards. On Musha, it said, magic is ever present. Visitors encounter magic that they can experience nowhere else on earth. I wondered whether he considered the island just another part of the show. Did everything he involved himself in have to be magical?

His monologue over, he leaned back against the wall, tilted his head downward, and stared up with a scrutiny calculated to decode my intentions. Wondering if he was hypnotizing me, reading my mind, shuffling through my innermost thoughts, I projected images of sandy tropical beaches onto my frontal cortex.

"So are you on the staff at *BestLife*?" he asked, wiping the sweat from his brow.

"No, just a freelancer."

"Have you written there before?"

"No," I admitted, realizing I needed to shuffle the deck. "But they really want feature stories about rejuvenation, so my editor loves this idea."

"There are lots of rejuvenating things down there." He raised his voice, excited. "The whole point of creating Musha was to help people feel like children again."

I nodded empathetically, thoughts trailing off into what it might mean to want people to feel like children rather than adults. He said something about how somebody I'd never heard of who'd vacationed in Musha had been on the cover of *BestLife* recently. Copperfield genuinely liked the magazine. He took the stack of postcards back for a moment and pointed a manicured finger at an image. "Here, I want you to check out that pink sand," he said, smiling benevolently at my companion. "Musha has pink-sand beaches the size of South Beach, in Miami. Sugar sand. That's what I'm talking about. If you come down, you'll see the sandbar where *Pirates of the Caribbean* was filmed. You can even watch it at my drive-in movie theater on Coconut Beach."

"It all sounds so wonderful," I said, steeling myself. "All these details will make the perfect backdrop for a story about the fountain of youth."

"The fountain?" He frowned, serious. "How did you hear about the fountain?"

"I read the newspaper reports." Martha probably hadn't told him anything about my angle.

"There won't be much of that." He smiled tightly, his fluttering fingers tracing arabesques of misdirection in the air. "You'll be immersed in the sense of wonder: there'll be amphibious catamarans, Jet Skis, and

Balinese daybeds. We're just not ready to show the fountain of youth yet."

"Is it real?"

"Yes, it's real, but we're not sure how real," he said, fingers stiffening. "We know it affects plants and insects, but we don't know if it's for humans. It's too early to discuss at this point. There won't be any of that."

"My editor needs there to be some of that." I was showing my cards. "They commissioned a story about Musha's antiaging qualities, including the fountain of youth." It wasn't me—it was *BestLife*.

"Musha is a place for people who have the wherewithal to go anywhere on earth," Copperfield continued, a hint of fluster creaking through. "It's about making billionaires feel like they're kids again. That's a story, isn't it?"

"It is." I nodded, with determination now. "But, like I said, *this* story needs the fountain. I don't need you to explain how it works. I just need to see it, to experience it in some way so that I can use it in the story."

Copperfield straightened up brusquely. "Were you promised the fountain of youth?"

"No," I answered, apprehending that Martha might get in trouble for this.

"So nobody promised you the fountain?" he reiterated, calming down, sensing a simple bureaucratic solution to the impasse.

"Nobody promised me anything. But there's a story behind the fountain of youth that hasn't been told yet, and *BestLife* is interested in publishing it."

He paused, balancing the odds. He looked at my date, and looked back at me. "Okay," he granted finally. "I might be able to give you a taste."

"That's all my editor wants," I exhaled gratefully.

"We'll continue this over the phone." He handed me a card. His number ended in five 5s, like a Hollywood pizza-delivery line. "I'm going to need to speak to you quite a bit before I agree to this."

"That's perfect!" I smiled.

Perplexed by my enthusiasm, he shook my hand, said good-bye, and faded into the inky recesses of the theater.

Following the Montreal performance, Copperfield and I spoke a number of times. If he called and I was out, my call display showed that he would keep calling, often every ten or fifteen minutes, until he got through. The first time he tried to reach me, I returned home from a short meeting to find three messages from him on my answering machine. When I called him back, he had some bad news for me. He'd been thinking about it and had decided I wouldn't be able to see the fountain.

That didn't mean I shouldn't visit, though. "I want to be very clear with you about what you'll see if you come down and what you won't see," he said. "I'm not in the bait-and-switch business. I'm in the billionaire business. I make billionaires happy. Is that a story we can do?"

"There are numerous possible stories—"

"The richest man in the world has been here four times. I'm not gonna name any names. . . . *Okay,* the Google guys got married here."

He repeatedly mentioned how he couldn't reveal who'd stayed on Musha in the past, even though his publicity materials referred to them, by name, frequently: "In the guest roster are the names of entertainment luminaries such as Oprah Winfrey and Robin Williams. . . . Musha's most recent guests include one of the world's wealthiest men, an internationally known computer software magnate." As mysterious as Copperfield tried to make it sound, there was no way I could even pretend I'd be focusing the story on Bill Gates's visit.

"The 'billionaire playground' angle is different than a feature about your fountain of youth, which is what I pitched," I said. "We're exploring the metaphor of rejuvenation, and the fact that there's a real fountain of youth on Musha. I can weave details about the billionaires into the story as long as we agree I'm there to write about the fountain."

"Musha is about being children again," he corrected me. "I'm redoing my childhood. My parents used to smuggle me into drive-ins. I'm haunting one of the islands. It's a metaphor about reliving and regaining things that we've lost."

"Yes, but you issued a press release about the fountain of youth, didn't you?"

"Yes, but it's *not* ready to be *seen* yet," he snapped, then immediately switched to a more patient tone. "The fountain-of-youth thing will reveal itself in the next few years, but not now."

"But isn't there a way to speak about it now? We *are* speaking about it now. And the way you speak about it suggests that it's very meaningful to you."

"It has a mystical quality, but it's real. If you want to think it's snake oil, fine."

"I don't need to bathe in it. I don't need an explanation of it or any full reveal. I can respect the mysteries of the fountain's ways, but I do need to write about it. I just need enough of a thread in order to bring the rejuvenation aspect to life."

"I really don't want to show it. If I turn water into wine, I'll get hammered. As soon as you give people something tangible, they'll rip it apart. It'll be dissected. I'll be crucified."

Had he just made a Christ analogy? I tried to assuage him, going on about how *BestLife* readers would relate to the story of the fountain.

"I want people to not think I'm . . . crazy," he replied. "But you're right: maybe we can *talk* about the fountain. On Musha. Yes. Would that be enough for your story?"

I wasn't sure. I didn't want to give in too easily. He told me to think about it and call him back.

I mulled it over and saw only one way forward. To not see it—but talk about it—would still be better than not going. I'd just have to sneak out at night and find it. When we spoke next, a few days later, I nevertheless pressed him to show me the fountain.

"You won't see my wrinkled hand go into a stream and come out young," he said firmly. "This is not a trick. But if you want to talk about the meaning of the fountain—that, we can do."

"A discussion of the fountain of youth, on Musha."

"You'll have a very passionate subject. I speak about the fountain with great verbal aplomb."

He suggested I speak to my editor about it. I told him that I was pretty sure it would fly. Stories invariably end up different from how they're conceived, I explained, reiterating my point about the numerous possible approaches to covering Musha. I was essentially agreeing to give him the story he wanted—"David's Magical Island for Supremely Rich People"—as a way of getting the story I wanted.

Sensing an opportunity to introduce my book into the conversation, I mentioned my intention to write more than one story about Musha. Even though he wasn't ready to show me the fountain for *BestLife,* would he be willing to show me the fountain before any other reporters once he was ready to unveil it?

He said he'd consider it.

"There's one last thing," he added. "It's important to bring someone.

To be there alone is going to suck. All the experiences on Musha are shared experiences. You should bring that girl you came to the show with, or someone else if you want, it doesn't matter. Do you think she'll be available?"

I decided not to invite her after coming across a leaked document, "Show Participation," containing instructions for his road team on bringing girls backstage. It revealed how Copperfield's onstage gestures indicated to his team which women he was interested in. Backstage, his assistants had to recite memorized lines, including a mention of Musha Cay: "Hi! How are you? Did you enjoy the show this evening? Good! . . . My name is (name). . . . I work for David. . . . We just wanted to thank you for your participation in the show, and David will be here in a few minutes to also thank you personally. Every time we see somebody in the audience or onstage with a look or personality that is special, we like to get more information so we can possibly contact you in the future . . . and so we can invite you and your family back to the show whenever you are in the city. Did you know that David has recently bought some islands in the Bahamas? Well they are BEAUTIFUL and we are doing a lot of project [*sic*] for these islands: ads, tv, radio and many other promotions. So we'd like to keep in touch with you in case there is a job in the future we think you would be interested in."

One section outlined methods of handling models, including important prospects David might want to speak to privately. When "pulling" the guests, assistants were advised to be smooth and discreet, especially with a "special." There were guidelines on extracting personal information from these female "prospects": "While they are filling out the form, you should try to chat with them and ask them the following questions: Where are you from? What hotel are you staying in? Who did you come to the show with? Husband? Boyfriend? Friend? Whenever it feels comfortable, you should take a Polaroid photo of them." If a boyfriend or husband started causing problems, assistants were instructed to do their best to calm things down by referring to the "What to Say" sheet for help.

The document noted that girls he'd bring onstage for a trick called "The Scorpion" were to be given extra attention backstage after the show. The routine needed two volunteers, but usually only one of the participants was brought backstage. "On occasion, David will have you pull in both scorpions, even if he is only interested in one of them, just for comfort."

The Scorpion. Hadn't he brought my date up to participate in his Scorpion routine?

I decided to invite my former bandmate Rafael Katigbak, an amateur magician who'd idolized David Copperfield as a child. Our band had been named We Are Molecules. When we'd still played music together, our mutual interest in magic had led us to cowrite and perform a live musical called *The Magic Idea Machine.** In the song cycle's plot, a supercomputer could turn any thought into reality. The Magic Idea Machine had only one glitch—an inability to understand the meaning of love. Our songs helped it learn to love.

"I feel like I've won the lottery!" Raf yelled when I called to ask if he'd come down to Copperfield's island with me.

"So you're in?" I double-checked.

He didn't answer. There were scuffling sounds in the background.

"What's going on over there?"

"I'm running around dancing," he hollered. I could just see him pirouetting through his perpetually cluttered apartment, trampling over the video games, karaoke equipment, musical gear, and vinyl LPs left out after yet another of his frequent parties. A short Filipino hipster with fifties nerd glasses, a swoosh of black hair, and permanently chapped lips, Raf was one of my oldest friends. Since our band had fallen apart a few years earlier, we hadn't seen much of each other, but this trip would be a way of catching up. "I can't believe I'm going to meet David *Copperfield*!" he said. "Every time I try to imagine him, I visualize Julio Iglesias with Coppertone skin. But then I remember what he really looks like: the geekiest guy ever."

Before Raf got too excited, I filled him in on details about the alleged rape, still pending.

"So we're going to Rape Island," Raf joked.

I reiterated that Copperfield hadn't been found guilty of any wrongdoing. The whole thing might have been an extortion attempt.

Raf didn't care. "God, I hope he rapes me," he sighed, faux-dreamily.

---

*We did so alongside the third member of our troupe, Liane Balaban, the date who'd accompanied me to David Copperfield's performance.

Perhaps some levity, courtesy of Raf, would cut through the situation's seriousness. I wanted him to come not just because of his comedic sensibility, or because he appreciated the strangeness of magic as much as I did, or that he'd help keep things grounded no matter how odd they got—or even to repay him for the time he'd helped me unclog the sump pump I backed up at our friend's country house—but because Raf had much in common with Copperfield.

They'd both started learning about magic as a way of getting members of the opposite sex to notice them. "I would always fall in love with the girl in the next seat, who would then ignore me and be in love with somebody else," Copperfield once explained. "But I knew that somehow, someday, I would capture her attention." He found a solution—by transforming himself into Davino: the Boy Magician.* "David Kotkin became Davino only to impress girls, for girls were his obsession," as Bill Zehme wrote in a 1994 *Esquire* profile.

In his tween years Raf had started frequenting toy stores to buy magic kits, for much the same reasons. "I was a skinny runt with bad skin and a voice that sounded like Steve Urkel with a cold," Raf once described himself, in an autobiographical column for the local weekly paper. When he'd first started dabbling in illusions, Raf wrote, he was so shy that if a girl spoke to him, he'd blush, get nauseous, and run to the toilet.

Magic didn't exactly help Raf's social standing in high school. At a regional public-speaking championship, his grand finale about "the light of peace" hinged on his pulling a napkin off his hand and making a two-foot-tall candle appear out of nowhere. For some reason, the spring didn't engage, and the candle came out only a few inches high. On top of that, his lighter didn't work. And the wick wouldn't light.

By the time we were adults, he'd overcome his timidity and mastered a few tricks. Raf once surprised our friend Seth by performing a card trick for his birthday. Raf fanned out a deck, asking Seth to choose a card at random and then rip it up. Seth gave all the torn pieces back, except one ripped bit. Raf made the rest of the card shards disappear. He then asked someone to go into the kitchen and bring back a green pepper. When our astonished friend sliced the pepper open, it contained the reconstituted card, intact except for the one torn piece, which slipped into place perfectly.

---

*At the age of fourteen, Davino became the youngest-ever inductee to the Society of American Magicians.

Raf, the most Peter Pan–like person I'd ever met, would be the perfect foil for a fountain-quest to Musha. "There *is* something similar between me and David Copperfield," he allowed. "It's gonna be so interesting to see him in person. I mean, what does a guy like that eat for breakfast?"

Visitors to Musha get to eat whatever they want for breakfast, as I learned when Martha e-mailed me a form about our favorite foods, our favorite drinks, and all sorts of other preferences they'd take into account while we were on the island. It asked us to rank different meats, poultry, and shellfish, by order of likability. One section asked if we had any special soft-drink requirements. "Write down Tab cola!" said Raf. "Just to see if they really take special soft-drink requirements seriously."

After filling out the form, I called Copperfield to tell him my date from his show wouldn't be able to make it. His response sounded distant. "Where are you?" I asked.

"I'm in the middle of nowhere," he said in garbled tones. "On the tour bus."

It took a moment for me to realize that I'd woken him up. Capitalizing, I quickly told him about Raf—"he's a total sweetheart"—emphasizing his interest in magic.

"Sounds good," Copperfield mumbled.

"Great!" I finalized it. "I'll let you get back to sleep."

Not long after that, Copperfield called to say he'd been thinking of bringing his girlfriend of three years to Musha at the same time. He asked if my journalistic code would require me to include personal details about her. "If you need to write about her, that's fine, I just won't bring her," he offered.

I assured him he could feel free to bring her, that I wouldn't need to reveal her identity as part of the story. It made no difference whether he'd been with her for three years or not. I reminded myself to focus on his feelings about magic and the fountain, to see if he could help shed light on the idea of belief.

I checked in with Martha to tell her how the two stories were evolving. She said she'd tell Copperfield a bit more about my first book to prep him. The *BestLife* team was thrilled it was all working out. Everything looked set. Our speedboat to Musha would be leaving in a few weeks.

13
Escapology

Robin Leach: What is the magic of magic? What is it that makes the average person so fascinated by magic?

David Copperfield: It is a need. We all possess the need to dream. . . . Dreams are illusions, and we can't let go of them because we would be dead.

—*Las Vegas* magazine

"The perfume of Eros!" he repeated and helped himself to wine. "I say, Orr, what the dickens is that?"

"Only the motor force of the universe."

"What?"

"Yes, indeed. It is the sublimate of love. And love is the source of human activity. It has no other."

—Edgar Saltus, *The Perfume of Eros*

IN HISTORY, magic's origins are inseparable from religious activity. "Magic is no other than the worship of the gods," explained Plato. The etymology of the word *magic* goes back to synonyms for "priest" in Proto-Indo-European (*magh*) and Old Persian (*magos*). The priestly magi of antiquity were sages and natural philosophers—early scientists. Like shamans, they healed, oversaw ceremonies, and performed sacrifices.

Both magic and religion have always been predicated on the belief in the existence of other realms or dimensions apart from the empirical. Today, we understand religion as that branch of experience focused on venerating and approaching the beyond, whereas magic aims at controlling it and harnessing its powers for personal or communal use. In the beginning, however, we didn't distinguish much between worshipping deities (religion) and trying to manipulate them to our ends (magic).

A systematized approach to the utilization of hidden natural forces took hold in ancient Egypt. The pharaoh's magicians were intermediaries between this world and the divine. Rumored to be capable of disentangling their souls from their bodies, they also had books of spells that could help others with everything from fertility problems to protection charms to harming enemies to ensuring posthumous survival. Words arranged in a particular fashion and pronounced in incantatory ways were thought to grant the spellcaster influence over the spiritual world. Peddling in secret knowledge, employing fetishes or talismans, they made offerings to the sacred while also performing rituals intended to make it obey their commands.

Water pops up repeatedly in these spells—not surprisingly, given the essential role it played in Egyptians' daily lives. Without the beneficial flooding of the Nile, their civilization wouldn't have flourished. Priests made invocations to imbue liquids with healing powers. In some rituals, water was sprinkled over dead bodies in hopes of pouring a regenerative force into the desiccated cadavers.

For the Egyptians, death wasn't the end; it was a magical transition that brought renewal and rejuvenation. Just as Ra dies with every sunset and descends into the underworld only to return with each new dawn, each person's life goes on after death. Immortality, at first the sole purview of the pharaohs, became a civilian birthright. "You sleep that you may wake; you die that you may live," explained the Pyramid Texts, a collection of magical spells from the third millennium BCE (and one of the oldest religious texts in existence). After dying, souls were ferried through the Nurse Canal, along a winding waterway, and over inundated fields to the place they originally came from, where they were "born again, new and young."

The hieroglyphic record reveals that, in both magical and religious thought, death for the Egyptians is a gateway. Destinations abound. Souls don't just float down paths of water to the land of the blessed; they also waft heavenward on smoke plumes of incense, or get sucked into the infernal regions, or get brought "to the house of darkness, where dust lies on door and bolt." In the Hall of Double Truth, disembodied spirits encounter Osiris, "he who springs from the returning waters." The foremost god of resurrection, Osiris blesses properly mummified souls with his draft of immortality, the cold water of everlasting life. This precipitates an apotheosis known as Osirification. "I am Osiris, Lord of Eternity!" declares each dead soul, wetly aware of its own immortality. Having become Osiris, souls enter a silent boundlessness, where they remain until everything reverts to its original state of primal floodwaters and starts all over again.

This circularity mirrors the recurring inundation of the Nile. And in ancient Egyptian cosmologies, there is a beginning to all these cycles. Everything started when the creator God Atum emerged from the watery darkness of initial chaos—a static ocean called Nun. Parentless Atum arose miraculously, self-born, auto-conceived, not there and then there, floating in the primordial liquid. Having generated himself, he promptly started masturbating. Atum, the original onanist, swallowed his own semen and spat out the first two gods, Shu (air) and Tefnut (moisture). Together they personified the sky and the sea, life and eternity, truth and infinity.

Whether ejaculating or expectorating, Atum exuded bodily fluids full of vitality (his sweat, too, played a theogonic role). His tears became the first humans. And his *ka,* or vital force, is the animating energy of all living matter. After spewing out the first gods, he hugged them so that his *ka* entered them. Every human also has a *ka,* or what we today might call a soul. It is our godlike component, the part that is reunited with its maker after death. This afterlife regeneration can only occur, notes Egyptologist Erik Hornung, "if what is old and worn becomes immersed in the boundless regions that surround creation—in the healing and dissolving powers of the primeval ocean Nun." The immortal *ka*s enter the otherworldly waters of timelessness, sloughing off physicality and taking on a life in nonexistence.

Even without articulating the idea of a fountain of youth, ancient Egyptians viewed water as having magical and spiritual capacities. They also formulated another central axiom of magical thinking: that a kind of energy exists in nature that can be tapped into. This intangible substance is in all things. Their word for magic was *heka,* meaning "activating the *ka*." (Heka was also the name of the God of Divine Magic.) To practice magic, they believed, adepts must learn first to discern this essence, and then to master it, to bend its iridescence, to make it do their bidding.

Perhaps because it is invisible and an article of magical belief, it's hard to put this force into words, let alone understand it. Historians of the occult describe it as an imponderable charge, a suprasensible cosmic medium, a nonluminous astral light that penetrates the tangible world, a fluidic emanation that suffuses all of creation. "The existence and possible employment of this force constitute the great secret of Practical Magic," noted Eliphas Lévi in 1860's *The History of Magic*. He spoke of this magical agent as a plastic mediator, something akin to the Imagination of Nature. He also called it "a universal life force."

This magical energy shows up throughout time, in a disparate range of places, under a variety of names. The Polynesians called it *mana*. In Malaysia, it was *pantang*. Native American tribes such as the Iroquois or the Sioux named this nonphysical spirit energy *wakanda,* or *orenda*. In the late eighteenth century, Franz Mesmer claimed to be capable of harnessing animal magnetism; he could heal patients, he said, by shooting magnetic fluid from his fingertips. Wilhelm Reich argued that sitting in orgone accumulators attracted this libidinal life force into our bodies. Occultists have long seen it as the basic ingredient in electricity and in our dreams.

According to the enigmatic *Emerald Tablet* of Hermes Trismegistus, ostensibly a summation of Egyptian magical thinking, this current is the strongest of all powers, the Force of all Forces, an energy that ascends and descends between earth and heaven to perform the operation of the magic of One. Anaximander, one of the earliest Greek philosophers, called it *apeiron*—the infinite, unlimited, boundless. Calling upon this constantly circulating metaphysical force remains a basic assumption of magic. One of David Copperfield's classic close-up illusions involves materializing water in an empty glass. While performing the trick, he speaks of finding the true source of wonder. "All you have to do is look straight up, because after the planets, and the moon, and the tides," he says, as live goldfish swarm from his hand into the water, "comes *life*."

Magic tricks aren't really magical, as anyone who's seen shows such as *Magic's Biggest Secrets Finally Revealed* can attest. "Real magic," oxymoronic though that may sound, suggests the possibility of transcending physical laws through secret wisdom or enlisting the spirit world's assistance. It operates under the assumption that, by performing certain actions, we can affect the outcome of events outside our control. This magick can be defined as the science and art of influencing nature, or causing change to occur in conjunction with will or intention. Lying somewhere between faith and knowledge, its potentialities can be benign or harmful. "Our method is science, our aim is religion," as one of the twentieth century's best-known ceremonial magicians, Aleister Crowley, once stated. Using powers from a certain source to alter the outside world and transform lives, real magic is results oriented. It can be akin to sorcery, spiritualism, or witchcraft. To attain physical immortality would be real magic.

Entertainment magic, on the other hand, or secular magic, as it is also known, means the phenomenon of card tricks, sleights of hand, optical illusions, misdirection, and simulated vanishing acts performed publicly. Magic shows only really became a popular form of spectacle in the nineteenth century, when secular magicians started pulling bunnies out of top hats and doves from rumpled tuxedos, rather than evoking supernatural beings. The word *magic* to describe legerdemains, performative tricks, and attention management came into currency in 1811. Its effect is to make us tantalized by the romance that what we're witnessing *may* be real. While it would seem that there should be a clear distinction between magic shows and real magic, the lines were always blurred, often intentionally, by magicians aiming to create a greater sense of mystique around their performances. The advent of stage magic paralleled the rise of spiritualism, and many magicians in the 1800s also performed as mediums.

In today's age of David Blaine's high-voltage stamina-stunts and Criss Angel's mascara-streaked mindfreakery, David Copperfield may simply have been channeling archaic aspects of magic history by saying he'd discovered the fountain, but something about it did seem beyond his usual repertoire. The "real magic" premise made me think of Doug Henning, the world's most famous magician before David Copperfield came on the scene, and the best-known illusionist after Houdini.

Fifty million viewers tuned in to NBC to watch Henning re-create Harry Houdini's Water Torture Escape in 1975. "Mr. Henning is beyond compare as an illusionist," raved the *New York Times*. "He believes in magic, and he makes us true believers, too." Henning was still a megastar a decade later when he abandoned his showbiz career,* which he characterized as "fake magic," to devote himself fully to the Transcendental Meditation (TM) movement.

"I have always believed in real magic, that there is more to life than the senses can perceive," Henning said at the time. He found what he was looking for by, as he explained, "studying the secrets of the yogis and the mechanics of unfoldment of creation with Maharishi. I realize now that even my wildest imaginings are but a fraction of what is really possible."

---

*He sold his stage illusions to fellow magicians, including David Copperfield.

His television specials always ended with him saying, "Anything the mind can conceive is possible. Nothing is impossible. All you have to do is look within, and you can realize your fondest dreams." He certainly believed it. Through TM's "real magic," he spoke of attaining such an exalted state of consciousness he could perform actual miracles. "You can disappear at a high state of consciousness because your body just stops reflecting light," he explained enigmatically. "When you reach your full potential, and you think, 'I want to levitate,' you can levitate."

No longer interested in delighting crowds with tricks that appeared magical, he wanted everyone to experience infinity just as he had. And if, as he said, the human nervous system could be cultured so as to control the laws of nature, it could certainly allow a body to live forever.

Too young to have witnessed Henning's time at the pinnacle of television magicdom firsthand, I'd first learned about Henning in high school, when he ran for Canadian office as a Natural Law Party candidate. News reports showed him yogic flying and joked about his making the national debt magically disappear. Laughed out of the political sphere, he threw himself into designing a meditation theme park in Niagara Falls called Veda Land. He and his partner in the project, the Maharishi Mahesh Yogi, drew up plans for a levitating building, a meditation university, and a Tower of Peace where world leaders could meet to settle disputes. A flying-carpet ride would carry up to 120 passengers onto a rose petal, plunge into its molecular and atomic structure, then finally come to rest in the flower's "pure consciousness." The plans were interrupted by Henning's battle with cancer, which he chose to fight without medication or chemotherapy. He died at the age of fifty-two in 2000. Until the end of his life, however, he maintained that illusion magic simply uses known laws of science and physics, whereas real magic involves laws that haven't yet been discovered.

No matter what happened, I felt sure I would be able to glean more about the intricacies of magic—real or otherwise—on the eleven Islands of Copperfield Bay.

14

The Sorcerer's Lair

What is this fountain, would'st thou rightly know?

—Thirteenth-century Persian poem

Under the fountain of truth we drank the questions down.

—Alden Penner, *Clues*

JUST AFTER noon on November 7, 2008, Raf Katigbak stood at an American Airlines gate in Miami, head craned toward a TV broadcast about an incipient hurricane swirling through the Caribbean. The National Weather Service warned that Paloma could hit the Florida coast within days. Flights to the Bahamas were still going ahead as scheduled, but they wouldn't be returning until the tropical cyclone wreaked herself out.

"We're going to be stranded down there," said Raf, high-fiving me. "I'm never going home." I hadn't seen him in weeks. He'd let his beard grow out. The greasy black hair spilling from his panama hat reached his chin. He wore banged-up Ray-Bans, rattan shoes, a cream-colored linen suit with ripped silk lining, and a paisley opera scarf.

"You look conspicuous," I told him.

"I've been getting a lot of funny looks. The only chink in my armor is the mismatched argyle socks."

"Is this your idea of dressing for the occasion?" I laughed.

"Actually, this is what I wore last night. I was at a bar until three in the morning. I went home, packed, and then left for the airport an hour later. The US immigration agent was convinced I was a Colombian import-export businessman. He was like, 'Do you have more than ten thousand dollars on you?' and looked at me all serious to gauge my answer."

"You do look like you have more than ten thousand dollars on you."

"I *feel* like I have more than ten thousand dollars on me. Come on—I'm going to the Bahamas to stay at David Copperfield's place!"

When we deplaned at Exuma International, two uniformed customs officials queried us about our Bahamian intentions. I explained that I'd come to write a story about the fountain of youth.

"I hope you tell the right story," said one of them, stamping my passport.

"Me, too," I replied.

Droplets of sapphire ocean spumed onto my face as we skimmed into the Out Islands. The waterway was a long liquid boulevard with private-island homes on either side. Sitting there in the back of the motorboat, I realized how impossible it would have been to sneak into Musha. "Pretty dreamy," Raf sighed, pointing at an isolated patch of ivory sand covered in driftwood formations and skeletal palm trees.

Island after island flashed by, each lovelier than the last. We passed a gray stone castle on a scrub-covered atoll floating in the aquamarine water. A few minutes later, our captain, a buff, thirtysomething, country-club type, pointed out Johnny Depp's island.

"I feel like I'm home," said Raf, putting his feet up and flipping through a copy of *Private Jet Lifestyle* magazine.

About forty-five minutes later, we coasted up to the dock on Musha. The stairs led to a large building called the Landings, a tasteful wooden affair painted in pastel green, blue, and yellow. Above it, perched atop the island's zenith, lay a dark mansion. As we prepared to disembark, the captain pointed out four or five sharks in the water, saying they lived under the quay.

"Are they pets?" I asked.

"Nah, they just live here."

"But do they belong to Musha?"

"No, they're wild sharks, but this is their home."

"So sharks just choose to come live here under the front porch?" interjected Raf.

"That's right," answered a slightly impatient, blond, bland, managerial-type woman standing on the pier, dressed in a Musha Cay golf shirt. "But they're not dangerous. You can even go down there and swim with them while you're here."

"Do they bite?"

"Not really," she said, smiling strainedly. "But don't stick out your fingers around them. Don't grab their tails, either. And don't creep up on them from behind."

As the various majordomos, butlers, and concierges introduced themselves and made sure we didn't lift any of our suitcases, a pair of skinny legs in gray Crocs and peachy-pink surfer's shorts strolled down the stairs. Copperfield's crisply ironed shirt was as black as his bushy eyebrows. His face was partially concealed by a small-domed black cap. As he approached, his deep-set eyes brightened, becoming big and glossy.

He was friendly, if formal, and appeared pressed for time. As soon as we shook hands, he looked at his watch and suggested we tour the island before dinner. He started by showing us a game room in the visitors' reception area. Houdini's personal billiard table was the centerpiece. When I asked what phrase Houdini promised to tell his wife if he could communicate from the beyond,* Copperfield answered immediately, "Rosabelle, believe." He showed off some of his other collectibles, including a creaky fortune-teller machine, an early motion-picture device called a Mutoscope, and a hundred-year-old claw-digger amusement device.

Several members of his team were seated around a television monitor watching video footage they'd shot a day or two earlier. Copperfield explained that he'd brought down some *Sports Illustrated* models and *Vogue* cover girls to do a shoot for a calendar he was working on. One evening, they'd all played an indigenous game called the Musha 500. We watched them go at it.

The bikini- and stiletto-clad girls were standing on the beach clustered around two shallow trenches filled with water. Each "racetrack," or aquatic corridor, was about four inches deep and four inches wide, and maybe ten feet long. Two girls each selected a goldfish from a central tank, then placed one fish in their respective trench. A whistle sounded. The models put straws in their mouths and started blowing bubbles

---

*Houdini, who didn't believe in "real magic," exposed numerous mediums and spiritualists as frauds.

into the water in order to make the fish swim forward. The freaked-out fish kept darting around, forward and backward, as the ultrathin models puffed furiously into their straws. One of them nearly got her goldfish to the end, but then it spotted the finish line, turned around, and zig-zagged back down the concourse. *"Merde!"* she cried.

Copperfield told us how much fun the models had while they were here. As he spoke, the staff would laugh in unison, even if he wasn't saying anything all that funny. Raf looked over at me and rolled his eyes. Copperfield then walked us out, explaining that we'd be able to check out the rest of the Landings later on, when we came back to eat dinner there together. Before the sun set, he wanted to show us the rest of the main island.

"Is the fountain on this island?" I asked, getting down to business.

"We can speak at length about the fountain tomorrow, after we go out and see the other islands," he explained tersely, leading us along a paved road.

"Are there actually cars down here?" Raf asked.

"There could be, but we prefer golf carts." Copperfield slid into a buggy's driver's seat, coolly indicating for me to sit next to him. Raf jumped into another cart driven by an assistant, and we all pulled out. "There used to be two limousines on Imagine Island," said Copperfield, explaining how drug smugglers used these islands as landing pads decades ago. "They'd bring in female accompaniment to inhabit it. The movie *Blow* really happened at Norman's Cay. A lot of cocaine went through Exuma."

The island was larger than I'd anticipated, and greener. Oleanders and other lush flowers pulsated in the subtropical warmth. The sky had been overcast on our arrival, but slanting daggers of sunlight were now carving through the clouds, illuminating the Listerine waves below. The water, beautifully translucent, shimmered with almost unreal blue-green radiance. I asked Copperfield what color he thought it was.

"I don't even try to describe the sea anymore," he answered. "You end up using adjectives like *cerulean*. After all this time in the Caribbean, I let the photographs do the talking. Scratch that—it's so many spectrums of blue, you can't even photograph it. You have to see it."

As we steered away from the ocean, he pointed out other Musha must-sees, such as a seventeenth-century head from Burma and a collection of royal thrones from Africa. "Here's a Sri Lankan god I found on my travels," he crowed, indicating a bejeweled, big-eared, mustachioed stone sculpture holding a conch in one hand, and what appeared to be a toilet plunger in the other.

"What's his name?" I asked.

"Super Mario."

Has sense of humor, I jotted into my notebook, and quickly flipped the page in case he was reading over my shoulder.

Our cart serpentined along a mazelike configuration of roads. I asked if it was possible to get lost on Musha. He stressed the importance of staying on the paths because there were holes all over the island. "If you fall in, you can go quite deep down. It's dangerous. Some holes stretch all the way through the island's core into the ocean."

The warning sounded genuine—but it could also have been a possible clue to the fountain's whereabouts. I started looking for any signs of life off the main path.

As we drove upward, toward the manor, he told me that he maintains a full-time staff of over thirty employees on the island, including a zookeeper. He pointed out some of his toucans. Toco toucans, he specified, "the Rolls-Royces of toucans." I wanted to ask him about the sharks, but he embarked on a long story about the herd of African giraffes he'd purchased that would soon be wandering free all over the island. "They'll eat off your plate," he said, "over there in the Valley of the Giants. I'm building them a whole compound with bedrooms for when the weather's bad." He was also putting the finishing touches on something called the Secret Village, a hidden passageway that opens into a three-acre replica of Angkor Wat with "mind-reading monkeys who crawl all over you."

As he spoke, a little bird scampered across the road. "Baby egret!" he exclaimed.

"Is a baby eagle called an egret?" I inquired, putting my pen down momentarily. "Or is it an egress? No, wait, an egress is an exit, a way out, an escape, right?"

"A baby eagle is an *eaglet*. We have a lot of crab-eating *egrets* down here." Copperfield glanced over at my notepad and suggested I transcribe the following sentence: "'As David Copperfield drove me to Highview, the highest point on Musha Cay, a crab-eating egret crossed my path.'"

Only when we walked into his mansion did it sink in that I was actually here, inside the magician's abode. He showed off more exotic collectibles: cobra sculptures rising from the ground, maharaja chairs, carved prayer beds from Afghanistan ("their heads point toward Mecca"). The downstairs suite contained an African room with idols, masks, headdresses, and figurines used in tribal ceremonies. He took us into the dining room and showed us a canoe attached to the ceiling. It doubled as a chandelier. "Check this out," he purred, pushing a button. The ship started slowly descending. "It levitates down from the ceiling on special occasions. That's cool, right?"

"Yeah!" exclaimed Raf.

"What happens once it comes all the way down?" I asked.

"Anything you want," he said, slightly miffed at my lack of imagination. "You can put dinner in it ahead of time, to impress your girlfriend, or place an engagement ring in it. Things like that."

In the master bedroom, he pointed out the neatly made bed, saying, "That's where I'm-not-allowed-to-name sleeps." Then he flipped another switch and a huge TV floated out of the ground in a wicker Indian chest.

"Whoa! Where does it come from?" Raf asked.

"It's magic." Copperfield grinned.

The other rooms, vaguely reminiscent of Graceland, were filled with the strangeness of peering in on a superstar magician's private domestic life. As I perused the rarities on display, I half expected to come across a genuinely magical item, like transparent wings or a cloak of levitation.

He wanted us to see the fitness center, so we jumped back on the golf carts. It turned out to be a basic corporate-hotel workout room. An antique carnival-strongman statue stood out front. On the wall inside, he pointed out a photograph of the strongman at the base of the Eiffel Tower, explaining that it dated back to the monument's unveiling at the 1889 World's Fair. The photo looked doctored. "It's amazing what you can do with Photoshop these days," I blurted out.

"That's not Photoshop!" Copperfield protested, almost hurt.

"No, of course not," I apologized. "Just a joke."

For our next stop, he took us down to Coconut Beach, one of the islands' main sand strips, neatly littered with Windsurfers, Yamaha WaveRunners, and the makings of Dave's Drive-In movie theater. He described Musha as his most important project. He oversees all the details, he said, from buying the board games himself ("like Clues") to designing the telephone users' manual.

He parked the golf cart on a bed of white sand meticulously raked into swirls and geometric patterns. "If it's not a hundred percent exactly how I want it to be, it's a waste of my time," he explained. The statement's underdrift pulled me in. His perfectionism suggested not only a desire to have things done properly, but that he saw himself as being somehow above reality's imperfections. It connected both to the boastfulness of that opening reel at his show ("greatest magician of all time") and to the very desire to own a private island, to create paradise, to be apart from the rest of humanity, to have no attachments, to live in a blissful state of narcissistic fulfillment. To have things be exactly how you want them to be. To no longer suffer. To be perfect.

Epic orchestral music reverberated from some hidden hi-fi system. The music sounded heroic: full of horn swells, harp swoops, flute trills, and tribal drums, *Braveheart* for a moment, *Lion King* the next. It heightened the dramatic intrigue of being on the island, creating a sense that something monumental could happen at any instant.

"What's this music?" I asked him.

"Magic music," he answered, straight-faced. "You'll find Klipsch speakers scattered among the palm trees."

Maybe I was reading too much into it. He certainly took pride in his possessions and clearly cared a great deal about the island. Sure, he had an obsessive side, but despite his ludicrous wealth, Copperfield seemed reasonably regular; tense, certainly, and highly sensitive, but also hospitable, sardonically humorous, and a bit goofy. Then his perfectionist tendencies came to the fore again when he demonstrated the precise way to arrange pillows on the chaise lounges. "Even the Balinese daybeds have secret compartments," he added, opening a chair to show us an instructional document on pillow placement.

Expounding on how much Musha meant to him, he said he wanted *BestLife* readers to understand the passion he felt for this place. It was where he came to "escape from the escapes." How strange it must be to need a break from escaping, I thought, to take time off from working as an escapologist.

He spoke about how, despite the prohibitive expense of running the island, nothing could prevent him from focusing all his energies here. I said I could tell how hands-on he was about it, from the time he'd taken to talk to me before the trip to the way he'd given this personal tour.

It wasn't just about the sugar-sand beaches, he went on, it was about imbuing them with stories. He explained his plans for a haunted island where it would snow on the beach, like in that short story he'd once written. Soon guests would be able to go on yeti quests, where Sherpas would make water magically appear. "What makes me happy is people going like this—" He dropped his jaw. "Everything comes back to that."

When he took a break to respond to a call on his walkie-talkie, Raf and I had a chance to speak openly. "This place is truly amazing," he whispered. "David is like a kid who just wants to show off all his stuff. And he's kind of tacky, no?"

"He has so many secret things. Secret cays, secret TV stands that rise out of the ground on hydraulics, secret passageways leading to secret monkey enclaves, secret underground chambers, secret daybeds. Everything has a secret compartment!"

"You never know what you need to hide, I guess."

The tour wound up with Copperfield showing us the various accommodation options, all of which were waterfront houses. In one of the buildings, he took a moment to make sure we noticed the laminated page he'd made explaining how the remote controls worked. He told us how guests who come to the island can spend the entire time letting it all hang out: "Some people just want to come here and be naked and play bongos. Musha is a place where you can be totally fucking naked because it's secluded and there's no paparazzi around."

As our golf cart hummed past an empty tennis court, Copperfield asked what I thought about the tour, and whether the things he was showing me would be good for the magazine story.

I told him that everything was useful, even though I wasn't sure what would make it into the final text—or texts. "You never know until you've done all the reporting," I confessed. "But I love the feeling of not knowing, of being lost in a story as it's coming together. I feel like I'm in the labyrinth right now."

As he dropped us off at our beachhouse, he mumbled something into the breeze about how much he appreciated being lost in illusions himself.

"What a shithole," said Raf, walking into Pier House, our hyperluxurious two-bedroom suite. The décor drew upon the Far East—large shadow-theater puppets, Buddhist masks, and a carved temple archway painted in fading primary colors. Raf threw open the blinds on the large windows, revealing a white-sand beach and green-gabled pier. On the coffee table, I found a metallic container embossed with an Egyptian scarab, that ancient symbol of immortality. Had Copperfield put it there intentionally?

Raf busied himself in the snack pantry, a larder closet stocked with chips, candy bars, cookies, crackers, nuts, popcorn, pretzels, Snapea Crisps, and any other junk food you could ever want. "No Tab cola though," he tsked. He grabbed a couple of coffee pods for the in-room espresso machine and started making macchiatos.

"That thing was so Photoshopped it's not even funny," he said, adjusting the steam wand's intensity.

"You really think so?" I asked, no longer sure. "He wouldn't be so audacious to pull a fast one like that."

"Like that?" Raf scoffed. "The guy walked through the Great Wall of China, he made the Statue of Liberty disappear—that's what he does, he pulls fast ones. That Photoshop image is small potatoes. *Audacious*. Ha!'"

After a final snort, Raf started singing the words "We're in David Copperfield's private island resort" to a childish melody. He laughed maniacally and looked to the heavens. Stopping suddenly, he threw me a cut eye. "You realize we're being filmed, right now, through this Laotian mask."

I immediately started talking loudly about how magical Musha felt, lauding David's passion, complimenting his taste, noting the exquisiteness of his many Oriental sculptures. I filled Raf in on what David had said about this place being his most important project. "He's a perfectionist," I concluded. "One hundred percent or nothing."

"That's David Copperfield," said Raf loudly. "An OCD Jewish nerd who can make you believe he levitated over the Grand Canyon. A magician surrounded by supermodels on his island. What a tough life."

Moments later, a black winged creature flew into the room. "It's a bat!" I cried, as the hairy dark thing fluttered about erratically. Transylvanian thoughts caromed through my mind.

"No—wait! It's a fucking *moth*!" Raf shouted, as it hurtled to a stop upside down in a corner where the ceiling met the wall. "Holy shit, David Copperfield just flew in here as a moth! Can he do that?"

We edged closer. Before us quivered an extremely large, velvety-black butterfly. Two striking dark brown eyes looked out at us, one on each wing. It had the wingspan of a sparrow. Raf started shooting photos from different angles. We made various shooing motions, to no avail. It didn't move again, so we decided to unpack. By the time we left the Pier House for dinner, the winged insect had flown away.

As we drove toward the Landings, I noticed a sign that said PETRIFIED LAKE. Had it been there earlier? Neither Raf nor I had seen it during the tour, and it seemed as if it would've caught our eye. I made a note to investigate it later. By the time we showed up in the Houdini billiard room, it was about ten minutes later than Copperfield had suggested. His vexation was palpable.

A couple of assistants were milling around, offering cocktails. His girlfriend, a gorgeous European model, sat next to him.* She spoke about how her supermodel friends from *Sports Illustrated* and *Vogue* had loved their time on the island. One of them had said how coming here felt like coming home.

"That's what Raf said when we were on the boat," I added, neglecting to mention that he had been joking.

"That's why I do this," Copperfield said. "To make you feel like you've gone back to being a child again. That, and getting this reaction—" His jaw dropped.

He directed our attention toward the television screen. It showed us a promotional clip about his home in New York, a three-story penthouse apartment on Fifty-Seventh and Lexington looking over Central Park. It was an urban palace. The camera panned through a playroom full of penny-arcade games, carnival muscle-strength challenges, and other antique funfair curiosities. A tracking shot of the living room revealed a number of nude wooden people nailed to the two-story-high wall in various contortions.

"What are *those*?" I asked. The mannequins pinned up there like human creepy-crawlies made me think of the dark butterfly in our room.

"They're incredibly rare life-size models," Copperfield explained.

"Models?" Raf asked, turning slowly to look at M.

"Around the turn of the last century," Copperfield answered, "it was illegal for artists to hire real-life models, so they used articulated lay figures like those. I even have one that belonged to Cézanne."

Watching the models dangling from the wall, I remembered reading something about his warehouse in Las Vegas, recently raided by the FBI during their investigation of the alleged rape: getting in requires tweaking a mannequin's nipple.

Other weird contraptions flitted by on the screen. "Initiation devices," he clarified.

---

*I didn't recognize her and promised not to reveal her identity, so I'll just call her M.

"Like for what sort of initiations?" I checked.

"You know, trick chairs, paddle machines that whack you in the butt, novelty electroshock games, kind of benign hazing things like that. I also have tons of ray guns. You've gotta come out to Vegas and see my warehouse. It's huge. I have a whole room full of ventriloquial dummies."

He spoke about the excitement of having acquired the Mulholland Library of Conjuring and the Allied Arts. He also owned the world's largest collection of Houdini memorabilia, including Houdini's Water Torture Cabinet and his Metamorphosis Trunk. During his own lifetime, Houdini had possessed the world's largest library of magical materials. Their respective era's greatest magicians, they also both specialized in straitjacket escapes, and they both had Jesus complexes. (Houdini is most often visualized, his biographer Kenneth Silverman wrote, "dangling upside down from a tall building, arms outstretched in a pose of inverted crucifixion.") They also shared a fascination with mutilated bodies: Copperfield's living-room collection paled when compared to Houdini's prized set of "revolting snuff snapshots that showed chunks of flesh being hacked from a woman tied to a stake." I recalled coming across an article somewhere about Copperfield's love of horror movies.* Afraid of finding out just how much he had in common with Houdini's dark side, I brought the conversation back to their joint interest in archiving the history of magic.

"Do you consider yourself a collector?" I asked.

"I don't really like that term," Copperfield said. "I'm not an accumulator. I love objects that carry with them amazing stories. But I don't want to be seen as a collector."

"Wouldn't you like to start collecting women's shoes, size nine and a half?" chided his girlfriend.

He pursed his lips and reached for a glass of water.

Raf leapt in. "Wow—nine and a half? You have *big* feet!"

"I know!" M groaned, growing self-conscious. "I'm so embarrassed about them."

Her feet seemed perfectly normal to me. I shot Raf a "behave yourself" look. He'd been reading a how-to book about picking up women, based on the teachings of a magician and master seducer named Mystery. The most important technique was "negging," a sleazy way of

---

* Although, that night, he elucidated, with great verbal aplomb, precisely why he considered *The Blair Witch Project* such a failure.

creating intimacy by jokingly pointing out someone's insecurities. Copperfield handled it deftly and politely, talking about how all people have parts of themselves they are sensitive about, and how we all deal with them as children. He spoke of his own complex about having big ears, which explained his affinity for Super Mario, that Sri Lankan god statue with the huge lobes. "Childhood is what shapes us," Copperfield said. "It's how you use your markers and devastations that counts."

He frequently dropped wisdom during our time together, sharing sayings about everything from forgiveness to decision making, such as "grudges hurt the grudger more than the grudgee," and "the more successful you become, the harder it is to focus on family," and "if you really want something to happen, you can force it to happen by your drive and your force, and that's a kind of supernatural effect."

As we moved over to the dining table, he told us a story about going to a camp in Warren, New Jersey, as a child. "At Camp Harmony, we spent two weeks searching for a guide who'd been kidnapped by Indians. It was just a game, but I was living it. That's what I do here on Musha Cay. My whole life goes back to that camp experience when I was three or four. The yeti quest I'm working on, where Sherpas will make it snow on the beach, it's just a variation on that. Everything is. Everything I do is about getting people's jaws to drop. The canoe is cool—but not as cool as having that canoe come down from the ceiling full of sushi. That's—*kaw,*" he said, dropping his jaw.

The kitchen staff served each of us a braised-lamb dish, except for Copperfield, who was brought a platter of breaded chicken fingers. He would eat the same thing each night, while the rest of us were treated to a variety of seafoods and other meats. Copperfield's fondness for chicken fingers goes way back. Shortly after he proposed to Claudia Schiffer, a journalist joined them on a limo ride to Planet Hollywood in Manhattan, where he watched them "feast" on chicken fingers.

Over dinner, David spoke of magic's illustrious past, mentioning how magicians had been kings' confidants, and how they'd always held high posts throughout history.

"So what position do you want in Obama's cabinet?" Raf nibbled.

"Well, Ronald Reagan did offer me a post after a show in Ford's Theatre," said Copperfield. "He wanted me to make things vanish."

"Like his wife," added his girlfriend.

"Now, now," chuckled Copperfield.

When the staff cleared our plates, he asked what we wanted to do.

"Should we people watch?" I joked.

"You can't do that here," he sniffed petulantly, and suggested we play board games or do some karaoke.

"Raf is incredible at karaoke," I jumped in, trying to get back to an upbeat place.

"We'll see about that," said Copperfield.

"I have my own machine at home," Raf retorted, unperturbed.

As Copperfield moved inside to set things up with a gaggle of helpers, Raf and I lingered on the dock, looking down at the sharks drifting through the waves below. Raf wondered what M thought of all those models crucified on the wall.

"She's so beautiful I can barely look at her," I said. "It's like watching the sun."

"Through diamonds," Raf added. "She's *too* beautiful. The whole island is. You almost need to turn your eyes away. Or talk about how big her feet are."

On the way into the karaoke room, I noticed an illustrated map of Musha. I couldn't locate the Petrified Lake, but did find a body of water marked *The Sanctuary* in its general vicinity. A sanctuary? A consecrated place where sacred objects are kept?

"Did we see the Sanctuary today?" I asked, walking into the room.

"No," Copperfield answered definitively. His seriousness made me reluctant to press the matter, while simultaneously affirming my hunch. Whether sanctuary or petrified lake, its liquids would be worth exploring.

The karaoke machine readied, a diminutive young male assistant tested out the system by singing a flamenco song in a pretty alto voice. Copperfield went up next, with a smooth rendition of Sinatra's "Love Isn't Just for the Young."

When he finished, he spoke about how much he admires Vegas crooners. I remembered Martha's telling me that he likes to sing Rat Pack songs. "I mean, he gets up there in front of guests and sings 'Candy Man.'* He's just not tuned in."

Raf went up next. He chose "I Believe I Can Fly" by R. Kelly. "If I just believe it . . . I can fly."

---

* "Who can take a sunrise, sprinkle it with dew, cover it with choc'late and a miracle or two?"

"He's pretty good," conceded Copperfield, mentioning his dislike of hip-hop.*

As Raf soared into an R&B falsetto, Copperfield mentioned that a famous rap producer had recently stayed on Musha.

"Which one?" I inquired, playing along.

"I can't tell." Copperfield smiled. "He watched *Scarface* at the Drive-In on the beach."

"Did he do any karaoke while he was here?"

"No," said M, "but he heard us covering an Eminem song he produced."

"Sshhhh," stage-whispered Copperfield.

That Dr. Dre had been here was precisely what Copperfield wanted me to be writing about. He took that opportunity to drop a few more names. The old guest book belonging to the previous owners, he said, was full of messages from comedians such as Billy Crystal ("The crackers were stale") and Steve Martin ("What? No tobogganing?").

My song came up next: "Just an Illusion," an old disco hit. The lyrics were simple and repetitive: "Is it really magic in the air? . . . It's just an illusion . . . in all this confusion?"

M followed with a heartfelt rendition of "I Will Always Love You."

"Will you really, though?" Raf whispered at one point.

Pretty soon, Raf and Copperfield were belting out duets. We took lots of photos, drank too much, and had a lot of fun. Some of us ended up wearing capes. As Copperfield belted out a Dean Martin number, Raf turned to me and said, "I can't not get a photo of me piggybacking him, then—*switch!*—him piggybacking me."

After an hour or so, it was time for bed. Copperfield and M slipped into a golf cart and headed up toward Highview. It took us a little longer to get going, as Raf couldn't remember how to start the cart's engine. He'd been drinking shot after shot of sambuca and ouzo ("to finally understand the difference").

"I guess it isn't drunk driving if it's just a golf cart," I said.

"And if there's no police," added Raf, "'cause you own the roads."

We caught another glimpse of Copperfield zipping up the hill.

"It's so funny watching him roll away in his golf cart," said Raf, fum-

---

*This despite (or possibly a consequence of) the fact that he himself had recorded rap verses in the past. For example: "I dress real cool, with an attitude; that's why I've lasted longer than that Henning dude."

bling with the ignition. "So undignified somehow." Raf didn't realize his foot was on the accelerator, and as soon as he'd turned the cart on, he sent us crashing straight into a dense thicket of shrubbery. "We're gravy," he cackled. He then backed into a Tibetan prayer pot, veered through a grove of palm trees, and finally managed to steer us back to the Pier House intact.

15

Sleights of Mind

You do yet taste some subtleties o' the isle, that will [not] let you believe things certain.

—Prospero, in *The Tempest*

One of the problems in writing magic history is that magicians rarely tell the truth.

—Simon During, *Modern Enchantments*

THE FOLLOWING morning, we toured the islands of Copperfield Bay, a seven-hundred-acre agglomeration of water and land. As our speedboat spanked over the waves, Copperfield pointed out each of the eleven islands, most of which were desolate, covered in twisted trees wind-battered into lopsidedness. I felt dismayed. The archipelago's sheer square footage meant it would be hopeless to attempt searching for his fountain.

He kept pointing out landmarks. This shoreline, he said, was where the Sherpas were going to be making it snow on that haunted beach. He directed our gaze toward a rock painted to resemble a mermaid. It crinkled under the waves. The boat took us to secluded islets he described as getaway spots, places to escape from escaping the escapes.

My notes were getting increasingly psychoanalytical. *Isolating ourselves from other humans is a desire to be separate from them,* I wrote, *to not be part of the human race, to be something more, godlike. Is this akin to wanting the Fountain of Youth?*

Raf kept snapping photos, including some of Copperfield, who quietly asked him not to. "I'm not David Copperfield right now," he said.

Back on the main island, we stopped to eat a quick lunch. Two attractive young girls sat in the dining room, each with a parental guardian. They had been recruited from his shows, they said, and were here doing modeling work for brochures about the island. That they happened to be on Musha at the same time as us made me wonder if it was an intentional PR move. He obviously knew about the leaked Scorpion-pulling docs. Or maybe he only invited pretty women to Musha for benign reasons? Either way, the two girls said they were having an amazing time.

After lunch, Copperfield handed Raf and me a one-page story to read before we went on a treasure hunt. "The Unknown Pirates" described a pirate who was David Copperfield's great-great-great-great-uncle. This poor pirate had wrongfully been accused of something and needed to have his image restituted—which we could help with by doing the treasure hunt.

As we read it, Raf looked at me. "What's up with asking us to clear his great-great-great-great-uncle's name?" he questioned, frowning. "That's so weird, isn't it?"

"Is it just a coincidence—or is it somehow related to him trying to fix his own image?" I asked. "Did he allow this trip to happen in hopes of being portrayed a certain way? The restitution theme is so Prospero."

The treasure hunt was a competition: Raf and I on one team, the two aspiring models on another. We split off, the winner being whichever team could solve the clues fastest, in order to clear the good pirate's name. We began in the woods, where a fully costumed pirate gave us the first set of clues.

The hunt consisted of puzzles and magical tricks scattered around Copperfield's domain. We had to enter a cove with skulls hanging from a hook and hike across a sharp, rocky terrain. Copperfield came along with us, making sure we noticed all the little details he felt made it somehow more authentic.

When our guide turned off the main road toward the Petrified Lake, I got out, pulled Raf aside, and told him to keep an eye out for clues to the fountain. We walked along a muddy path in the woods, which opened onto a sizable body of water surrounded by a large copse of bone-white trees. The skeletal tangle of lifelessness stretched back in for a hundred yards or more.

"What is this place?" Raf gasped. "Fucking eerie."

We'd come to the Petrified Lake. The unmoving water, framed by that mangled landscape, seemed more like a bay of death than a fountain of youth. Copperfield kept a close watch over us as we walked down to examine the forsaken cove. The water, brown and brackish, stood completely still. The trees appeared to have been bleached with salt.

Leaving the Petrified Lake, I told Raf we'd need to go back, later, to see if we could make brown leaves turn green again.

"You know the only way you're going to be able to prove this is by killing me and throwing me in there," he answered.

The pursuit took us to another secret beach nearby. "Look at the rock formations, all the different shapes," Copperfield said. "Mother Nature's just sick amazing."

Forging ahead, he led us down a path strewn with dead palm fronds. When we emerged onto yet another beach, he pointed to a sign with an arrow that said BLUE SPRINGS THIS WAY. "Follow your destiny," he added, smiling.

We walked forward, past more and more signs for the Blue Springs. They led into a mini-forest, and for a moment I felt certain we'd be coming to one of those holes he'd warned me about. But, after a few twists and turns, the path spit us out into a clearing in the woods, where we nearly tripped over a dozen knee-high metal coils. It took a moment for the gag to sink in. The Blue Springs weren't some magical waters—they were just a bunch of blue-painted metal springs spiraling out of the ground.

Copperfield gleefully told us how much the people who develop the computers we use and the people who program our book clubs love that joke. "I'm trying to create wonders a billionaire can appreciate, and that's a tough crowd. 'Tough' in quotes, please. The people who come here have seen it all. They own it all. They've had every experience possible, so if I can make them come back, then I've done my job."

Raf and I lost the treasure hunt by a long shot. It didn't matter; we ended up sipping champagne on another beach, where the pirate did a magic performance. (The pretty girls who were here doing promotional work didn't drink any of the champagne; they weren't old enough.) At the climax of the show, a young female assistant levitated on the point of his sword. He also unearthed a box of pirate's booty from under the sand. For one trick, involving a guillotine, he called Raf up as a volunteer. Before decapitating him, he demonstrated the blade's sharpness by positioning a stubby carrot in front of his pelvic region and chopping it in two. "Castration humor, eh?" mused Raf.

"It's amazing how you don't feel or sense yourself transforming with age," Copperfield exclaimed, back at the Landings, where the two of us had embarked upon the promised discussion about his fountain. "Even when I'm eighty years old, I wonder if my point of view will be different than when I was eighteen. There are certainly reminders that you've changed physically, but the sensibility remains the same."

"In what way?" I inquired.

"You stay attracted to the same things. What's fun never changes. For example, we all karaoked last night, and none of us acted our age or cared how we looked or sounded. That's just like a twelve-year-old's get-together with friends: you throw on some records and have a license to be ridiculous. When you're here, you give yourself permission to have the feelings you felt when you were still discovering things. It's very life-enhancing."

"'Very life-enhancing'?"

"You mentioned how you like being in the labyrinth and not knowing, and still discovering things about yourself. There was a time, when you were still young, when you were learning the rules of how to act. If you can take those away, you can feel the fountain of youth."

"But that's not the fountain of youth you told the media about."

"No, but it's as important. As communicators we do affect people's lives, we do lengthen their lives. The medical properties of enjoying being transported are well documented. As artists, that's why we're celebrated. There's a thirst for our work. The responsibility we have is to youthen people. We prolong life. We parse verbs to get an emotion. To dream, to exhale, to have adventures—that's almost as important as politics. As a victim of a lot of stress, I can tell you it may not be as important as medicine, but it's right up there. Everybody has a need for the curative properties of storytelling."

I sensed an opening. "Speaking of stories, tell me about the real fountain."

"As reported in the press, I'm doing experimentations with the natural elements that are here, elements that have amazing qualities." He spoke deliberately, choosing his words carefully. "Will that translate to a discovery that benefits mankind? Time will tell."

"Yes, but is this just another Copperfield illusion?"

"I don't want it to be brushed aside as a PR stunt or something that can be explained as a magic trick. Magicians can replicate everything in the Bible. In this case, the things that I'm talking about could have practical applications for people. The fortunate part is that I'm a magician, and it's the unfortunate part as well. The hard part is for people to believe it."

"It won't be hard for them to believe it when they see it."

"That'll be in the future."

"But surely you realize people assume the fountain isn't real?"

"It's okay to have puffery, but this is entirely different—"

"What do you mean by 'puffery'?"

"Puffery is yeti quests, the Blue Springs, the snowing on the beach, all of that stuff. But when you're talking about a discovery that can seriously affect people, I'd rather have all the backup done before demonstrating it."

"So you see a distinction between the puffery and the fountain."

"Absolutely. There's a distinction between things that we wink at, like the treasure hunt, and something like the fountain. When you come and see my show, you know that I have permission to lie; you suspend disbelief. But when you get into things that can affect people's health—meaning liquids that can reverse things—I want to be really prepared."

"Liquids that can reverse things?"

"Liquids that can reverse genes."

"Can you explain how it works?"

"I want to have my research done," he parried, "and have all my ducks in a row before doing so."

I pressed him, again, to explain how this wasn't puffery.

"Look, here's the difference," he clarified. "Allowing people here, giving them a new outlook—it's rejuvenating. But that's distinctly different from a liquid that can turn a brown leaf into a green leaf."

"Have you seen that happen?"

"Yes."

"Can I?"

"We've discussed this. We have a deal. People will think it's a sleight of hand."

"Has anybody else seen it?"

"Yes."

"Who?"

"Me and a couple of . . . biology people."

"How did you feel when you first encountered it?"

"That there are amazing possibilities here."

"Why did you tell the media about it if you weren't prepared to show it?"

"When you discover something exciting, you want to share it. It's like when you fall in love and you want to tell everybody about it—but then you realize it was maybe a mistake to do so. If it doesn't turn out the way you'd hoped it would, next time you'll be more cautious. So you only go into details when you're sure it'll last and be real. It's the same here. There are things happening, little things that have bigger potential. But then I'm an illusionist, so if I talk about it, people will shrug. But it could be very, very cool."

"What is the timeline?"

"Things happen very slowly down here in terms of research. It takes a long time. Certain things need to be combined with other things, and that's a slow process. In my work I normally use science to my benefit, not the other way around. I use technology to do magic all the time, and I learn about existing science to do illusions, but the idea of doing new science that's never been done before is new to me."

"So is this a new technology?"

"Yes. We're experimenting with learning certain things and combining them with other things. It's very wild, but slower than I'm used to. This type of research and possibility can't be rushed, so I'm learning to be patient and learning not to make promises. . . . Did I fuck up your story?"

"I think this *is* the story. Tell me about the scientists."

"I'd rather not. I've made promises to them that I wouldn't talk about it."

"How seriously do you take this?"

"I don't understand the question."

"How important will the fountain be to you in coming years?"

"I'm not sure what the end result will be. It's a big project. Normally,

I can fast-track things in my work. I can expedite things for my show. But not in this case."

"Let's go back to the beginning. Did you know about the fountain before you came down here?"

"No."

"How did you find it?"

"I spent a lot of time walking on all the islands, exploring with a machete—I mean a cutlass. At one point, I saw things that were particularly . . . *vibrant.*"

"What did you do when you saw these vibrant things?"

"I said, 'We should take a closer look.'"

"But you won't let any reporters take a closer look."

"No."

"So how did this story go out to the media?"

"Poorly. I'm guilty about getting very excited about something. Normally, I would have everything more detailed before showing it to people, but I got very excited. It goes back to the love analogy."

"What's the ETA?"

"We're not talking months, we're talking years."

The rest of the interview continued in this vein, with me asking the same questions in different ways and him reiterating his answers. As we spoke, I thought of something the NDE expert Raymond Moody had said about the overlap between parapsychology and entertainment, how both evoke realities and have consolatory powers. Just as Father Gervais described watching films as a form of prayer, Moody also felt that art—or entertainment, or what Copperfield was speaking of as storytelling and magic—can plumb the depths of our spiritual concerns. "Entertainment has this really profound function," Moody said. "It can evoke feelings of wonder and make us realize how little we know and even though it can't prove anything, at least it can leave the crack of the door open for hope."

In the end, despite all my attempts, Copperfield insisted that I wouldn't be able to witness the fountain. Still, even without any first-hand evidence, he'd almost persuaded me that it might be real. He'd held up his end of the bargain: he had indeed spoken about it with great verbal aplomb. I again felt as though precious gemstones were falling out of his mouth as he spoke. "Patter" is considered a key form of attention management in magician know-how. And I'd fully succumbed to his stream of conversation.

Yet as we left the Landings, I was already planning to sneak out to the Petrified Lake to do some experimenting of my own. I asked him if he had anything else he wanted to add.

"I'm a one-trick pony, in a beautiful way, I hope." He shrugged. "Here, watch my face"—his jaw dropped—"that's what it's all about. It all goes back to the 'whoa' factor."

The idea of a one-trick pony who keeps blowing your mind returned to me a few hours later when, perusing the bookshelves in my room, I came upon a copy of *Beyond Imagination,* the sequel to his first anthology of short stories. This one, which focused on "the magic of the human condition," contained another of Copperfield's own stories. As with the previous one, the hero's name was Adam. This Adam loved collecting sun-bleached driftwood. The plot revolved around Adam's discovery of an invisible baby eagle. He decides to tell people about his pet eagle, hoping it'll make others like him, but instead, when they ask to see it and he refuses, they start mocking him. In the end, however, he makes the impossible happen, amazes everybody, and is rewarded handsomely for his magical abilities.

As I lay there, drifting to sleep, I thought of the egret that crossed our path as soon as we landed on the island, how we'd spoken about eaglets and egrets. Fiction and reality seemed to be overlapping in an almost destabilizing way. A Shakespearean hero who orchestrates an elaborate setup on his island, using "trumpery" to reinstate himself as king. A storybook Adam in a world of bleached-white wood, baby eagles, and real magic. A famous magician who'd found a liquid that no one could see that reverses genes.

That night, I dreamed that, if I stared long enough at water, I could fly. Exhilarated, I soared through the air, gravity-free, feeling an unforgettable sense of limitless potential. Then, somehow, I became David Copperfield. I fell back to land, addled with obligations. I had so many responsibilities, so much to take care of, oversee, fix.

"You have to do it all," I told myself.

"No, I don't," I replied in the dream. "I'm not David Copperfield."

On waking, I sensed the fountain must really be out here, somewhere, just waiting to be found. I opened my laptop and watched videos on YouTube of Copperfield levitating. Despite his cheesy faux-romantic touches, the trick remained as striking as it had been in my youth. He simply rose off the ground and flew through the air. Magic.

At breakfast, I told him about the flying part of the dream, neglecting to mention the creepy part where I morphed into him.

Copperfield listened intently. "So you could fly just by looking at the water?"

"Yes, that's how it worked."

"Isn't that amazing?" he exclaimed, genuinely. "That's exactly like my fortune-telling act! The way it works is that I just need to look at the lights." He told me that people's dreams often found their way into his performances. Over the years, he added, he'd done at least four different illusions that involved his becoming young.

"But those were just puffery, right?"

"Right."

The sun shone, having nothing else to do, and Raf and I spent the rest of the day on the beach and racing around on WaveRunners. We also went swimming with the nurse sharks. As I approached them, snorkel on, they seemed nice enough. Then one of them flipped over, swishing its tail. It headed straight at me, fin quivering menacingly. Hearts racing, we scrambled up the ladder and onto the dock.

An assistant stood waiting for us at Coconut Beach, walkie-talkie in hand. Copperfield was on the other end. He wanted us to choose a film for the Drive-In theater that evening. Raf suggested *Thunderball*. Copperfield hemmed and hawed, saying he wanted a comedy backup just in case.

"Okay, how 'bout *Scarface*?" I said.

"Uh, that's not a comedy," Copperfield pointed out.

"Any comedy suggestions?" I asked Raf, putting my hand over the mouthpiece.

"How about some Mel Brooks?"

"No," Copperfield retorted, "not that."

I asked Raf to pick something else.

"Are we choosing a film, or is he?" Raf laughed, mock-exasperated.

"Welcome to our world," said Copperfield's assistant quietly.

"Well, why don't you just choose something?" I suggested. After a little more back-and-forth, he said he'd find something appropriate, like an Adam Sandler film.

I handed the walkie-talkie back to the assistant, who explained that Copperfield probably wanted something new, to show off the resolution. That way we'd see how high quality his theater system was.

"Oh, so he wanted us to not pick an old movie?" I puzzled. "Why didn't he just say that in the first place?"

She shrugged. Having established a connection, I asked if he'd ever spoken to her about the fountain of youth.

"He never talks about it with anybody." She smiled uncomfortably. "I don't even know which island it's on. I've never asked him about it, but I know he wouldn't tell me. He doesn't tell anybody about it. He's a very mysterious man. He's so charismatic, but also very shy at the same time."

"Yeah, for sure," I said amicably.

"Well then, enjoy the rest of your trip!" She waved and walked away.

Diving back into the ocean, I resolved to investigate the Petrified Lake before the end of the day.

Shortly before sunset, Raf lay zoned out on the couch, watching a Cindy Crawford infomercial about antiaging cream made from French cantaloupes.

"I'm going to go turn some brown leaves green," I told Raf. "Come put a leaf in with me, so you can back me up."

"No," he said, unmoving and unmoved.

"What? Why not?"

"Because I'm too lazy."

"Come on!"

"No." He shifted sides, chewing on some pizza-flavored Combos. "Really."

"It'll be weird if I do it by myself."

"It'll be weird if we do it together. This whole island is weird."

"Of course I know it will be 'weird.' But it might be frightening."

"Nope. Don't care. Not coming. You gotta fly this mission solo, my friend."

Settling into the golf buggy alone, I hung a right on Imagine Avenue. Around the zookeeper's area, I started getting even more paranoid than

I'd been since arriving. Some assistants looked up from their landscaping work. Did I seem suspicious? If they questioned me, I'd just say I was exploring. *This is totally normal,* I told myself. Except the sun was setting, it was grimly overcast, the wind had come untethered with the approaching hurricane, and I was sneaking into the master's fountain of youth.

Another assistant drove toward me in a buggy and motioned for me to stop. It felt like being pulled over by the police. I smiled broadly, to cover the guilt.

"Just want to know if I can get you anything at all," she said.

"No, I'm fine actually," I said, exaggeratedly calm. "Just exploring."

"Awesome—have fun!" She zipped away.

A few moments later, I came to a sign saying SANCTUARY with an arrow pointing away from the water. This was where the PETRIFIED LAKE sign had been yesterday. It was gone now. Were the signs' comings and goings just a part of the treasure hunt, or did they have something to do with the fountain?

Fingering the dead leaves in my pocket, I walked through the brambles. The dirt path led to the postapocalyptic pond, surrounded by all those gnarled, white trees lying jagged in their boneyard heap. They'd been salt-whipped, sun-bleached, battered by the elements. The rest of Musha was so lush, but this neck of the woods was broken, brittle. Deathly. It felt like a setting from the darker pages of *Titus Andronicus*.

A big, crowlike bird with bright green feet flapped clumsily away. The wind howled through the spindly, brittle trees. They were frozen into place, like petrified formations. I felt a bit petrified myself, especially when I noticed Copperfield's black aerie atop the hill. *He could be watching me right now,* I thought. A giddy, childish fear pulsated through my nervous system at the thought of being caught.

I took a few brown leaves out of my pocket, breathed deeply, and submerged them in the water, holding on to their stems. Umber muck swirled up through the coppery brine. I swished the leaves around a bit, then took them out. Nothing. Still brown. I tried stirring the leaves into the mud at the bottom of the lake, coating them with brown, mossy sediment. One after one, I drenched them in various ways, to no avail.

Although I'd been half-hoping that something magical might happen, I turned away with a strange sense of relief. The fountain remained a possibility, unrealized.

That evening, the storm we'd heard about in Miami intensified. Raf and I sat in the Landings, sipping tea. The captain came over to discuss transportation logistics with us. The hurricane had intensified from a tropical depression to a Category 4, he informed us. "That's a monster. If it hit the island, it would destroy everything. But she's supposed to be downgraded to a tropical storm by the time she gets here."

He told us we could either spend three more days on the island or leave in the morning on a private jet. We thanked him for the update and said we'd think it over. Perusing the cloud banks, Raf made an aside about its "really gusting out there."

"What did you say?" asked David Copperfield testily. He'd approached without us noticing, sensitive that even part of the world *disgusting* might be used in conjunction with anything to do with Musha.

"Um, it's really *gusting* out there, like it's really windy," Raf explained, and turned away.

Copperfield relaxed somewhat. He told us we wouldn't be able to go to the Drive-In unless the weather changed radically, but that we'd instead be able to watch something on the flatscreen across the room. (Later that night, he popped in Adam Sandler's *You Don't Mess with the Zohan,* which he'd already seen, but as he watched, he laughed heartily, freely, like a kid, until, about a quarter of the way through, he fell asleep on the couch.)

I asked him about the bleached-white mangroves.

"They're not mangroves," he said with finality. He didn't elaborate. We'd had our interview about the fountain. That was it.

Over dinner, he told us stories about being the shy kid in the corner at Andy Warhol parties: "I didn't interact with people, I just sat there and watched." He talked about his regret at choosing the name David Copperfield. "It was the wrong idea," he lamented. "Now I have to share it on Google with a Dickens character, who deserves a lot more credit. And, also, it's too big for a marquee." He showed us stamps with his face on them, from the Dominican Republic, from Grenada, and from St. Vincent. "All sorts of people lick my head when they send someone a letter," he mused.

"The back of your head," clarified Raf.

"How weird is that?" Copperfield asked, then answered himself: "Pretty fucking weird."

He spoke about how following the path to accomplishment means drifting away from your family. He wasn't that close to his relatives, he admitted, explaining how he'd been trying to reconnect with them. The road to stardom had been full of sacrifices. It really is lonely at the top, he added, lonelier than any fridge-magnet platitudes can express. It seemed an earnest attempt to open up, to provide me with material he didn't normally give journalists, yet I still sensed how guarded he was, how controlling.

In his experience, he sighed, the more money one has, the more complicated things get. He had everything so many people desire—wealth, fame, success, babes—yet as he spoke, we could hear how tightly wound he was. He'd hoped this island would set him free. But one's desire, as Freud said, is always in excess of the object's capacity to satisfy it. Escaping from the escapes, into an illusionist's illusion. He'd invested more than a hundred million dollars here, but now, even as the leaves tremored in the hurricane's approach, his refuge—his place of unfettered freedom, his utopia—remained chained to the one thing he most wanted to escape: reality.

16

Technical Interlude: Magick, Eros, Symbolism

> The rain, fallen from the amorous heaven, impregnates the earth...
> and from that moist marriage-rite the woods put on their bloom.
> Of all these things I [*Eros*] am the cause.
>
> —Aeschylus, *Fragments*

> I strove to seize the inmost form . . .
> But burst the Crystal Cabinet,
> And like a weeping Babe became—
>
> —William Blake, *The Crystal Cabinet*

MYTHOLOGIES OF all times speak of a magical liquid of life, be it chi, rasa, haoma, or amrita. This imperishable substance, the myths tell us, is the essence of immortality. Religions teach us to allow this flowing substance into our lives, to view ourselves as sustained by it, and to realize that we are actually a physical embodiment of it. All of this, of course, takes place on a symbolic level. When we view this as a literal possibility, we are thinking magically. The symbol is not the same thing as the thing it refers to.

The history of immortality and our quest for it is inherently bound up with the symbolic. Symbols are tools we use to grapple with the impossibility of understanding our impending demise. And stories that attempt to speak about what can't be known are called mythologies. Today the word *myth* is synonymous with untruthfulness or falsity.

But mythologies, in their original context, were symbolic stories whose truths couldn't be conveyed any other way. "A myth is above all a story that is *believed,* believed to be true, and that people continue to believe despite sometimes massive evidence that it is, in fact, a lie," writes Wendy Doniger in *The Implied Spider*. Myths, being metaphorical, are both real and unreal, simultaneously true and untrue. They point toward some greater truth. It's not their veracity that counts; their implicit meanings are what matter. "Only when a mythical story is considered as factually false is it of any use," explained the literary critic Northrop Frye. "The need to see it as 'the truth' factually is an infection."

Myths are concerned with something insoluble and ineffable. To discuss that which words cannot discuss, myths use symbolism. A symbol is a go-between, something that stands for something else. Symbols are multivalent; they have multiple meanings, many values, various possible consequences. We can imbue them with our own significance. By insisting that there is one real, or superior, sense to a symbol is to restrict its very nature, to neuter its power.

Using symbolism is akin to describing a dream. We can never make another person experience our dream (let alone remember everything that happened ourselves). We can only attempt to convey aspects of what happened in the dream. The words we use when speaking in dream language, or in mythologies, are symbolic.

Symbols encapsulate a fleeting sublimity that can never be fully articulated. They are incipient transitions. They float between thoughts and feelings. They connect the conscious mind with the collective unconscious, the within with the external, the above with the below. The word itself derives from a Greek term meaning "to bring together, to combine, to integrate." It is at least two things in one.

Symbols live in our hearts, affecting us on profound, inner levels. They can also inspire us to venture further into a mystery, deeper into the wisdom concealed behind the symbol, a truth whose luminescence outshines any attempt at demystifying it. Symbols transport us past the limitations of language into inexplicable territory. Those with a scientific-mechanistic worldview think that if something cannot be defined outright or coherently, then it probably isn't anything. This contrasts with those who consider the spirit world to be a reality full of potential.

Anything spiritual cannot be directly articulated. We can only allude to it obliquely or use suggestive descriptors. Symbols nod toward the revelatory, the transcendental. They are attempts "to translate the truth of that world into the beauty of this," wrote Evelyn Underhill. Symbols don't tell us what True Reality really is; they carry us beyond the expressible and ascertainable into the unseen. They link us with the cosmic. In Coleridge's words, a true symbol is characterized "above all by the translucence of the eternal through and in the temporal. It always partakes of the reality which it renders intelligible; and while it enunciates the whole, abides itself as a living part in that unity of which it is the representative."

As Coleridge knew, the symbolic language of metaphysics is also the language of visionary poetry. As Kathleen Raine put it, in her remarkable work *Blake and Tradition*: "The schools do not teach this learning, but the poets find it out." Without a road map through their symbolic wildlands, the transcendental and Romantic poets can at times seem incomprehensible, which is part of the reason they are occasionally perceived as madmen rather than sages. But consider their plight: whether Dante or Shelley, they could only employ terrestrial words to approximate divine constructs. To bring us close to the heart of things, they had to use symbols. Raine's contention is that the same key unlocks all their texts: "Each poet may have his chosen symbolic themes, but all speak one language." It is possible, she argued, to unpack a symbol by exploring its common and traditional usages.

The trope of water surfaces repeatedly, and homogeneously, in symbolist poetry. In William Blake's works, for example, water represented the mystery of souls entering—and then departing—human bodies. W. B. Yeats put it succinctly: "What's water but the generated soul?" Walt Whitman saw in seashores the theme of liquidity marrying solidity, an image of the soul becoming a body.

The key to such thought, as Raine demonstrated, can be found in ancient Greek philosophy, Orphism, and Neoplatonism. A foundational work of Western literary criticism is Porphyry's *On the Cave of the Nymphs,* from the third or fourth century CE. In his deconstruction of a Homeric allegory from the *Odyssey,* Porphyry also explores the symbolic meaning of water. He relates it to the Greek term γένεσις, meaning *genesis*: coming into being, being born, the act of generation, but also ensoulment, the act of souls entering into bodies. Back then, the sea symbolized *bios,* or the life of the flesh. The soul was referred to as a "life-fluid." It arrived into this watery world and took on a body itself primarily composed of water. As Heraclitus noted, "Souls descending into generation fly to moisture." To be alive was to be "wet," awash in the sea of matter. And like Odysseus, we all had to spend years sailing the wine-dark, storm-strewn waters of existence. At death, our soul would reawaken into eternity. Our final destination was a place beyond all wave crash, a tearless place where "oceans are unknown," as Homer wrote, where everyone is ignorant of the sea, where we can finally forget these salty waters.

Porphyry and other metaphysicians believed that, just as Narcissus drowns by confusing his reflection for an actual substance, souls die into life through an attraction to the wateriness of materiality. In myths about the process of genesis, your soul begins as light. It is not yet a spirit, just a kind of ethereal luminescence sleeping peacefully, dreaming pure thought.

One day the moon glides through your realm. She comes in silk and gauze, wearing a tiara of rays. Lifting her gossamer veil, she looks right through you. She's so captivating you stir into motion. As you come together, for an instant, you feel close to everything, immersed in the sensation of complete love. But this lunar bliss gets complicated fast. Where before you were intangible radiance, just a glow, you now have an outer layer. The encasing material hardens, trapping you within a flexible diamond. Your union consummated, the Moon Goddess continues along her way.

You have entered the Chrystall Castle, becoming a spirit in its receptacle. You are in a chrysalid state, "confined and imprisoned by lawful Magick in this Liquid Chrystall," as Thomas Vaughan put it in *Aula Lucis*. The container filled with your energy is an unborn soul, an entelechy, a "cocoon-like Integument." Clothed in this hylic envelope, you are now ready to enter human life. Gravity pulls you down. It's a shredding sensation. Breaking apart, you dissolve into flesh. Bits of you are dispersed throughout the cells of a newly fertilized ovary. You are caged into a body.

As an infant, fascinated by light, you recall fragments of how life took shape. These memories fade as you age, until you die and begin the journey back to the shores of that wave-washed coast whence you sprang. The moon bathes the clouds off you. You liquefy in Lethe's waters. Forgetting everything, you sink back into the source.

This Moon Goddess is variously called Persephone, Kore, Diana, Luna, Hecate. She has so many names, explained philologist Károly Kerényi, "because the real one was not allowed to be uttered." The symbol of the moon is used for obvious reasons. Both luminous and dark, bright and then gone, she shines and extinguishes, dies and comes back. But the moon is also something greater than the moon: a goddess, a deity, a metaphor, a force within love, within each of us. In the mystery cults, she is also called the Honied. Like honeybees, wrote Raine, "souls that descend into generation will make their way home to the eternal world."

Today, such esoteric thought has to be spelled out for us to even begin grasping its meaning. But for those in whom this pictorial language lived, the meaning was clear, indisputable. Such myths were reminders. We die and then live on, returning again and again—like bees to the hive, like the moon in the sky, like water evaporating up and raining down. Another dimension to this mythology is hinted at in the actual encounter between the light-spirit and the Moon Goddess. Love is sacred, it suggests, divine in nature—and the feeling of love is what makes souls become human. Variations on this concept are found in mystical teachings of all times.

According to the Swedish scientist and theologian Emanuel Swedenborg (1688–1772), the sensation of love corresponds physically to the spiritual experience of God. Beginning in his mid-fifties, Swedenborg experienced a protracted series of religious revelations in which he spoke with angels. These conversations led him to a realization of the connections between this world and the spiritual world. His theory of correspondences states that everything on earth refers to something above. And we can only approach the things of the spiritual realm indirectly, he said, through correspondences, which take the form of symbols. To decipher the symbols around us is to come into contact with the sacred.

Swedenborg composed several encyclopedic works on the nature of symbols. These books, especially his treatise on the afterlife called *Heaven and Hell,* became sourcebooks for artists such as Borges, Dostoyevsky, Baudelaire, Whitman, Strindberg, Balzac, and Goethe. As William Blake put it, "The works of this visionary are well worthy the attention of Painters and Poets; they are foundations for grand things."

Swedenborg wrote at length on the symbolic meaning of fountains and water. At their most elemental level, both represent the possibility of coming into contact with divine truths. When the Bible speaks of a "pure river of water of life, bright as crystal, proceeding out of the throne of God" or a "fountain of water springing up into everlasting life," he wrote, it should be clear that something spiritual is meant, not something material—hence the use of correspondences. "It is manifest to everyone that the 'waters' here do not signify waters," he explained, "but that spiritual waters are meant, that is, spiritual things which are of truth; otherwise this would be a heap of empty words."

The ancient Greeks also spoke of correspondences in their doctrine of the Forms. They considered Forms to be eternal, unchanging archetypes of true Reality. All life in this universe, they said, spills forth from the Form of Beauty. Through love, we set eyes on the limitless ocean of Beauty, the fountain itself. Those witnessing the final object of all seeing cannot remain mere watchers, Plotinus wrote. Filled with the fountain's overflow, they become molten into oneness with the vision. They become Beauty. The soul that beholds beauty, as Father Gervais often said in class, itself becomes beautiful.

In ancient Greece, Eros played a role in this process. Truth, explained Socrates, can only be attained through the inspiration of love. And all lovers desire both beauty as well as immortality. In the *Phaedrus,* love sent from above is described as "the greatest benefit that heaven can confer on us." Why? Because it shows us that the soul is immortal and indestructible.

At that time, Eros was considered an "all-begetting and all-uniting life force." It brought things together, had preservative powers, and tempted us to bring forth art and life. All creative efforts, what the Greeks called *poiesis,* or poetry, is done with immortality in mind. Plato distinguished between three main forms of *poiesis.* The first is sexual reproduction, a creative act which provides immortality in the sense that a genetic lineage will survive the parent's own bodily existence. The second category of *poiesis* is the attainment of fame, which leaves a legacy after death. The third, and highest, expression of *poiesis* occurs when one's Eros-inspired pursuit of wisdom results in an experience of the soul's indestructibility.

Socrates's speech in the *Symposium* is more than a discourse on the possible meanings of love. It also serves as a guide on how Eros can launch us into an awareness of the sacred. Socrates begins by explaining that Eros is not a god but rather a spirit, half-human and half-divine. Eros, he says, is a kind of magical force that transmits messages and energy between mortals and immortals.

When Eros enters our life, a metaphorical ladder appears before us. Love is an invitation to climb. On the first rung of the ladder, we see beauty in our lover. On the next rung, we find ourselves appreciating the beauty in everybody. The following step is the realization of inner beauty, of mental beauty—that a beautiful soul is of higher desirability than a beautiful exterior. Once this stage has been attained, the lover starts noticing the beauty in daily habits, in everyday surroundings, in life as it is. This in turn leads to the erotic love of wisdom, an ascent that grants an inner vision of ultimate beauty, the breathtaking Form of Beauty itself, that limitless ocean of beauty. It is beyond time, eternal and uncreated, an everlasting loveliness from which all other manifestations of beauty stem.

Those falling in love can find themselves glimpsing into what lies beyond the grave. Having done so, assured of something greater than themselves, they feel ready to die. Socrates considered this to be the profoundest experience of immortality available to humankind.

For many pagans, the power of Eros was felt in a simpler way: through orgasms. In that little death, they felt the same thing mystics felt in their Eleusinian trance. Believers in the mystery cults considered orgasms to be proof of human indestructibility, spasms emanating from and connecting us to the bright world of the Forms. Perhaps they thought sexual climaxing was what everything would feel like in the never-ending perfection of the afterlife.

In 1927, the Nobel Prize–winning French writer Romain Rolland coined the term "oceanic experience" to describe a spontaneous, mystical experience of oneness with the world. He called it a "simple and direct experience of eternity." It's the ineffable limitlessness felt in religious experiences, and in certain heightened moments of Eros, a sensation of an indissoluble bond with the divine. These oceanic moments transport us into what Rolland called a *sur-vie,* meaning something more than life: a blissful, exalted sense of omnipotence and elation.

Wondering what they signified, in psychological terms, Rolland sent a letter to Sigmund Freud, asking for his thoughts on such oceanic presentiments of eternity. Freud, who'd never been much of a mystic, replied that the oceanic feeling might be a vestige of the primordial unity everyone feels in their mother's womb. Perhaps all mystical experiences, he speculated, are flashbacks to the tranquillity preceding birth.

During the intrauterine period, floating weightless in amniotic fluid, we really are one with our universe. We are the world. As newborns, we're in a state of pure narcissism: we can't distinguish between ourselves and the outside world. Gradually, we come to realize that our mother is not us, that we are not her, that we are the child but we are not, in fact, the world. We've left the prenatal garden of Eden. The trauma of separation makes us want to return to the state preceding it. The desire for the womb, to revert to a state of longed-for unity, remains with us in adulthood, if unconsciously so.

Freud felt that religious experiences confused the wish to reconnect with the womb with the wish to reconnect with the divine creator of all existence. This instinct to return to our literal maker—our mother, in whose belly we lived, in living water where all our needs were met—got its wires crossed with the idea of a God in a carefree heaven where pure fountains gurgle through the meadows of paradise. Regardless of Freud's opinions on religion, the womb-connection may explain why so many creation epics involve watery beginnings, why so many religions speak of aquatic miracles, even why the fantasy of a fountain of youth has persisted in our imaginations for so long.

Indeed, various hunter-gatherer tribes actually spoke of death as a watery world akin to the womb. Semang pygmies believed that immortal souls end up in Belet, but they can only eat the island's miraculous breast-milk fruits once they have broken every one of their bones and flipped their eyes around in the sockets, reversing them so that they look inward. Papuans told of spirits ending up in Hiyoyoa, an idyllic land in the underwater ocean. Bulbous fruits seeping mother's milk may or may not be there; we'll find out when we die. Other Pacific Islanders spoke of the undersea paradise called Tsiabiloum, an enclosure without sickness or sorrow or any want whatsoever.

Freud's contemporary Sándor Ferenczi developed his own psychoanalytical theories about "thalassal regression." (*Thalassa* is a Greek word meaning "the sea.") Our desire to climb back into the womb goes even deeper than infancy, he argued: it's actually a wish to return to our oceanic origins, to the primal evolutionary sea from which humanity emerged eons ago. He felt that we all yearn, on some deep, amphibian level, to rebecome the aquatic creatures we once were.

It is essential, if complicated, to distinguish between a symbolic image and the ulterior Reality of which it is suggestive. We sometimes lose sight of this important difference. Just as the image of something is not the thing itself and a map is not a territory, a symbol is not the object it represents. It's common to identify too closely with a symbol. Physical immortalists do it all the time. No life-form has ever been proven capable of existing indefinitely, of living for millions upon millions of years, yet radical longevists think we'll get there. Perhaps I'd made the same error by trying to see Copperfield's fountain. Or perhaps it really existed. I didn't know.

Symbols have a special power over us. We may even come to feel that a symbol can permit us to transcend the human condition. But that is not within the scope of a symbol's attributes. We all need myths. When old myths wear out, we create new ones, such as the myth of physical immortality. When we enter into new mythologies, we should be careful about mistaking a symbol for what it represents. It's so simple to slip into magical thinking.

Spiritual immortality is *not* literal immortality. And the fountain of youth isn't real. But humans have always been particularly adept at literalizing the symbolic, at confusing divine constructs with actual reality. It's part of our innate tendency to attempt to make sense of our ever-shifting surroundings and interior states by dipping our fingers into something intangible and then triumphantly declaring that it's actually tangible.

Back at home after my trip to Musha Cay, I still harbored hopes that the fountain might be real. My *BestLife* story wouldn't be able to resolve the mystery of his gene-reversing liquid, but I wanted to believe that writing "David's Magical Island" would incite him to invite me back so I could actually see it. I sent off an e-mail, telling him that the magazine piece would essentially discuss what it's like to visit Musha, with its treasure hunts and that incredible assortment of artifacts, while also incorporating David's vision of its future, from the valley of the giants to the yetis. "This means that we're going to veer away from the fountain-of-youth angle for *BestLife*," I concluded. "That said, I seriously plan to pursue that aspect for another story in the future (as we've discussed, David)."

But before I even had a chance to write the puff piece, my editor e-mailed, saying that *BestLife* had folded.* Martha got in touch shortly thereafter to express her condolences. Musha Cay was doing fine, she

* Shortly before this book went to press, Rodale relaunched *BestLife* as a quarterly.

wrote. "David has had some interesting bookings and he keeps relentlessly building his world. He has added another Treasure Hunt. He really does live in a world of fantasy!!"

I asked if his giraffes had arrived yet.

"No giraffes yet," replied Martha. "Let me see if I can get an ETA."

Later that day, the phone rang while I was in the bathtub. Recognizing Copperfield's number, I picked up and suggested we speak after my bath. He quickly mentioned that the giraffes weren't there yet; he was still designing their barn. But he'd just bought a flock of eighty flamingos. "That's pretty cool, right?" he asked, uncertain.

I reminded him to contact me once he was ready to show the fountain. He said he would be sure to. He never did.

Six months after my trip, Lacey Carroll, his accuser in the rape case, called Washington State law enforcement after blacking out and waking up in a hotel room in the company of a man assailing her. Video footage showed the two of them at the front desk of the Bellevue Club Hotel together. Carroll seemed in full control of her faculties. But that's not how she remembered it. She told 911 she couldn't remember checking in at all. The man, whoever he may have been, had a very different story. He told police that she asked him to "put $2,000 in my purse and you can have it all." He refused to pay her. She then got upset, he alleged, and stormed out of the room. He found her in the lobby, surrounded by hotel staff, saying that he had taken advantage of her.

The police charged her with prostitution. Copperfield's lawyers pounced, vociferously denouncing Carroll as a fraud. At her trial, Carroll's prostitution and false-statement charges were dismissed in exchange for her pleading guilty to charges of obstructing a police officer. The bargain cost her $953 and thirty hours of community service. "She is not a prostitute or a liar," said her attorney, Robert Flennaugh, in a statement to media. But she was done with the court system. "She just wants to put this behind her and get on with her life."

As a result, the assistant US attorney heading the federal grand jury investigation announced that they were dropping the Copperfield investigation "based on jurisdictional grounds." Whatever happened in the Bahamas, it was staying in the Bahamas.

"I know this much for sure: Copperfield never woulda raped anybody," Rick Marcelli told me over the phone. I'd called Marcelli, Copperfield's former personal manager (and the image-maker who gave him his trademark romantic-rocker matinee-idol look), to ask his opinion about the fountain. "Fountain of youth my ass—tell his hair that," Marcelli responded, laughing. "It looks like he has a dead rat up there."

I did recall being puzzled about Copperfield's hair. On the island, he always wore a baseball cap, but at times he'd angle his hat up slightly to wipe his forehead, revealing wispy, sparse hairs matted there. "What—does he have hair plugs or something?" I asked Marcelli.

"I have no idea what's going on with that. And I should know! But to me, this whole 'fountain' scheme is brilliant marketing for the island. Listen, he's a magician. It's not the Bible business. It's not the truth business. It's the art of misdirection. It's deception. It's *lying*."

Bill Zehme opened his 1994 *Esquire* profile of David Copperfield with the line "Nothing he does is real." After reading the article, the magician faxed off a three-page letter expressing his discontent: "I was, I will admit, excited to have an in-depth article about my career. . . . I will also admit, I wasn't quite prepared to learn that my soul was primarily about Pringles, diarrhea, trivia, nipples, and the importance of having a tan." Gossip columnist Liz Smith, commenting on the dustup, noted, "Reading about oneself is a very subjective matter indeed. What strikes others as fascinating may make the subject's own skin crawl."

A few weeks after Lacey Carroll's case imploded, an e-mail arrived from Martha, asking if I'd be able to write about his having been "officially exonerated." Even though the "David's Magical Island" angle died when *BestLife* folded, I assured her I would certainly tell the story of my island experience. He may have withheld the fountain, but Copperfield had been generous enough to give me a story. If it wasn't the story either of us had hoped for, there was still a kind of magical complicity in what transpired.

One moment in particular will always stay with me. On the speedboat leaving Musha, smiling uncontrollably, childishly, I looked at Raf and realized I felt so *happy*. Whether or not the fountain existed, Copperfield had made me believe it might. And that was enough.

17

Transmuting Magic into Science

There is in this business more than nature.

—Shakespeare, *The Tempest*

Then it seemed like falling into a labyrinth; we thought we were at the finish, but our way bent round and we found ourselves as it were back at the beginning and just as far from that which we were seeking at first.

—Socrates, in Plato's *Euthydemos*

THE *TEMPEST* ends with Prospero breaking his wand in two. Ridding himself of enchantments, he rejects magic, finds forgiveness, and accepts reality with all its limitations and infirmities. In this way, just as Prospero acts as a stand-in for an aging Shakespeare, Prospero's story is our story. We're all betrayed by growing old.

We're also all born with an epistemophilic instinct, an urge to know, to understand where we come from and how we got here. This natural inclination can never be fulfilled. We'll never understand why life is or what death means. As we age, the magical explanations we concocted in childhood either fall away in disillusionment or evolve into increasingly complex illusions. Our ravening hunger to know more leads us into the desire to unearth hidden arcana, to compute the suprasensible, to solve existence.

We can only do that, as Foucault wrote, "at the unattainable end of an endless journey." Still, we overburden our scientific enterprises, appointing them to fix all of humanity's intractable problems. What we often expect from knowledge is the same thing we demand from fairy-tale love: to be exempt from the human condition, to fuse into a oneness with the Other, to attain permanent perfection. But hearts do get broken. And our lust for knowledge entails a similar peril, as Adam and Eve found out.

Despite the myriad accomplishments of our technological society, there are limits to what we can understand. We can never completely integrate all knowledge into ourselves as knowledge. To know the truth of the universe would be to experience every speck of agitation and indifference occurring over the vastness of every galaxy and black hole and nebula at every single moment always.

The hope that accruing knowledge will lead to omniscience is akin to imagining that creativity will grant us supernatural powers. It's magical thinking. Scientific breakthroughs are real and will continue changing our lives, but the longing to unveil all mysteries is an impossible fantasy. Science isn't a technique for solving the insoluble; only magic is, and magic by definition isn't real. The more we penetrate into nature's secrets, the clearer it becomes that we really will die. We stop conjuring. We abide.

Prospero exemplifies the human journey from magical thinking to adulthood. Once he gives up on the dark arts, he grasps just how fragile life is. From there on in, he says, "Every third thought shall be my grave." This anticipation of death is as exaggerated as our denial of death is in youth. Still, freed from self-enchantment, Prospero finally understands what he really is: mortal. It's what we all are. Realizing it, we can sleep in peace.

Magic is like avoiding reality. Inattention and misdirection keep us from noticing the way things really are. But we can move beyond illusions. W. H. Auden, writing on *The Tempest,* described art as the alternative to magic. Rather than something entertaining or consoling (i.e., magic) that distracts people from their true condition, art allows others to "become conscious of what their own feelings really are: its proper effect, in fact, is disenchanting." Art can be a mirror that shows us reality. Under magic's distorting influence, Auden felt, death is inconceivable, an intellectual fiction.

Enlightenment thinkers believed that art ought to liberate us from our painful involvement in reality. But now we don't expect art to lift us into some higher state of nonbeing. Its purpose is to help us feel, to help us find meaning, to help us confront the details of life as they truly are, composed of grief, suffering, and loss—as well as contentment, relaxation, and fulfillment.

Goethe wrote at length on the duality of pleasure and anguish, on the necessity of creative struggle. His Faust begins as a young alchemist who calls on spirits to grant him a vision of nature's innermost force. He wants to see the fountainhead. Conjuring, he glimpses the fundamental machinery of existence, a kind of surging liquid energy. A spirit appears. Faust, being mortal, cannot comprehend him. The entity vanishes, leaving Faust suicidal.

Not long after, Mephistopheles materializes and offers to provide Faust with a moment so wonderful—one in which he comes to an understanding of the elemental mysteries of the universe, perhaps?—he'll want it to last forever. In exchange, should such a moment occur, he'll then spend eternity in hell with Mephistopheles. Faust agrees, but says he doesn't want simple pleasure; he wants joyous pain, refreshing frustration, loving hate. He wants to be able to suffer and bear the suffering.

At the end of his life, however, having created homunculi, traveled through time, and come as close to divine creation as a human can get, Faust concludes that it's not for us to be gods. "There is no view to the Beyond from here," he realizes. The sacred remains hidden in a flash of the unknowable. Glimmers are all we're granted.* Going blind, finally capable of accepting his fallibility, Faust renounces sorceries. "I must clear magic from my path," he vows. "Forget all magic conjurations." He abjures supernatural aid and comes to an acceptance of this world's imperfection, of humanity's mortality. He breaks the spell. "Illusions are such fun," Mephistopheles mutters. "If only they would stay with one." An illusion is something that promises to set our souls free. That's magic. We can't "escape" from life. We can't return from adulthood to childhood.

Nor can we understand. We've been thrust into a world we didn't make, a world whose ultimate explanation will always elude us. Heidegger

---

*"The truth, being identical with the divine, can never be perceived directly by us, we only behold it in the reflection, in the example, the symbol," as Goethe wrote in "An Essay on Meteorology."

calls this sensation *thrownness.* As we tumble, we can only know that we can't know. Alchemy won't help; neither will science. Just as Prospero and Faust turn their backs on the deceptions and misapprehensions of magic, we all need to be vigilant about succumbing to the lure of a magical version of science, a science many believe is capable of improving everything indefinitely.

The notion that physical immortality might be something we can engineer goes back to a time when magic and science were indistinguishable. Throughout the first millennium CE, esoteric Daoist and Tantric longevity potions were concocted from mercury, cinnabar, and sulfur in a practice known as *kim* or *chin*. Transliterated into Arabic as *al-khymia,** the combining of mineral elements and turning base metals into gold became known in Europe around the twelfth century as *alchemy,* which gradually morphed into what we today call *chemistry*. One of its goals was the discovery of an elixir that could make a person live forever.

Science, as we now conceive it, did not exist at that point (the word *scientist* only became commonplace in the mid-1800s). But in thirteenth-century Europe, systematic and experimental approaches to nature began to emerge from the medieval darkness. British philosopher Roger Bacon (born c. 1214 CE) put forth a distinction between magic aimed at commanding or appeasing spirits, and magic that dealt with the factual mysteries of nature, such as rainbows, waves, and clouds. He called this second approach "natural magic." Unlike occult, otherworldly magic, which dealt in miracles and was the rightful domain of religion, natural magic had practical applications. Using the experimental method, the secrets of this world could be explored—and, more important, explained. Science, Bacon posited, was magic that actually worked.

A Franciscan friar and "excellent wise man," Bacon was known as Dr. Mirabilis, the "wondrous doctor." He believed that the study of natural philosophy didn't simply confirm the facts of nature—it also complemented a devoutly religious life. Bacon depicted science as something

---

* *Khymia* may also refer to the Arabic word for "substance" or "art." Alternately, the term *chem* could have derived from the Egyptian term for "black," from their oxidized silver jewelry. Yet another possibility is that *khym* sounds exactly like the Greek word for "mixing," as in "mixing fire with mercury to make an immortality potion."

that reinforced one's faith in God, as opposed to the damnable magic of conjuring spirits, which, beyond being irrational and heretical, he noted, was altogether impossible.

Over the past few decades, we've grown accustomed to seeing science and religion as incompatible, but in Bacon's time, they were deeply entwined. Most researchers until modern times found nothing strange in worshipping God and doing experimental studies. Copernicus viewed his galactic investigations as a religious activity that filled him with awe. Newton's discoveries were predicated on the existence of God, the all-powerful "Mechanick" of the cosmos. (Newton's formulas established that absolute space is equidistant to the Creator's omnipresence.) Darwin firmly believed in the soul's immortality. Einstein spoke of a "God who reveals Himself in the harmony of all that exists," of a spirit manifest in the laws of the universe.

Science began as tests seeking to prove the reality of the sacred, and Bacon's natural magic was intended as such. The knowledge of our material surroundings leads to divine knowledge, he argued; and everything knowable stems from God. His book *The Cure of Old Age and Preservation of Youth* portrays aging as a curable disease. Its cause? Insalubrious living. With the correct use of healthsome ingredients, everyone should be able to live at least as long as biblical patriarchs such as Methuselah, who died at the age of 969. For Bacon, science was a way to perfect ourselves, even to attain physical immortality. According to his scientific calculations, maintaining our innate moisture entailed ingesting precise dosages of seven types of medicine, each containing various vital principles.

He prescribed powders made from gold, pearl, or coral. The connective tissue from a stag's heart works wonders for energy levels, he counseled. Eating vipers and serpents was also highly recommended, as they are reborn by shedding their skins. (Dragons contain the most vital spirit, but they are hard to come across, Bacon granted, unless you happened to be in Ethiopia.) The best remedy of all is breathing in the exhalations of virgin girls. The word *gerocomy* refers to the idea that a certain "vital principle" can be absorbed by aged men through proximity to young maidens. Bacon swore that it rejuvenates even the dustiest old fogey. Despite being charged with "suspected novelties" by the Vatican, Bacon himself lived almost to eighty, far longer than most of his contemporaries.

Bacon's work was indebted to that of another important medical pioneer, Jābir ibn Hayyān of Persia. He famously sought a rejuvenating tincture called *al-iksir*—the elixir. He never found it, but he did invent fireproof paper and develop glow-in-the-dark ink. And he was already emphasizing the importance of experimentation in the ninth century CE: "The first essential in chemistry is that thou shouldest perform practical work and conduct experiments, for he who performs not practical work nor makes experiments will never attain to the least degree of mastery." Unfortunately, it's hard to glean anything concrete from Jābir's experiments, as his writing is prolix to the point of impenetrability—an intentional maneuver. "The purpose is to baffle and lead into error everyone except those whom God loves and provides for," he wrote in *The Book of Stones*.

Baffling though they may be, works of his such as *Investigation of Perfection* and *The Book of Eastern Mercury* were translated and made available to Europeans like Bacon by the twelfth century. As alchemy rose in prominence, disputes about its limitations were raised, but no one knew its capacities with any certainty. The esteemed Persian physician Avicenna, also known as Ibn-Sina, decreed that it simply wasn't possible to transmute base metals into gold, but that panaceas of eternal youth would eventually be uncovered by chemicomedical researchers.

Both ideas captivated European alchemists,* who pursued *ars aurifera* alongside seeking the *arcanum universale*: the philosopher's stone, the preserver of the macrocosm, the miraculous stone that is no stone. Its liquefied version could make one live forever, they thought. Most alchemists from the thirteenth to the seventeenth centuries hoped to drink the quintessence of the universe, possibly made from water of gold (whatever that may have been).

Misguided though such efforts seem in retrospect, they were an important step toward what we now consider scientific chemistry. Medieval alchemists learned how to alloy copper with zinc (yielding goldlike brass); they figured out how to make porcelain; they invented distilling equipment; and they harkened the scientific method's approach by performing experiments over and over.

During the sixteenth and seventeenth centuries, every major European town had its own "laborer of the fire" toiling away in a shady apothecary. Capitalizing on the hope that alchemy could find a way around

* As they did for Chinese and Indian alchemists. (See the next chapter.)

death, countless swindlers rooked the upper classes with bogus medicaments for eternal life.

Researchers also harmed *themselves* in their chimerical fossicking, ending up half-asphyxiated by arsenic fumes or going blind from noxious vapors. Many physicians from that epoch who are well respected today (such as Newton) also performed alchemical research, which was essentially their term for "science." Newton died of mercury poisoning, as did countless gilders. Mercurous nitrate causes erratic, flamboyant behavior; its use by milliners in felting hats gave rise to the expression *mad as a hatter*.

In the late 1700s, a Scottish quack named James Graham, Servant of the Lord, O.W.L. (Oh, Wonderful Love), became the talk of London for claiming anyone could live to 150 simply by making regular visits to his private clinic, the Temple of Health. He encouraged valetudinarians to rub themselves with his patented aethereal balsam. He also advocated earth baths, in which naked patients climbed into holes in the ground and were covered neck deep in mud. He spoke of the salutary effects of thoroughly washing one's genitals in cold water or, even better, in ice-cold champagne. His most in-demand device, however, was the celestial bed, a massive stallion-hair-filled mattress supported by forty glass pillars that administered mild shocks of electrical current. Graham's clients hoped the effects of "holding venereal congress" in the bed would help them live longer, if not forever.

During that same era, the German physician Franz Anton Mesmer made a fortune convincing people that the secret to health and longevity resides in a special fluid coursing through our circulatory system. This invisible electrical fluid, he said, is an internal ether subject to blockages. Fortunately, the cure lay in using "animal magnetism" to unblock the inner fluid. Brandishing magnets and a wand, Mesmer would stand with one foot in a bucket of water and stare deeply into patients' eyes (whence the term *mesmerizing*). Sometimes his methods worked. But, of course, we didn't yet understand the vagaries of the placebo effect.

Eighteenth-century Parisian aristocrats were easily swindled by colorful crooks peddling elixirs of eternal life. One miracle *docteur* sold a concoction consisting almost entirely of tap water. The Count of Saint Germain was a self-promoting "wonderman" who "never dies and knows everything." He manufactured immortality potions for wealthy patrons and claimed to be capable of astral travel: "For quite a long time I rolled through space. . . . I saw globes revolve around me and earths gravitate at my feet." Rather than a real count, Saint Germain was simply a dazzling scoundrel. "As a conversationalist he was unequaled," marveled Casanova, in his memoirs. Only one other courtier-impostor even came close, Count Alessandro di Cagliostro (real name: Giuseppe Balsamo). Like all hucksters worth their salt, Cagliostro pushed amulets and nostrums. Living forever was merely a matter of payment. Not immortal himself, he perished during the Inquisition.

Graham, Mesmer, Saint Germain, Casanova, and Cagliostro all lived during times not entirely different from our own. "They were received into the centre of a small, skeptical and libertine world that had, in principle, rid itself of prejudice," explains the Romanian historian of longevity Lucian Boia. "These people who pretended to believe in nothing at all, except, to some extent, in philosophy and science, were ripe to be caught in any trap that a person of speculative intelligence could set. Because they believed in nothing, they were ready to believe anything."

When we think (or pretend) that we don't believe, we're setting ourselves up. It's human nature. Death demands it of us. We're so unconsciously desperate to believe in the possibility of immortality that we can be suckered easily—if we don't have a belief system in place already.

It can be hard to distinguish between science and folly. The well-respected empiricist Francis Bacon (1561–1626) was one of the earliest advocates of the scientific method. He also considered youthfulness to be a vital moisture in the body that could be replenished in various ways. He recommended wearing scarlet waistcoats and self-medicating with anything redolent of fresh earth or newly turned-up soil, such as strawberry leaves, raw cucumbers, vine leaves, and violets.

The sixteenth-century Swiss alchemist Paracelsus innovated and introduced the concept of chemicals as medication. Before him, all the way back until Galen, healing remedies in the Western world were primarily plant-based, rather than chemically derived drugs. An impetuous, pudgy warlock with scalpel eyes and a no-bullshit attitude, Paracelsus argued that alchemy's aim was medicine rather than gold. His discovery that mercury helped cure syphilis yielded a salve used until 1910. Alongside his medical innovations, he also is credited as being the first to articulate (however gropingly) the concept of the unconscious. Jung considered him a pioneer in the study of psychology.

Interestingly, Paracelsus had no problem advocating the use of sorcery in helping patients. "Magic has power to experience and fathom things which are inaccessible to human reason," he wrote. "For magic is a great secret wisdom, just as reason is a great public folly." An example of this secret wisdom, according to his treatise on longevity called *De Vita Longa,* is learning how to live for a thousand years, even forever. Like Jābir and Bacon, he spoke of attaining a life without end, although he himself died in his late forties, without ever finding that mercury-based potion of immortality.

Paracelcus's medical practice was predicated upon the existence of a physical thing called *iliaster* that keeps bodies alive. *Iliaster* is a compound word from two Greek morphemes: *hyle,* meaning physical matter, and *astron,* meaning of the cosmos. In Paracelsus's teaching, *iliaster* is a substance from the heavens animating life here on earth, a kind of astral secretion produced naturally within the body. This celestial-cum-terrestrial stuff is our vital spirit, the essence of existences, "the true spirit in man, which pervades all his limbs." It originates within the Great Mysterie, the source of all created things. Paracelsus used words like *breathlike* or *vaporous* when describing *iliaster,* what we today might call energy—or the life force.

This energy dissipates as we age—but Paracelsus believed it could be replenished. He considered mercury to be a key ingredient. Just as bodily *iliaster* consists of internally transmuted mercury, alchemical *iliaster* can also be crafted through an elusive blend of mercury, sulfur, and salt. (Mummified remains were thought to contain traces of *iliaster,* so they were also incorporated into the mélange.)

For Paracelsus, mercury had abstract qualities beyond its slippery chemical properties. As a growth stimulator, he said, it allows humans to develop from children to adults, at which point it starts to run out. If prepared quicksilver (what Paracelsus called the "eternal liquor") were added to the bloodstream in adulthood, one could conceivably live forever. Mercury also allowed the lower, physical world to come into contact with—or be transformed into—the higher, spiritual world.

In Greek religion, Mercury was Hermes, the God of Going Back and Forth, an intergalactic go-between, a means of communication between humans and divinities, between life and death. In pre-Hellenic times, Egyptians called him Thoth. During the early Christian era, Hermes-

Thoth evolved into a new figure called Hermes Trismegistus, or Thrice-Great Hermes, known for bestowing a body of esoteric knowledge called the *Corpus Hermetica*. (Purporting to be texts from ancient Egypt, many of them actually appear to date from around 200 CE.) One of the most important documents in the Hermetic canon is *The Emerald Tablet,* which contains the proclamation "that which is below is like that which is above, and that which is above is like that which is below."

One commentary on *The Emerald Tablet* contends that the tablet's meaning is that "the secret of everything and the life of everything is water." This text, *The Book of the Silvery Water and the Starry Earth,* describes something called upper water—a divine spiritual water that represents the soul. There's also lower water, the earthly, everyday water that nourishes the body. These two can be brought into contact with each other through the translating medium of mercury.

Paracelsus latched onto this idea. In his books, mercury is spoken of as the spirit of water, living water, *mare nostrum*. Alongside the *iliaster,* he wrote of a psychical (rather than physical) principle called *aquaster*—meaning liquid of the cosmos, or star-water. This invisible watery force has quasimaterial qualities and is the wellspring and birthplace of *iliaster,* that "which animates and preserves the liquids in the body." One way of thinking about it is as the fountain of the vital spirit. Paracelsus saw this as the basic element of creation, a radical moisture whose nontangible wetness nourished the universe with life force. If we could just tap into it, we'd live forever.

Magical though it all sounds, Paracelsus was a late-Renaissance visionary who helped push us forward into that uncharted terrain we now call the Enlightenment. Although conflicted in many ways, wrote Jung, Paracelsus "was spared that agonizing split between knowledge and faith that has riven the later epochs"—a split we're still agonizing over to this day.

As chemistry emerged out of alchemy, it started the long, slow process of concealing its religious implications. It kept its practical side, letting its mystical and cryptic aspects fall into obscurity. But the magical facet of science hasn't disappeared entirely. Most immortalists today still feel that, as English philosopher John Gray has argued, "if only they are able to penetrate the secret order of things, humans can overleap natural laws." In time, perhaps they (or "we," as my friend Billy Mavreas would say) will be able to.

For now, true scientific achievements involve nonglamorous technicalities and incremental advances that rarely correlate with the sensationalist way they're depicted in the press, where they often have more in common with magical thinking than with science itself. Can technology really find a way to make us live forever? Or is that very idea a mythology that helps us to avoid thinking about a truth we can't bear, that we have to die? To find out, I would need to speak with physical immortalists about how exactly they see us attaining eternal life. I'd need to ask real scientists to clarify the extent of their discoveries. And I'd need to learn more about the ways our ancestors attempted to live forever—none of which worked, of course. But exploring their failures sheds light on the reasons we continue to pursue immortality and helps explain why, despite all the evidence to the contrary, we still believe we will be able to live forever.

Part 3

Science

Magic, religion, and science are nothing but theories of thought; and as science has supplanted its predecessors, so it may hereafter be itself superseded by some more perfect hypothesis, perhaps by some totally different way of looking at the phenomena—of registering the shadows on the screen—of which we in this generation can form no idea.

—J. G. Frazer, *The Golden Bough*

18

Mercurial Times

When all the ties of the heart are severed here on earth, then the mortal becomes immortal—here ends the teaching.

—*Katha Upanishad* (or, *Death as Teacher*)

I knew at last that on the plane of Assembled Occasions
One cannot escape from the secret laws of predestination

—Bai Juyi, 772–846 CE

UNTIL THE Neolithic revolution, bodies from different cultures around the world were painted red when buried in the earth. Decorating cadavers with vermilion signified the hope that even in death there is life, a returning to and from the earth. Powdered pigment of cinnabar—a hardened, red, metallic ore resembling the dried blood of the earth—was a typical dye used in these funerary rites.

The discovery that heated cinnabar's vapors condense into silvery, liquid mercury perpetuated the belief in its transformative powers. Just as cinnabar's death becomes mercury's life, ingesting mercury seemed to be a means of bringing shiny, new vitality into the eater. This notion captivated the elite of imperial China, where the science of chemistry began with efforts to manufacture physiological-immortality potions.

As early as the Warring States Period (476–221 BCE), mineral-based philtres ostensibly bestowing longevity, if not eternal life, were being manufactured. Cinnabar and mercury were the main active ingredients. Medical treatises contained directives for grinding burnt minerals to the fine, ashy consistency of dust motes swirling in a sunbeam falling through a window. Aspirants to immortality were instructed to imbibe cinnabar formularies mixed with sow lard, vinegar, saltpeter, orpiment, realgar, sal ammoniac, and wine. The result? The twenty-five-year old Emperor Ai of Jin died in 365 CE, after overdosing on longevity drugs. He wasn't the last leader to die trying to live forever. The fascination with chemical immortality reached an ironic apogee centuries later, during the T'ang dynasty (618–907 CE), when elixirs poisoned those hoping for precisely the opposite effect.*

In that Golden Age of China, the capital, Chang'an, was the most important city in the world. The T'ang nobility's insatiable desire for the fantastical fueled the importation of countless marvels. Among them, writes Eliot Weinberger in the *New York Review of Books,* were grains that could make you light enough to fly, crystal pillows that gave you hallucinations of faraway lands, and heat-emitting rhinoceros tusks powerful enough to warm entire castles. Little wonder then that the availability of mercury-based alchemical elixirs intended to make one live forever skyrocketed.

One T'ang cinnabar votary, Emperor Ming (also known as Hsüan Tsung), managed to reign for forty-three years, living to the wizened age of seventy-seven. A few rulers later, however, Emperor Xianzong wasn't as lucky. Alchemical pharmacists sold him golden concoctions that made him increasingly paranoid and demented. At age forty-two he was assassinated by palace eunuchs. Xianzong's successor, Emperor Muzong, executed the mountebanks who'd formulated the poisonous preparations by forcing them to take their own medications. But then Muzong himself became an elixir convert, repeating the same mineral mistake. His tenure lasted four short, pain-filled years. Sixteen years later, horror upon horror's head, Emperor Wuzong, too, started popping alchemical pills, becoming severely manic and "very irritable, losing all normal self-control in joy or anger." Then he died.

How could they have been so ignorant? For a simple reason: because they were human. *Errare* is our style. But more precisely, in this case, emperors and aristocrats "considered themselves eminently suitable for survival," as sinologist Joseph Needham wrote. Common valedictions

---

* Of the various sources consulted for this chapter, Joseph Needham's multivolume *Science and Civilisation in China* played such a pivotal role it deserves to be singled out here. For more, please see the sources section.

at the time included "May you have a never-ending span, long life that lasts for ten thousand years." Beyond fluffing high-ranking officials' already inflated self-esteem, alchemists were canny salesmen, hinting at side effects such as prolonged stamina and increased sexual prowess. Their drugs could make you live—and love—longer. Like all psychotropics, these chemical preparations bestowed an initial sense of euphoric disorientation. This transient inebriation foreshadowed a breakthrough to true immortality, the pushers explained, enticing their marks further. Imperial elixir junkies perished seeking life, impelled by the same phototropism that draws moths to flames. The more they paid, the sicker they became. Overdoses were portrayed as premonitions of eternity, "temporary deaths" reframed as portals into everlasting life.

Pharmaceutical tomes insisted that preparations be taken slowly, over time, unless one wanted to attain the unseen world immediately. "If you should desire to ascend to the heavens with all speed, the elixir is to be swallowed in a single dose; you will fall prostrate and die immediately. But if you wish to prolong your stay among men, you are to consume it little by little, and when at last it has all been taken, you will then find yourself an immortal."

Given the pain of prolonged mercury consumption, though, it was preferable to die fast. One official who fell ill after taking mercury described the pain as being akin to having a red-hot iron rod piercing him from head to toe. He felt flames billowing out of every orifice and joint. He vomited blood for ten years. He woke from sweat-drenched fevers with mercury pooled in his mattress. Cinnabar addicts ended up with ulcerous lesions all over their chests. They'd begin as wounds the size of rice grains, then gradually swell. The users' necks would expand until the chin and the chest appeared continuous. After every bath, cinnabar dust would seep out of the sores and collect visibly at the bottom of the washbasin. By the ninth century CE, so many people were suffering from immortality potions that alexipharmics for elixir poisoning started being circulated. The *Mysterious Antidotarium* (*Hsüan Chieh Lu*) recommended a mixture of fruits, herbs, nuts, and deer glue stirred vigorously with a willow-wood comb.

"Elixir mania" hit during the medieval Chinese centuries, but it began much earlier. As early as the eighth century BCE, thinkers were discussing methods of "becoming an immortal." In ancient Chinese thought, souls of the dead were believed to end up in the Yellow Springs, a shadowy underworld just below the earth's surface.

Lore had it that not everyone ended up in the Yellow Springs: some jackpot winners sprouted feathers and became immortal, ethereal, purified beings. They were humans in levitative, aerostatic, subtle bodies, the "spirits of just men made perfect." They were able to roam interminably through the countryside and the Milky Way, freed from all physical wants, endlessly enjoying and meditating upon Nature's bounty. By the fourth century BCE, a certainty arose that the technical means of attaining immortality—"not somewhere else out of this world, nor in the underworld of the Yellow Springs, but among the mountains and forests here and for ever"—was within reach. The only thing needed was to find and consume the elixir of immortality.

As nonsensical as the concept may seem to us today, it grew wings with the emergence of Taoism somewhere around the third or fourth century BCE. Almost immediately, the Taoist movement branched in two directions. Classical, or philosophical, Taoism concerned itself with a contemplative, mystical approach to the ineffable, immaterial, ever-present and everlasting Tao, that which transcends everything the mind can conceptualize. Religious Taoism, on the other hand, offered a more practical and materialistic system focusing on health, medicine, and longevity. Both approaches had the same goal: to attain the great Tao. Spiritual and physical, they were two sides of the same coin. A more apt metaphor is the central symbol of Taoism: yin and yang. Laozi's formative text, the *Tao Te Ching,* emphasized these interconnected, polarizing forces as the definitive enigma of Taoist cosmology ("Indeed, truth sounds like its opposite").

Both forms of Taoism recommended yogic postures, breathing techniques, and gymnastic exercises, but where philosophical Taoism aimed at a rigorous, meditative serenity in which the adept became okay with dying, religious Taoists believed that a combination of physical exertion, diet, and herbalism could extend physical life indefinitely. This second model, eventually responsible for the deaths of all those emperors and nobility, became known as *hsien* Taoism (*hsien* is variously translated as "immortal" or "living angel" or "enlightened and eternal sage"). The cult of the *hsien* promised, by definition, material and bodily immortality. "The conviction crystallised that there were many men who had liberated themselves from death, and were continuing in perpetual life," wrote Needham. "This became a fixed belief in Qin and Han times, taken immensely seriously by emperor after emperor."

Chinese alchemy began with *hsien* hopes of prolonging adherents' lives. It started as benign natural products—concoctions made from roots, herbs, berries, and animal parts—but soon began incorporating minerals, chemicals, and precious metals. To supply impatient Tao-hungry rulers who lacked the discipline and isolation needed for philosophical Taoist training, pharmacists sprung up offering soups of powdered gold and mercury: druggy elixirs of never-ending life.

The alchemists were known as *fangshi*—chemical intuitives, recipe gentlemen, masters of methods. Sometimes these medical magicians died imbibing their own mineral whimsicalities, but usually they lived cushy lives sponsored by wealthy, easily blandished patricians. It wasn't hard to find buyers for something like "the subtle potion of the Flying Springs that brings one to the land of deathlessness, where the feathered ones are." An early reference tells of one such itinerant mendicant during the Warring States Period who convinced a prince of Yen that he knew the techniques of deathlessness. The prince sent emissaries to learn the full teachings. Unfortunately, the sage died before he could fully transmit his secrets. The prince, furious, punished his men for their dilatoriness. It never occurred to him that if the old shaman couldn't make *himself* live forever, then he certainly couldn't help anyone else live forever.

By the second century BCE, charlatans abounded, as contemporaneous reports testify: "There were thousands of magician-technicians in the regions of Yen and Chhi who glared around and slapped their thighs, swearing that they were the real experts in the arts of achieving the life of the holy immortals." These bug-eyed leg-slappers found a ready sucker in Emperor Wu of Han—so receptive to becoming immortal that he'd built a high-elevation copper tower atop a mountain in hopes of harvesting heavenly dew, the nectar or manna he believed would ensure eternal life.

Around 130 BCE, a thaumaturge called Li Shao-chun, who came highly recommended by the Marquis of Shen-Tse, told Emperor Wu to make a certain sacrifice to the spirits of the furnace, which would cause primordial supernatural beings to appear. These ancients would help transmute cinnabar dust into yellow gold plates, utensils, and drinking vessels. Meals consumed with these would grant prolonged longevity.

In time, the emperor would be able to visit the blessed immortals on holy islands to the east. Appropriate oblations were to be made to these eternal, perfected ones. *"Then,"* the intermediary whispered, *"you, too, will never die."*

These holy islands, located somewhere in the Bohai Sea, were thought to lack roots, so they just floated around aimlessly on the backs of fifteen tortoises. The emperor's *fangshi* Li Shao-chun swore that once, on a boat, he'd met an immortal inhabitant of these isles who'd found the flower of deathlessness. Named Anqi Sheng (Master Aṇqi), this entity remains hidden to all unless they've followed the prescribed measures for materializing him. Once successfully evoked, Anqi Sheng would divulge all the secrets of transcending the human condition. Another wizard named Luan Ta backed up Li Shao-chun's account, saying that he, too, had encountered Anqi Sheng.

Venerable precedents had been set for such stories, as the first emperor of unified China, Qin Shi Huang, himself claimed to have met and discoursed with Anqi Sheng for three straight days. Subsequently, he endeavored to reconnect with the immortals on their home turf in the blessed isles. On three separate occasions he sailed into the Bohai Sea in search of the elusive archipelago of eternal life. He skulked along beaches, gazing out at the ocean in hopes of glimpsing the islands. The older he got, the more frantically he searched. He ended up dispatching his official court sorcerer, Xú Fú, alongside hundreds of male and female stewards, into the unknown. They set out on ornate, silk-sailed ships with orders not to return without the elixir of eternal life. They suspected that Anqi Sheng and his fellow immortals lived on Zhifu Island (today a peninsula connected to mainland China). The hunch proved incorrect. Emperor Qin Shi Huang's doctors ended up poisoning him with a mixture of powdered jade and rare-earth elements.

He was buried in a jewel-and-pearl-encrusted subterranean tomb protected by the now-famous Terracotta Army. More than seven hundred thousand men labored on his necropolis, an underground microcosm of the entire universe. Geomancers selected an auspicious spot near Mount Li. Entire rooms were carved from jade. The ceilings were painted with a zodiacal firmament, traces of stars, cosmic dots, and heavenly bodies. The mausoleum's pièce de résistance was a mechanically operated stream of flowing mercury. They filled the inner sanctum with statues of singing musicians, pirouetting dancers, and acrobats frozen midflip. After the emperor's liver failed from the toxic elixir, his cadaver was carried into the hypogeum, behind gates that locked themselves automatically. Numerous *Goonies*-style booby traps were installed to deter grave robbers. Tons of soil concealed the entranceway. A forest sprung up over the tumulus, and almost two thousand years went by before some peasants stumbled onto a life-size clay soldier while digging a well in 1974.

Given that backstory, it's easy to see how the Han emperor Wu might've felt justified in hunting for the abode of the blessed as well. After a time, however, his patience ran out, and he came to wonder whether the immortals' dwelling place lay not to the east but rather to the west, in the vicinity of the Kunlun Mountains. Inscriptions from the region described timeless beings flying around in carriages yoked to scaly dragons, sipping from springs of jade, nibbling on jujubes, plucking the Herb of Life alongside white tigers.

King Mu of Zhou, who'd ruled China eight hundred years earlier, also traveled to Kunlun, where he met a goddess next to her turquoise pool. According to the third century CE *Monograph on Broad Phenomena,* Emperor Wu met her as well. Called the Queen Mother of the West, she arrived on purple clouds, wearing a robe of blue pneuma and a crown of stars. Her retinue included a nine-tailed fox and a moon-dwelling hare who mixed holy drugs with a pestle and mortar. At their powwow, she gave Wu a gift of magical peaches. A later document (*Esoteric Transmissions Concerning the Martial Thearch of the Han*) suggests that the Queen Mother also bestowed certain recipes for elixirs and other sacred texts upon the emperor. She lectured him on the importance of relaxation techniques and Taoist self-cultivation, which may explain how the emperor lived into his late sixties. Unfortunately, the texts were all lost when his library burned to the ground.

Shortly after Emperor Wu's death, philosophers started cautioning against immortality potions, explaining, "There is no such thing as the Tao of the immortals, 'tis but a fable of those who like to talk about weird things." In the second century CE, however, a significant alchemical treatise called *The Kinship of the Three* was published (it is still considered the world's oldest book of alchemy). In 366 CE, a volume of documents called *Confidential Instructions for the Ascent to Perfected Immortality* appeared. A philosopher of that time, claiming to know the recipe for walking on water (eat cinnamon, onion extract, and bamboo sap mixed with turtle brains for seven straight years), shared his own formula for an immortality drug: "Take three pounds of genuine cinnabar, and one pound of white honey. Mix them. Dry the mixture in the sun. Then roast it over a fire until it can be shaped into pills. Take ten pills the size of a hemp seed every morning. Inside of a year, white hair will turn black, decayed teeth will grow again, and the body will become sleek and glistening. If an old man takes this medicine for a long period of time, he will develop into a young man. The one who takes it constantly will enjoy eternal life and will not die."

Concurrently, other nonchemically derived yet equally out-there anti-aging methods emerged. Medics started investigating whether mercury's beneficial effects might be attained through exercises. Just as scientists today have identified dopamine-producing regions in the hypothalamus, specialists in the Three Kingdoms period wrote about cinnabar fields in the body and the brain. Activating our cinnabar-based embryo of immortality requires doing handstands or hanging upside down, so that our sexual fluids flow into the head.

Sex played a vital role in Chinese yoga.* As a Jin-dynasty official wrote, "One branch of Taoists seeks solely by means of the art of intercourse to achieve immortality." Numerous how-to sex texts outlined the genital gymnastics initiates had to master to reach eternal life. A fundamental tenet of male Chinese sexological magic entailed penetrating a female lover without reaching climax. Ejaculating not only prevents immortality—or so the reasoning went—it also upsets the Five Viscera, does damage to *ching* channels, and causes the hundred illnesses. And just as sperm beget children, they devised, *injaculating* could only lead to inner birth and then eternal life—especially if the nonejaculating penis in question is able to absorb the woman's *chi,* or energy, by "drinking" up her orgasm. As Sun Ssu-miao's *Priceless Prescriptions* explains, "If one is able to mount just twelve women without ejaculating, this makes a man youthful and handsome. If one can have intercourse with ninety-three women and remain locked, one will live for ten thousand years."

This form of *coitus reservatus* was intended to reconnect practitioners with their prenatal paradise (*wu-chi*). But it wasn't all just oceans of tin-

---

*Those desirous of pointers are directed toward Douglas Wile's *Art of the Bedchamber: The Chinese Sexual Yoga Classics Including Women's Solo Meditation Texts* (1992), which recounts the intricacies of how women can shoot golden mystery elixir into a man's urethra, thereby "completing the work of immortality."

gly restraint. After doing three thousand practices and perfecting eight hundred exercises, aspirants had to stroke the zither without strings, blow slowly on the flute without holes, and spend 5,048 days nurturing the mysterious pearl that hovers above the void. Then, after eating the pearl, they'd sprout wings and soar away to a feathery eternity.

Women seeking perpetual life through sexuality had their own precepts, grouped together under the banner of "solo meditation." Various manuals of feminine self-cultivation promised immortality in a mere three years, using a combination of techniques such as gripping the stork, straddling the crane, directing sweet dew to the crimson palace, letting the splashing stream of true juice overflow from the fountain of the heavenly gate, and massaging breasts until they take on the consistency of walnuts and the solo meditator feels "suspended like a cascade, white as silk."

Dissenting Taoist leaders believed all these sexual and chemical activities were futile, akin to clutching a piece of jade and hurling oneself into a fire. Their assessment gradually became the consensus,* but it took centuries and countless deaths for interest in immortality blends to slowly peter out. A high-society fad for cinnabar flared up again in the twelfth century. Around that time, Song-dynasty medics started testing philtres on death-row inmates. Despite a success rate of zero percent, alchemists still argued that they simply hadn't perfected the recipe yet: "We cannot deny the possibility of the existence of methods for transforming people into feathered immortals." The finest peculiarity of belief is that believers do not recognize themselves as believers.

India experienced its own alchemical heyday in the Middle Ages. Experiments with mercury-based potions were being described as early as the fourth century BCE, synchronously with the earliest documented Chinese forays. Like Taoism, Indian religion is best known in the West for its contemplative paths, but it also had a pragmatic side. Numerous meditative, ritualistic, and physico-sexual disciplines were developed in hopes of attaining physical immortality—of transforming oneself into a diamond body or a sound-emitting crystal capable of sparkling forever.

Beginning in the sixth century CE, various salvific systems grouped under the heading of Tantra spread across India. Some forms of the practice were more yogic and spiritual; others emphasized the use of

* By the Ming period Chinese alchemy found itself in an irreversible decline.

elixirs. Marco Polo, in his travels, reported encountering *ciugi* (yogis or gurus) on the Indian coast who could live to two hundred by drinking potions made of quicksilver and sulfur mixed with water.

The most important book of Indian herbo-metallic alchemy is the *Rasarnava* (*Flow of Mercury*), an anonymous treatise dating from the eleventh or twelfth century CE. It describes procedures for manufacturing globules of mercurial ash that, when kept in the mouth for an entire month, bestow a life span of 4,320,000 years. Pills made from smoke-sight-*vedha* mercury held under the tongue for three-quarters of a year render one both omniscient and omnipotent. Hatha-yogic breathing exercises were another means of becoming God: if you could just hold your breath long enough (meaning at least twenty-four years), you'd absorb sufficient pranic juju to gain complete mastery over the entire universe.

As with esoteric Chinese Tantra, bodily fluids played an important role in the Siddha tradition.* Sperm was seen as having vivifying effects, especially if raised along the spinal cord toward the crown chakra. Other life-extending liquids best expressed themselves if hydraulically drawn up into the genitalia. Through urethral suction, mercury, water, menstrual blood, and other liquids were sucked *into* the yogi's penis hole. This permitted fluids to then be sublimated into a sacred nectar of immortality. To expedite that process, initiates occasionally swallowed their yogis' saliva. Some gurus would even dab droplets of their semen onto the tongues of followers in the belief that it would be cranially transmuted into a death-defying balm.

Such arcane notions go back to the earliest Indian Vedas, which described a mysterious immortalizing liquid called *soma*. One famous passage states, "We have drunk the soma; we have become immortal; we have gone to the light; we have found the gods." The Holy Grail for medieval Indian alchemists was to materialize that *soma,* the cosmic fluid of life. *Soma* was also sometimes spoken of as *soma-rasa,* or *soma* juice.

The word *rasa* (as in the alchemical handbook the *Rasarnava*) refers to the flowing, liquid elements of life. Water, juice, rain, syrup, sap, plant resins, tears, and other bodily liquids—all fluids are manifestations of divine *rasa,* the symbolic source of existence, the basic mythological element of creation.

---

*Caveat emptor: My primary source on Indian tantrikas is David Gordon White, a professor at the University of California in Santa Barbara who is open to the possible existence of "bodies immortalized through the use of mercurial preparations."

Any mythology relies on ritual, on the repeated practice of a certain behavior. Today, we all worship technology constantly: checking e-mail, turning on machines, flicking light switches. We're all trying as hard as possible to cram ourselves into our computers: making art on them, being entertained and informed by them, feeding them information, meeting friends with them, even falling in love through them. We are teaching machines who we are, telling them everything there is to know, programming them with intelligence in the hopes that they will develop consciousness. And if not, we'll scan the contents of our brains into hard drives and speed things along. Soon, modern immortalists assure us, we'll *become* computers.

19

Preservation's Particulars: Longevity and Longing

We don't crave immortality, but we must reach out to the limits of what is possible for mankind.

—Pindar, *Odes*

You have to get old. Don't cry, don't clasp your hands in prayer, don't rebel; you have to get old. Repeat the words to yourself, not as a howl of despair but as the boarding call to a necessary departure.

—Colette, *Les Vrilles de la Vigne*

IN 1912, the year he won the Nobel Prize in Physiology or Medicine for his breakthroughs into sutures and vascular surgery, medical biologist Alexis Carrel started a tissue culture of fibroblast cells from the heart of a chicken embryo. He placed the muscle cells in a stoppered flask and had lab assistants replenish the culture medium regularly. The culture proliferated, as he expected it would. Regularly nurtured with fresh poultry extract, the cells stayed alive for another three and a half decades. His findings proved—or so he thought—that cells are naturally immortal. "Death is not necessary," Carrel wrote, it is "merely a contingent phenomenon." Soon, he predicted, we would be engineering human tissues from cell clusters and growing replacement organs in vitro. Growing old would be a thing of the past. We, like cells, are meant to live forever.

He was wrong about cells living forever. Fifteen years after Carrel's death, scientists realized that cells, like us, senesce and then die. But his predictions regarding regenerative medicine may now be getting closer to reality. In recent years, scientists studying tissue engineering managed to print out a fully beating, three-dimensional, two-chamber mouse heart using a modified desktop, ink-jet printer. By filling the ink cartridge with cells, they've been able to "publish" functional human kidneys.

This is a time when we can grow human ears on the backs of mice and implant culture-grown lungs into rats. In the near future, specialists say, whenever we need replacement body parts, from blood vessels to bladders, we'll use rejection-proof artificial organs grown in laboratories using our own cells. "By putting in the parts you need, you'll be able to extend life by several decades," explains Anthony Atala, director of the Wake Forest Institute for Regenerative Medicine. "We may even push that up to 120, 130 years."

Bolstered by such promising discoveries, our understanding of aging is changing rapidly. Outside the field of organ regeneration, other genuine life-extending breakthroughs are being made in model test species. In 2011, *Nature* reported that dying worms yellow with a pigment called Thioflavin T (or Basic Yellow 1) makes them live 60 to 70 percent longer than the norm. There's more. Researchers are currently finding clues to longevity everywhere from Texan bat caves (where biochemists are investigating the role of misfolding proteins in long-lived bats) to the soil of Easter Island (where antifungal microbes known as rapamycin can raise the life expectancy of mice by 30 percent or more). Spermidine, a molecular compound found in human semen as well as grapefruit, has also been proven to significantly prolong the life span of worms, fruit flies, and yeast.

These strange-sounding experiments are yielding findings that could affect our lives. Will longevity research yield breakthroughs leading to immortality? Tinkering with the genes in yeast or roundworms has real effects on longevity in those species; that doesn't mean those genes will perform similarly in humans. And experiments on human cells *in vitro* do not guarantee similar functioning *in vivo*. So dying yourself golden yellow will be useless—unless you plan on standing really, really still in an urban center's touristic thoroughfare. It won't help you live longer, but sightseers will likely throw consolatory pennies at you.

There's no documented validity to any life-extension strategy, but that hasn't deterred the making, selling, and buying of countless longevity creams, potions, and pills. For $36.95 you get a one-month supply of micronized Longevinex™ capsules, "designed to help Americans live longer." For $39.95 you get 120 ml. of Clustered Water™, a solution of water organized into clustered structures that ostensibly rejuvenates interstructural cells. For $159 you get 0.7 fluid ounces of Rejuvity's Ageless Renewal Serum™, containing a specially formulated concentration of Repair-Plex™. Unfortunately, once you start using it, you shouldn't stop: "If discontinued, then the aging process of the skin simply continues," explains the website. The fine print on such products can be a great way of pushing more product.

We'll believe anything—and belief is our most powerful panacea. As scientists conducting clinical drug trials know, placebos are often as potent as medication. If we simply believe we are taking a drug that fixes our problem, our problem can end up fixed—even if we're just taking a sugar pill. The way belief affects healing is called the expectancy effect. In studies, placebos have proven effective in treating everything from minor headaches and depression to sore joints, irritable bowel syndrome, and skin conditions. Dermatologists can dab inert water onto patients' warts while explaining that it's a treatment that eliminates unwanted viral lumps. In 48 percent of such cases, the warts disappear.

Placebos may not cure all illnesses, but inactive substances do alleviate symptoms and offer therapeutic relief in those who believe that they will work. Faith heals. It isn't sufficient for patients to simply take a placebo without knowing what its effects are: they have to trust that the prescribed remedy will help in order for it to work. If the medical specialist who administers the sham pill explicitly tells patients that it will resolve their complaint, the condition can be cured by a placebo. But if the patient doesn't believe that the treatment will work, it definitely won't. (In fact, if we believe the treatment is bad for us, a totally unharmful dummy drug can aggravate a patient's condition and cause other injurious effects. This is called a nocebo.) Belief, it seems, has the possibility to be as curative as many drugs or treatments.

The placebo effect isn't fully understood but it appears to be a result of the way beliefs interact with endorphins, the body's self-made opioids. "The poppy fields of the mind" spring into bloom when the body experiences extended physical activity (as in runner's high), spicy foods, or intense pain; they are also linked to religiosity. In fMRI scans of test subjects exposed to low-level electric shocks, the pain receptors in their brains light up. When they are given a placebo ointment, however, their brains behave differently. Even though they are being administered the exact same shocks, their pain receptors remain inactive. Instead, the endorphin-manufacturing quadrants of the brain become engaged. "Our brain really is on drugs when we get a placebo" is how the scientists behind these tests explain it. A similar thing occurs to religious believers. When administered shocks while being asked to look at an image of the Virgin Mary, Catholics don't feel the pain. Their pain receptors shut off, and instead their internal apothecary kicks in.

Belief and placebo are deeply linked, which is why we're so open to suggestion. All marketers need to do is make consumers *trust* in them, and their products will have certain positive effects—as long as they aren't actually harmful. Unfortunately, some antiaging remedies *are* hazardous to our health. With the rise in availability of modern-day youth elixirs at the beginning of the twenty-first century, the US government formed a Special Committee on Aging to investigate these potions' supposed benefits. The congressional report's title summarized their findings: *"Anti-Aging" Products Pose Potential for Physical and Economic Harm.*

Unstudied though these products often are, their performance is akin to that of any placebo—it *might* have an effect. Stem-cell activating and resveratrol-laden age-management nostrums may not help prevent visible signs of aging, but it doesn't really matter if they work. What matters is that consumers *believe* they will work. Hopes for creaselessness are bolstered into faith by hypnotic ad copy in fashion magazines: hydroxy acids have exfoliatory antioxidant powers. Kinetin causes humectant agents to remain on the skin longer. Hyaluronic acid's renewing properties are legion. Copper peptides are where it's at. You mean you aren't yet taking the age-decelerating coenzyme Q10?

Marketers often use scientific-sounding jargon, but a 2011 study of wrinkle-reduction creams found that "at best the products had a small effect, and not on everyone." In other tests, those that sell for inflated prices were outperformed by no-name moisturizers. Double-blind trials show that about 20 percent of participants' wrinkles can be improved slightly to moderately after six months of serums, while about 12 percent of participants' wrinkles improve simply when taking placebos.

That hasn't stopped women from risking delirious sums on the off chance they might shrink a wrinkle. The cost of a 16.5 oz. jar of Crème de la Mer (which bioferments the curative powers of the sea into a light-and-sound-wave-treated elixir users are encouraged to "apply day and night—for a lifetime") is $1,900. Carita's "infinitely rare antiaging formula" Diamant de Beauté, containing pulverized diamonds (which obviously makes facial skin last forever), costs $600 for 1.7 ounces. Ads for SK-II, one of the priciest beauty brands in the world, tell of a Japanese monk crafting the cream's secret formula, a nutrient-rich fluid called Pitera: "After many experiments, he discovered a liquid that seemed to defy aging."

But even if it "seems" to, that doesn't mean it "does." Clinique Youth Surge Night cream claims to suspend age and interrupt time by building on sirtuin technology to "virtually slow the signs of aging." But sirtuin technology, as I would soon learn, remains uncertain at best.

"Younger-acting skin leads to younger-looking skin," claim ads for Olay Professional Pro-X Age Repair Lotion, which signals the moisture barrier to perform "more like it did when it was younger." Pro-X research is based on "one of the great scientific achievements," explains Olay's website: the sequencing of the human genome. Its efficacy is "proven," explain advertisements. But according to some consumer testers, using it on half of one's face while using a vastly cheaper cream on the other leads to no noticeable difference. Other testers, though, report favorable results.

Beyond their logic-stretching promotional slogans, nothing much supports the claims made by most antiwrinkle eye-cream manufacturers. Some evidence suggests that compounds called retinoids may have an effect on photo-aged skin, which has suffered prolonged exposure to UV radiation. A study funded by UK chemist Boots found that their own No7 Protect & Perfect Intense Beauty Serum benefits users with sun-damaged skin who apply it for at least a year. (Since the study was made public, sales have skyrocketed: a bottle of the product is bought every eight seconds at Target.) Regardless of efficacy, we spend billions on cosmetic wrinkle-reduction treatments every year. Even young people are getting in on it: in 2011, Walmart announced a new line of cosmetics with antiaging properties aimed at eight- to twelve-year-olds.

While cosmetic skin creams can contain heavy metals and other toxins, they aren't as dangerous as the use of human growth hormone (HGH), whose use as an antiaging remedy or for age-related problems is not authorized by the FDA. Covertly used by pro athletes and bodybuilders, there's a false perception that HGH can make people younger. On the contrary: the misuse of growth hormone has been shown to cause organ malfunctions and tumor formation in test species, as well as an "increased probability of early death," the most perfectly ironic side effect to a remedy hyped as having life-extending qualities. (Things haven't changed much since the time of the Han dynasty.)

We don't know exactly how hormones affect us. Preliminary results of a six-year, $45 million National Institutes of Health study of testosterone therapy among elderly men will be available in 2015. In the meantime, HGH and other supposedly age-defying hormones remain procurable online, as do books such as *Grow Young with HGH: The Amazing Medically Proven Plan to Reverse Aging,* by Dr. Ronald Klatz, MD, DO, "a world recognized authority on preventive medicine and advanced biotechnologies." Klatz is the president of the American Academy of Anti-Aging Medicine (A4M), an organization whose position statement is that "Aging is no longer inevitable."

We've tried to explain the causes of growing old in many ways. One of the twentieth century's most discussed theories of aging is the oxidative model. In the post–World War II years, a chemist and biogerontologist named Denham Harman suggested that aging is caused by cellular exposure to oxygen. This basic concept has obvious appeal based on a causal truth: air makes things age. Just as it causes a cut apple to brown or uncorked wine to oxidize, it weakens our mitochondria. Hence the notion that if we could only prevent oxygen from invading our molecular structure, we could prevent aging.

Antioxidants—the phytonutrients found in colorful fruits and vegetables—seem to protect the body from free radicals, the intercellular equivalent of terrorists. But nobody knows for sure. Numerous other experiments have given rise to doubts that oxidative damage is the main cause of aging. Antioxidant-based vitamin supplements may not limit bodily oxidative damage or even influence aging whatsoever. We can't say. The answer is blowing in the mitochondrial wind. But the dearth of evidence hasn't prevented this idea from being accepted as a proven fact, especially among canny marketers.

"In simple language, we don't get old, we rust from oxygen," announced Dr. Harry B. Demopoulos in 1989. An occasional actor with a baroque comb-over, Demopoulos was among the first to theorize that consuming antioxidants might slow aging. His company, Health Maintenance, Inc., supplied Hollywood celebrities with a patented blend of vitamins, generating an estimated $10 million in 1990. Since then, sales of antioxidant superfruits such as açaí and blueberries have soared.

In 2002, UC Berkeley's Dr. Bruce Ames, a winner of the US National Medal of Science, launched a product called Juvenon whose antioxidant supplements have the tagline "Stay Young." Another company, Eukarion, joined the fray, licensing its discoveries to Estée Lauder. The excitement over Eukarion slowed when the mania for free radicals started fading.

Today antioxidant superfoods are available at most supermarkets, in a multitude of forms. Consuming them won't make us live longer, although they may make us healthier. "Antioxidants haven't extended life—that whole idea is out the window," one senior researcher at the US Department of Health & Human Services' National Institute on Aging told me. "Things we thought we understood we realized we don't. The simple ideas just don't work. There is so much we don't know."

Cynthia Kenyon is a molecular biologist at UC San Francisco whose findings appear to contradict what we assume we know about aging. Her genetic work on roundworms called *C. elegans* has demonstrated that they can be made to live four times longer than normal. "That's not immortal," she concedes. "That's not to say that you couldn't. But we haven't." She feels confident that her discoveries will translate into human applications. On her website, she has posted a special section for nonscientists in which she explains her views: "We don't know yet, but to me it seems possible that a fountain of youth, made of molecules and not simply dreams, will someday be a reality."

Her position is precisely the opposite of another UCSF scientist, Leonard Hayflick, an evolutionary biologist and anatomist who has spent his career working with cells and debunking false scientific claims about immortality. He's best known for positing the Hayflick limit—that somatic cells can only divide a certain number of times, after which they senesce and die. (Previous to his demonstration of this phenomenon, cell populations were thought to be immortal.) The biological cause of aging, Hayflick says, "is the same as the cause of nonbiological aging—it's the second law of thermodynamics." Everything, he asserts, eventually breaks down, collapses, and falls apart. That's the nature of entropy; and entropy is a fact of nature. "Let's take something infinitely simpler than your body and mine: automobiles," says Hayflick. "Even if you put the car in a garage and don't use it, it won't stand there forever. Eventually, it will age and disintegrate. This is an inevitable law of physics."

Kenyon, on the other hand, starts speeches by rejecting such reasoning as old-fashioned. "In the past," she states, "we thought you wear out, like an old car." Her colleague the MIT molecular biologist Lenny Guarente, who studied stress in yeast and found that certain genes express themselves in those that live longest, puts it this way: "The wear-and-tear theory is best viewed as a laudable initial attempt to come to grips with the problem, but is not a serious scientific theory; the problem is that people turn out to be more complex than Chevys." The difference between inanimate objects and biological organisms, the two point out, is that living systems have self-repairing mechanisms. True, Hayflick and his peers concede, but our body can repair itself only for so long. "Aging occurs because the complex biological molecules of which we are all composed become dysfunctional over time as the energy necessary to keep them structurally sound diminishes," explains Hayflick. As we age, the defense and maintenance programs that protect us when young gradually stop working.

What if we could find a way of activating such genetic pathways later in life? Scientists have done so with worms, yeast, fruit flies, and mice. The names for these longevity genes are usually technical: SIR2, InR, p66Shc, fos, chico, age-I. (One exception is the recently discovered fruit-fly gene INDY, which stands for "I'm Not Dead Yet.") The results are indisputable, but interpreting them is another issue altogether. "When single genes are changed, animals that should be old stay young," Guarente and Kenyon summarized in a *Nature* article about genetic perturbations that increase life span in simple animals. "On this basis we begin to think of aging as a disease that can be cured, or at least postponed." Of course, not everyone agrees with their conclusion. It's certainly a leap to go from genuine genetic findings to speculation that aging in humans is a curable disease.

If anything certain can be gleaned from the current public argument between evolutionary and experimental biologists, it's how limited the scientific understanding of our aging actually is. The deepest researchers in the world today still don't know whether there's a universal biological process behind aging. One side, composed mainly of experimental genetic biologists with ties to pharmaceutical companies, believes that we will be able to manipulate longevity genes in all species. They hope to translate the findings they've made on a molecular level into human-ready medicines. On the other side, evolutionary biologists don't think we'll ever have the ability to intervene in fundamental human aging. For the rest of us, seeing aging as a disease or not is a personal choice.

Ending aging is for now as elusive as ending time, but scientists have found aging research to be fantastically rewarding financially. In 2008, GlaxoSmithKline paid $720 million for the rights to exploit pharmaceutical drugs based on the genetic pathways uncovered by Guarente and his graduate students. "What we're working toward is a drug that gives the benefit of exercise and diet without having to exercise and diet," explained Sirtris's cofounder David Sinclair, in a statement that neatly encapsulates the contemporary American dream.

Their findings remain inconclusive, at least in human medicine. GSK executives defended their expenditure as high risk—a shot in the dark with the possibility of a high return. Their inferences may prove to be useful, but an inference is not something meant to be believed; it's meant to be tested. Once the testing is further along, we'll know more about whether aging genes can be affected in humans.

"I doubt that aging can be reversed," says Hayflick. "Aging is a random, stochastic process that occurs after reproductive maturation and results from the loss of molecular fidelity." Guarente, Kenyon, and fellow experimental researchers contend that the genes they've isolated have the power to keep an animal's "natural defense and repair activities going strong regardless of age." As they wrote in *Scientific American,* these genes, in lab species, can "dramatically enhance the organism's health and extend its life span. In essence, they represent the opposite of aging genes—longevity genes." Heady, but such hopes always have been.

In both biology and reality, the end begins at the cellular level. Without exception, flora and fauna consist of cells. Our bodies are made up of trillions of cells doing what cells do best, which is making replicas of themselves. Then they die.

For most of history, life consisted of unicellular organisms popping around. To this day, the majority of creatures on earth are single-celled. When a single-celled organism such as a bacterium or a protozoan reproduces through binary fission, what was one becomes two. In some cases, such as division in a yeast or *E. coli,* clearly a younger cell and an older cell result from each fission. The young one gets all new parts, and the older one gradually senesces. And on it goes.

In other cases, there appears to be no distinction between the progeny and the original cell. The two fission products are effectively clones. There are two ways of describing what happens here: the parent cell either dies while becoming two identical offspring, or it gives birth to an identical replica of itself that will also divide again at the same moment as the parent. Whether the initial cell dies into twins or gives birth to itself, both definitions are valid, and they both showcase the limits of language to explain natural processes.

When cells started getting specialized in order to aggregate into more complex organisms, they became two distinct types of cells: germinal cells and somatic cells. Our soma cells constitute the majority of our body. Our germ cells are those in our gametes—the female ova and the male spermatozoa. Somatic cells age and can replicate a finite number of times (known as the Hayflick limit). Germinal, or reproductive, cells, however, have the capacity to keep on dividing more or less continually. They contain the hereditary information in DNA that is passed on through subsequent generations.

A benefit of ovigerous sexual reproduction is greater complexity for the entire species. Each descendant obtains a bit of both parents' germ cells. This form of ever-radiating genetic diversity is a defense against potential threats. Even if a large swath of the species gets wiped out by a genetic invasion, some part of the population is likely to be immune. But when species do go extinct, their genetic code disappears. Species are not immortal; nor can what's lost be re-created.

The mysteries of the cell may one day reveal something about aging that we don't as yet understand. For example, just as certain single-celled organisms are capable of regenerating themselves seemingly endlessly, so can cancerous cells. The replicative capacity of a normal cell is finite; cancer cells are simply normal cells that mutate and, for some reason, begin to proliferate continually. The cell buildup is what's called a tumor. Within each of us, stem cells also have the ability to make apparently limitless copies of themselves, but they are usually quiescent, active only in youth or in a medical emergency.

Cancer cells, stem cells, and unicellular organisms that experience unstoppable growth are spoken of as having cellular immortality. This does not mean they are indestructible. They are not undying. When a host body dies of cancer, the uncontrollable cancer cells die with it. So-called immortal cells can keep dividing in the correct medium (as with the cells of Henrietta Lacks, known as HeLa cells), but obviously they perish if taken out of that culture. Cellular immortality doesn't mean we can orchestrate the immortalization of life-forms.

In the nineteenth century, August Weismann, author of *Upon the Eternal Duration of Life,* wrote of how immense numbers of organisms "do not die." Our somatic cells perish, he conceded, but germ cells are "potentially immortal" as they can transfer themselves into a new individual. Even though we die, if we have children, our genetic information outlives us.

Just as cells make copies of themselves, our bodies evolved to live long enough to reproduce. We reach our peak in our twenties and start going downhill in our thirties and forties. In our postreproductive period, many of the exact same genetic processes that contributed important and valuable functions in our youth start to have harmful effects. This phenomenon, technically known as antagonistic pleiotropy, is one of the biggest obstacles to solving the aging riddle. Genes involved in cancer or Alzheimer's are also those that promote healthy growth earlier on. Aging-related illnesses are the price we pay for living to eighty.

Immortalists make much of how certain jellyfish, sponges, corals, and deep-sea creatures appear not to senesce. That these creatures lack nervous systems or memory isn't an issue. They "don't age," prolongevists say. They "may be practically immortal." Yet as slowly or imperceptibly as they age, they die when killed. Some of them can sprout new limblets if the conditions are favorable or bud body parts, but that doesn't mean they are eternal.

The freshwater hydra is a tiny, brainless organism with fascinating regenerative capabilities. Even if much of it is killed, it can spring back to health, giving the impression of imperishability. Imagine a three-millimeter-long tube topped with dreadlocks. The hydra is often compared to a fountain, its body a jet of water that shoots into a splash of tentacles. The tube is made up of cells that seem to be both germinal and somatic. As the tube's old cells age, die, and are sloughed off, new ones are generated. Being hermaphroditic, a hydra can bud off some of those cells into entirely new hydra. It can also rebuild itself from almost any bit of its body. Still, as amazing as it is, it isn't immortal. It's a simple organism whose body cells are also germinal cells. Take it out of water and it perishes.

Whether or not we consider immortality a feasible aim, specialists have made insights into the lengthening of lives—although there's nothing simple about their discoveries. One of the earliest scientifically replicable means of delaying aging is known as caloric restriction (CR). Eighty years ago, scientists realized that limiting the diets of lab rodents increased their longevity by 40 percent.

Feeding model organisms just enough calories to fulfill their minimum nutritional needs does appear to extend lives in everything from fish to apes. Protozoans that normally live for a maximum of thirteen days can live for up to twenty-five days on CR. In rhesus monkeys studied for twenty years, CR also delays the onset of age-related diseases. Not surprisingly, this has led to rampant speculation that CR might work higher up the phylogenetic ladder as well.

We've suspected as much for centuries. The pioneering proto-gerontological researcher Luigi Cornaro (1467–1566) was a Venetian nobleman who lived to his late nineties by eating only twelve ounces of food a day. He outlined his precise regimen in *Discorsi della Vita Sobria*. He spoke of being healthy, satisfied, and full of joy at a time when most were "sad, sick, and bored." But modern research into CR suggests that the diet isn't all smiles and satisfactions.

CR is currently being studied in humans under a multiyear program called CALERIE (Comprehensive Assessment of Long-term Effects of Reducing Intake of Energy), funded by the National Institute on Aging. As of 2013, CR hasn't been proven effective in humans, but even if it were, there are numerous catches.

To begin with, few people want to be hungry all the time in order to live longer. That doesn't mean there aren't practitioners out there. They've been dubbed the Skinnies by the media, and they eat just enough to stay alive. To follow the regimen requires establishing your daily caloric usage; then you need to eat around 20 to 30 percent less. The diet allows approximately 1,500 to 1,700 calories a day for women, and 1,800 to 2,000 for men. If overseen by a certified nutritionist, the program is not malnutrition—in fact, eating tomato soup with celery sticks for dinner is optimally healthy. It's just hard to do. And there are—as there always are—side effects. Not only do you lose weight, you end up looking gaunt and sallow. Those in CALERIE complain of often feeling cold—the same phenomenon occurs with underfed rodents, who shiver away hungrily in their cages. Caloric austerity also carries a heightened risk of bone-mineral deficiencies, lowered blood pressure, and anemia. If undertaken without medical supervision, it can also lead to anorexia—and one out of five anorexics ends up dying of complications from the disease. Perhaps the greatest deterrent is the resultant infertility: mice on CR completely lose their sex drive. Their human counterparts can also experience a plummeting libido.

Sexual malfunction has a murky relationship with longevity. Certain mutant strains of yeast that live 50 percent longer than average also forsake the ability to reproduce. The same mutation that causes their sterility also causes longevity, explains Leonard Guarente, whose team first noticed the coincidence. There is more evidence of this link. Decades-old studies from mental institutions demonstrate that castrated men live an average of fourteen years longer than noncastrati. Worms who've had their gonads excised also experience extraordinary longevity.

Cynthia Kenyon, who has used laser beams to zap the sexual organs of roundworms, found that hormones produced by their reproductive systems appear to have an effect on aging. A single mutation in a gene called age-I, linked to fertility, can extend life spans in worms. No one knows what might happen if these results were transposed onto sentient beings. Kenyon has focused on a suite of genes known as DAFs, and she argues that there may be a way of kick-starting their human equivalents, creating her hoped-for fountain of youth made of molecules rather than dreams. In other words, she believes that we will soon find a pill that activates age-influencing genes without having to practice caloric restriction or to castrate ourselves.

The goal of finding a pharmaceutical means of reproducing the health benefits of CR has long tantalized scientists. In 2002, one promising CR mimetic called 2DG made a splash until it was found to have a major flaw: in only slightly higher than recommended dosages, it is toxic and can lead to cardiac mishaps. A company called BioMarker Pharmaceuticals, Inc., has been exploring the CR-like effects of an antidiabetic medication called metformin, which may have potential but has also been shown to cause death from lactic acidosis. By far the greatest source of optimism is sirtuins, the family of genes discovered in yeast by Guarente and his team. If GlaxoSmithKline's billion-dollar investments prove correct, we may soon be able to take a capsule that triggers sirtuin pathways, allowing us to eat whatever we like, as David Sinclair suggests, while still receiving the biochemical effects of CR.

Some scientists studying other facets of CR express doubts that there will ever be any pharmacological means of triggering the complex interaction of the various genes involved in longevity. "My perception right now is the effects of calorie restriction are multiple," says Dr. Luigi Fontana of the Washington University School of Medicine's Center for Human Nutrition, "so I think it's highly difficult to find one or two or three drugs that will mimic such a complex effect." It remains to be seen whether a pill or pills will be made available for human usage. Even if a cocktail of age-defying drugs were released, nothing suggests it would be riskproof.

For now, sirtuins and CR appear to have something to do with aging, but that doesn't mean they are the aging process itself. "Sirtuins are a sexy narrative, that's for sure," one source close to the research, who asked not to be named, told me. "Imagine: an antique, archaic gene that came bubbling out of primordial sludge—it's beautiful." Almost as beautiful as another sexy narrative that captured the media's interest in the 1990s.

Telomeres are snippets of DNA at the ends of chromosomes. The lay description of a telomere is that it's the cellular equivalent of that plastic tip or cap on the end of a shoelace. Telomeres are devices that prevent genetic data from fraying. As a cell divides repeatedly, its telomeres eventually grind down. The wearing out of a telomere brings with it the death of a cell.

Imagine there was a way of preventing telomeres from shortening. In the 1990s, the same scientists who discovered telomeres also discovered an enzyme named telomerase, which, when added to cells in culture, allowed telomeres to maintain their structural integrity. As the telomeres stay long, cells can divide endlessly. The life span of such telomerase-enhanced cells can be prolonged indefinitely. This was an amazing discovery; but just because cells in vitro can be manipulated does not mean that telomerase can make *humans* live forever.

It's a fundamental difference, but one the media had trouble understanding or explaining. Journalists declared that scientists had finally found the microchemical fountain of youth in the form of an "immortalizing enzyme." It was widely reported that the length of one's telomeres directly determined one's life span. Aging would soon be a thing of the past.

Alas, the idea that telomeres are the sole cause of senescence is a simple misapprehension. Just because they play a role in cell aging doesn't mean they are *the* mechanism of aging. Telomerase does not eliminate the disease of aging—but it does play a role in the uncontrollable proliferation of cells afflicted with cancer.

The scientists who made the discovery, Elizabeth H. Blackburn, Carol W. Greider, and Jack W. Szostak (who together won the Nobel Prize in Physiology or Medicine in 2009), are still trying to understand the role of telomeres in aging, as well as the effects of telomerase. "Everybody *wants* to find that there's a great simplifying principle," explains Blackburn—but telomeres and aging are anything but simple. To counter all the hype, her colleague Greider wrote a paper explaining that telomere length "is clearly not directly correlated" with longevity. Telomere research has the potential to offer insights into how we age, and into fighting cancer, but for now what it really reveals is the mechanism whereby chromosomes deteriorate as cells divide.

Whether or not the elimination of aging and death are possible end points, many researchers have formed companies that court or are funded by the multinational pharmaceutical industry. While the idea of Harvard and MIT biologists in bed with giant corporations may seem upsetting to nonacademics, it's been a growing reality in the field for the past two decades. "In the early 1980s, it was nearly unthinkable for academic scientists to found for-profit corporations," explains George Washington University professor of law Lewis D. Solomon. But by the 2000s, the situation had changed completely. As Nobel laureate Eric Kandel stated in 2003, "These days, it's hard to think of a really good biologist who isn't involved with a company." Guarente portrays the division this way: "My lab tries to learn more and more about the basic biology underlying aging and survival. The company, meanwhile, is trying to translate this knowledge into drug development."

Adding to the complications of academics-cum-businessmen, scientists at major institutions occasionally make unverifiable, potentially irresponsible statements to garner publicity and grant funding. Attracting investments from the phalanx of wealthy older Americans looking to sink their private fortunes into the war on aging takes snazzy sales pitches, not dry displays of data collection. Professors with connections to Big Pharma may honestly desire to help civilization, but the skewed gold-rush aspect of longevity research means even the most scrupulous academic biologists stand to benefit personally from misunderstood discoveries. Because the potential for personal gain taints most aspects of aging science, trustworthiness can be hard to gauge.

This is nothing new. In the nineteenth century, John Stuart Mill said that there is no scientific evidence against the immortality of the soul except for negative evidence, meaning the absence of evidence that it exists. In earlier times, promises of immortality were a way for the powerful to maintain power over other people. By hanging the threat of eternal damnation over their heads, the ruling classes could keep civilians in line. For this reason, Mill's predecessor Hume considered the notion of an afterlife to be a barbarous deceit. "There arise, indeed, in some minds, some unaccountable terrors with regard to futurity," Hume wrote, "but these would quickly vanish, were they not artificially fostered by precept and education. And those, who foster them: what is their motive? Only to gain a livelihood, and to acquire power and riches in this world. Their very zeal and industry, therefore, are an argument against them." Exploiting the fear of death is a venerable, often lucrative, tradition.

Longevity studies still occupy a windswept limbo-land where the distinction between pseudoscience and verifiable research is often intentionally obfuscated. A measured approach is usually an indicator of reliability in a researcher, but many well-regarded aging scientists demonstrate a lack of reticence when making sweeping declarations about ending the disease of growing old. Take this twenty-five-year-old example: "We absolutely have within our hands the technology to manipulate and reverse ageing in every tissue and system." The intervening decades would seem to refute that claim, but it hasn't stopped such proclamations. In a 2011 open letter, biotech company Sierra Sciences wrote of an urgent need for investors when they lost funding during the recession: "It is no exaggeration to say that we are on the brink of actually curing the disease we call aging!" Actually, most rational humans would agree that statement *is* an exaggeration. However, Sierra Sciences was unarguably on the brink of something else: bankruptcy. The letter, circulated widely on antiaging websites, appealed for $200,000 a month to keep the lab operational; if no one stepped up, they wrote, it "would be a tragedy for humanity, as well as a missed opportunity to create a multi-billion dollar industry."

To understand how scientists appear to get ahead of the themselves in seeking investors, it's worth glancing at the saga of a corporation called Geron, whose stock soared when the telomerase story broke, in much the same way Sirtris's did when sirtuins had their moment in the spotlight. One of the company's founders, Michael D. West, was exceedingly vocal about his optimism that telomeres are the solution to human aging. In March 1990, West sent a breathless newsletter to investors about "the spectacular events unfolding. . . . We can take senescent (old) cells and make them immortal. . . . We have found the genes that regulate aging."

Within ten years, he wrote, the technology would be engineered into a form that could be administered to people. Ten years later, no longer with Geron, West boasted to media that he was "close to transferring the immortal characteristics of germ cells to our bodies and essentially eliminating aging." Ten years after that, more interested in stem cells than telomeres, he gave a lecture, "The Practical Uses of Immortality," in which his predictions hadn't really changed: "The potential applications in age-related degenerative diseases are the demographic trend of our time. It's going to be *the* story of the decade. So I think whenever you see opportunity like that, it's imperative to seriously consider it from a business standpoint."

But those who consider business opportunities for a living caution against taking him too seriously. A feature in *BusinessWeek* magazine questioned West's credibility: "Mention West's name to fellow scientists and they either sigh, cringe, or ask, 'What has he done now?'" As Carol Greider, one of the Nobel winners for telomere research, put it, "I never saw any science that he did that was that interesting. He was always just promoting ideas." The head of Sierra Sciences, however, considers West a guru.

A spectrum of people are engaged in cutting-edge antiaging science, from salaried researchers to self-professed bio-authorities to autodidact hobbyists to passionate longevists eager to sample any life-prolonging drugs. The greatest hype-man of this era, however, is indubitably Aubrey de Grey. His book *Ending Aging* notes that aging kills one hundred thousand people a day—"many old people, yes, but old people are people too."

An amateur gerontologist, de Grey gained mainstream credibility everywhere from *60 Minutes* to TED Talks, often leaving the impression that he was a biology "professor," when in actuality he only had a part-time job as a computer programmer for the University of Cambridge. Nevertheless, he has been depicted in conference bios and in the media as a "professor at Cambridge's Department of Genetics." It's the sort of blurred logic that typifies much of his work—the semblance of factuality trumps actual factuality.

De Grey's former supervisor, Michael Ashburner, a genuine professor of biology at the University of Cambridge, reports that he reprimanded him several times for misrepresenting his position. When I contacted Ashburner to sort out the truth, he explained that de Grey was never a professor employed by the university. He also called de Grey's gerontological activities "nonsense."

One of de Grey's contributions to the field of aging research is a concept he calls WILT: Whole-body Interdiction of Lengthening of Telomeres. WILT is his way of curing cancer. The idea came to him in an Italian café: if we could only excise the gene that produces telomerase from the human body, then we could eliminate cancer. Telomerase, which is active in cancerous cell growth, isn't usually turned on in normal cells. WILT would do away with it entirely. The problem is that, even if we *could* do away with the gene behind telomerase, it is vital in the creation of new blood cells.

De Grey's solution? We'll just inject ourselves with genetically engineered stem cells whenever we need to.

The trope of scientists striving to attain immortality, which makes them deranged and evil, is such a cliché it occupies its own subgenre in science-fiction fantasies, from *Frankenstein* to the *Highlander* movies. The gamut of unsavory immortals includes vampires, ghosts, zombies, werewolves, aliens, and witches. Infinite tales describe the downside of unendingness, the boredom, disinterest, and despair of being unable to die.

In *Gulliver's Travels,* Swift tells of the Struldbruggs, immortals whom Gulliver assumes must be the happiest people ever. But, it turns out the immortals hate being condemned to perpetual continuance. They are peevish, friendless, morose, and have terrible memories. Because language evolves, they speak weird, old dialects nobody understands anymore, and therefore they barely ever talk. When they do, it's to spout envy toward people who can die. They're bitter, crotchety grumps whose message for the rest of us is that living forever might be the harshest curse imaginable. We only imagine immortality being a good thing, Swift concludes, because of the imbecility of human nature.

But immortalists today would say that Swift committed a newbie mistake called the Tithonus error—the presumption that extending life would also extend those difficult years at the tail end of most elderly lives. (When the goddess Aurora beseeched Zeus to immortalize her lover, Tithonus, she neglected to request eternal youth as well; decrepitude ensued.) On the contrary, contemporary longevists explain, defeating aging will mean eliminating that entire period. As de Grey sees it, "There will, quite simply, cease to be a portion of the population that is frail and infirm as a result of their age."

The only simple truth in the aging world is that nobody understands exactly how aging works. Exhilarating theories abound claiming to explain it. Whenever a good story combines with a whiff of scientific respectability, pills are readied to capitalize on the hype. But what would happen if we could really live forever?

•

To find out, I decided to meet with modern-day eternalists, to attend their gatherings, to spend time with them and discuss their worldview. Looking for an insider connection, I tried to interview the Japanese architect Arakawa. He lived in Manhattan and designed homes that could "counteract the usual human destiny of having to die." Unfortunately, it turned out that he'd died a few months earlier. Then I set about tracking down the astrologer Linda Goodman, whose step-by-step guide to never dying includes visualizing your cells spiraling in reverse. She, too, had passed on. I phoned Dr. Daniel Rudman, who'd told the *New England Journal of Medicine* in 1990 that "growing old is not inevitable." But the inevitable had taken him, too.

I considered paying a visit to People Unlimited, a group of "physical immortals" in Arizona focused on living now and forever. Their motto is "It's time to end death and it takes a community of like-minded individuals to do it." They are united in their belief that every human deserves deathlessness. "Why do we need to die? It makes no sense," writes one recruit, Caleb Escobar, in the site's comments section. He then adds, "I'm not living to die, I'm living for a major reason—TO LIVE!" Followers speak of how they experience "zero struggle" and feel totally freed of all limits. Dean Moriki's testimonial explains that he joined PU after his wife suffered a stroke and became disabled: "It was a major change from a normal life to a daily living with a wheelchair involved. We divorced soon after." Since then, he'd become convinced of his noncorruptibility and now lived without suffering or disease. *Poof!* Magic. The denial was so astounding—and also disconcerting—that I decided against infiltrating their sanctuary.

It didn't take long to find a more appropriate group, one counting doctors, scientists, and philosophers among its members. The Immortality Institute is a 501(c)(3) nonprofit educational organization whose mission "is to conquer the blight of involuntary death." Members of the institute don't see aging as a necessary reality: they see it as a disease. A curable disease. Or, to be more precise, "a sexually transmitted terminal disease that can be defined as a number of time-dependent changes in the body that lead to discomfort, pain, and eventually death." To support the institute is to support research that will inevitably lead to eternal life. They all believe that science will ultimately orchestrate the defeat of death, and they're trying to accelerate the process. Their forum boasts intricate debates about how exactly science will accomplish physical immortality. While surfing the site's "events" section, I came across an open invitation from Dave Kekich, a registered imminst.org* user.

> If you want to celebrate the future, then come to my 125th birthday party on Saturday evening, May 23rd in Huntington Beach, CA. It's a "Come as You Will Be Party." Here's the deal:
>
> 1. First, even though some people think I look it, I'm not nearly 125 . . . yet.
> 2. We are going to project ourselves to 2068 at the party.

---

*The website's name was subsequently changed to longecity.org.

3. You will act and speak as though it is 59 years from now. In fact, if you don't agree to spend at least two hours "in character," then stay home. There will be no 2009 discussions here unless you're reminiscing. We will talk about what we've done and accomplished from 2009–2068, where we've been, what you've become, what changes the world saw in the past 59 years.
4. If you violate #3, you will be deposited into a worm hole in the basement which will immediately transport you back to where you came from.

The possibility of interviewing members of the institute in the flesh was tempting enough that I RSVP'd via e-mail: "Your 125th birthday party sounds like a wonderful idea! I'm a writer and I'd love to come. (I feel like it's 2068 already)."

Soon, a response arrived from Kekich, with directions and elaborations about the theme:

> Everything will be quite different 59 years from now. Since the power of our technology is doubling every year now with no end in sight (at least over the next 50 years or so), our tools will be over a thousand times more powerful in only ten years. If this doubling holds up over the next 59 years, that means our technologies, particularly our computational power, will be a billion billion times more powerful. Yes, that's a quintillion, or 1 followed by 18 zeros. . . . Imagine what America will be like, the world, the solar system. Aging? In my humble opinion, that solution will be a slam dunk. I believe we will have solved aging well before then.

Guests were asked to arrive in 2068-appropriate attire. It sounded perfect. I couldn't wait to see how they lived—and partied.

Looking for advice on how to blend in with a crowd of extreme life-extensionists, I forwarded the invitation to Jenna Wright, a good friend and talented costume designer in Hollywood. "Are you into coming to an immortality dress-up party?" I wrote. "It takes place in the year 2068."

"Are we supposed to be ourselves in 59 years?" she asked. "Or just anybody?"

"The theme is 'come as you will be,' but we can come however you'd like," I responded, CC-ing Clay Weiner, a director who often collaborates with Wright.

"Silver bodysuits seem like a YES," he replied.

My two wingmen offered to stop by a costume house to choose outfits for the three of us. "I'm thinking octogenarian leisure wear," e-mailed Wright. "Possibly some new biomechanically engineered organs on display?"

20

Biological Calculus

All or nothing! . . . Eternity, Eternity!—that is the supreme desire! The thirst for eternity is what is called love among human beings, and whosoever loves another wishes to eternalize himself in the other. Nothing is real that is not eternal.

—Miguel de Unamuno, *The Tragic Sense of Life*

What mad pursuit? What struggle to escape?
What pipes and timbrels? What wild ecstasy?

—John Keats, "Ode on a Grecian Urn"

"WHERE TO?" asked the airport customs official, absent-mindedly scanning my passport.

"Huntington Beach," I answered.

"And what are you planning on doing down there?" He had a crew cut, piercing eyes, a testosterone jaw.

"I'm going to a party with people who consider themselves immortal." It's important to always be honest and succinct when crossing international borders.

The immigration agent sized me up. "Immortal religiously?"

"No, physically. They really think they are going to live forever."

He'd never heard this one before. "They *consider* themselves immortal," he said slowly, acknowledging my use of the term, then considering it himself. *"Consider,"* he repeated, thinking out loud.

"That's right." I smiled.

He seemed unsure how to proceed, taken aback by the idea that anyone would even desire eternal life. Deciding to interrogate me, he asked me what I do for a living.

"I'm a writer," I told him.

"Are you going to write about this party?"

"Yes."

"Okay. So it's fiction?" Relief crept over his face.

"No: nonfiction," I answered. "Science nonfiction."

He considered it for another moment. On the verge of asking me something else, he caught himself, shook his head, and brusquely stamped my passport. "Good luck," he grumbled, waving me through.

On board the plane, wanting to understand the mind-set of those wanting to live forever, I flipped through *The Scientific Conquest of Death: Essays on Infinite Lifespans,* an anthology published by the Immortality Institute. It offers a solid introduction to contemporary immortalist beliefs. It is also one of the weirdest, most entertaining books I've ever read.

We'll soon be delivered from the limitations of biology, one of the institute's advocates of artificial intelligence tells us, through diamondoid nanobots assembled from billions of precisely arranged structural atoms. Computers will update our bodies' Stone Age software so that we can eliminate and even reverse aging. Once we've figured out how to digitize memory by frying a mind's salient details onto computational substrates, the Singularity will have been attained. There will be little distinction between RAM and brain. We will be able to live forever, but only if we diligently maintain our mind file by making regular backups and porting to updated data-storage formats. In the highly unlikely event that overpopulation ever becomes a problem, we'll simply move off-planet. There's an infinity of unspoiled stars out there just waiting to be colonized.

It all sounds so easy. Reading the book, it occurred to me that humanity's willingness to believe in immortal life hasn't changed much since the days of the T'ang emperors. The essays depict life as an intermediate-level puzzle to be solved through technological advance. But illusions don't account for the interminable complexities of reality. Fleets of tiny medical nanorobots that communicate with one another through Wi-Fi signals is a catchy idea, but if the problems we have simply connecting to the Internet are an indication, its real-world application would be a disaster. Every technological appliance I've ever owned has glitches. They don't always work properly. They break. Do we really want tiny robots malfunctioning in our bodies? Computers are fragile, not foolproof. Imagine having to fix an intracellular motherboard crash? What about computer viruses infiltrating our bloodstream? They can already be programmed to contaminate chips in pacemakers, defibrillators, and cochlear implants.

"Nothing made or engineered works perfectly or lasts forever," explains Henry Petroski, a professor of civil engineering at Duke University who specializes in man-made structures. It's naïve to imagine that nanobots would suddenly become the exception. Yet members of the institute are already speaking seriously of mind uploading. We can't cure AIDS, Alzheimer's, or autism—and science still can't explicate how consciousness works, why it evolved, or whether it even exists—but somehow we're going to find a way of transferring the entire bundle of synaptic connections in the human brain onto computers? The brain isn't just a computing device; it's a thinking, feeling, remembering, fallible receptacle of consciousness. We rely on vast subpopulations of neurons—will those need to be backed up as well?

Reading the Immortality Institute's collection of essays, I learned that many immortalists align themselves with a movement called extropianism (meaning those *without* entropy). Their guru, Max More, speaks of exploiting nanotechnologies in order to achieve complete control of matter at the molecular level. His wife, plastic-surgery poster child Natasha Vita-More, is a turbocharged optimist and aspiring posthuman who writes of her longing to transcend all limitations and live forever in an endlessly upgradable body. They and other extropians are disciples of the techno-prophet Ray Kurzweil, an immortality luminary who balances actual scientific legitimacy with arguable predictions about the promises of computing. He's an undeniably brilliant computing expert with pioneering accomplishments in fields such as speech-recognition technology and electronic keyboard synthesizers. He is also convinced that informatics will get faster and faster until we attain immortality in the year 2029, or 2045 at the latest. If he dies before then, his corpse will be cryonically preserved in the hope that technology will catch up with his dreams.

The bedrock of the physical immortalist worldview is a supposition Kurzweil calls the Law of Accelerating Returns. According to his calculations, an inviolable fact of reality is that "technology and evolutionary processes in general progress in an exponential fashion." This is an article of faith. Even if technology has grown dramatically over the past few centuries, we can't know whether it will forever proliferate at an exponential rate, let alone demonstrate that it has done so over the past five years. But Kurzweil has no doubts. Those who criticize him, he says, are merely Luddites who don't understand the exponential nature of progress.

Darwinian theory doesn't accept his assertion that "evolutionary processes in general progress" exponentially. The theory of evolution can be interpreted in various ways, but it does not offer proof that nature always improves. Instead, it suggests that nature constantly changes without direction at all, sometimes toward increasing complexity, sometimes not. There is no progress toward any goal in evolution—there is random drift, competition, chance extinctions, survival of the luckiest and often cruelest.

The truth is that there's no consensus on progress, let alone on exponential progress. "We are not seeing exponential results from the exponential gains in computing power," says William S. Bainbridge of the National Science Foundation, who disputes Kurzweil's position. Bainbridge cautions against basing our "ideas of the world on simplistic extrapolations of what has happened in the past."

Kurzweil has been making predictions for long enough now that anyone can verify how they've panned out. While he's been right on some counts (people do read books and magazines on screens these days), others are worded in ways that make it hard to say whether they even transpired or not. Google's driverless cars exist, as he predicted, but most people aren't yet driving them from LA to SF while napping in the backseat. Some of his forecasts were undeniably mistaken. By 2010, he wrote, "computers will disappear as distinct physical objects." They would instead be woven into our clothing and embedded in our furniture. Are our retinas synched to a computer feed? The average household is supposed to have over one hundred computers. It doesn't. And most people today still don't want to become cyborgs, whether or not the Singularity ever happens.

When his own followers question why many of his predictions have not come to pass, he lambastes their inquiries as "biased, incorrect, and misleading in many different ways." Again, a peculiarity of belief is that believers don't see their story as a mere belief, but rather as the Truth. There's no contradicting a believer because evidence has no place in the realm of belief. No matter how many charts and graphs Kurzweil deploys, his opinions remain just that—beliefs cloaked in factual-sounding garb.

When he was a child, Ray Kurzweil's hobby was magic. In his teen years, he replaced his parlor tricks with technology projects. He's gone on to be phenomenally successful as an inventor, innovator, and entrepreneur. His command of computerized technology is vast, and his many books contain pithy observations such as "intelligence selectively destroys information to create knowledge." His hope for the future is "to infuse our solar system with our intelligence through self-replicating non-biological intelligence. It will then spread out to the rest of the universe."

Singularity activists speak of algorithms as "incantations," as though they were magic spells. Reality for them *is* magical. They like to quote sci-fi author Arthur C. Clarke's third law: "Any sufficiently advanced technology is indistinguishable from magic." Who needs spiritual salvation, the death solace traditionally provided by religions, when you can have physical immortality? Kurzweil disputes the notion that his views are a substitute or alternative to customary belief systems: "Being a singularitarian is not a matter of faith but of understanding," he declares.

But can one prove the existence of software-based humans?

"Everything I do actually has a lot of evidence," he has said.

To prove this evidence, his book *The Singularity Is Near* drowns out doubts with a tellingly excessive hundred-plus pages of backup material. The footnote to a typically nonsubstantiable sentence—"we have the means right now to live long enough to live forever"—leads only to another of his books. He frequently cites other immortality "authorities" whose books also cite him in a feedback loop of back-patting "proof."

Although Kurzweil is ostensibly discussing science, he seems to me more like a high priest of digital materialism. "Science wrought to its uttermost becomes myth," wrote William Irwin Thompson in *The Time Falling Bodies Take to Light*. That's certainly true in Kurzweil's case. He ingests hundreds of supplements daily, many of which are sold through his company, Ray and Terry's Longevity Products. "The vast majority of our baby boomer peers are unaware of the fact that they do not have to suffer and die in the 'normal' course of life, as previous generations have done—*if* they take aggressive action," writes Kurzweil. What's the most effective approach? Buying his pills, naturally.

Anyone entering the world of radical life extensionists quickly notices that many "scientific" immortalists also happen to be in the longevity business, so they stand to profit from gullible consumers looking for a stay-young fix. Many legitimate antiaging researchers are working on finding ways to extend healthy lives, but those who sell products and self-promote with talk of attaining infinite life spans are invariably out for personal gain. "Every single generation before us has worked under the assumption that they possessed all the major tools for understanding the universe, and they were all wrong, without exception," writes neuroscientist David Eagleman. "Does it seem reasonable that we are the first ones lucky enough to be born in the perfect generation, the one in which the assumption of a comprehensive science is finally true?" Kurzweil's books answer in the affirmative.

As all those Chinese emperors and European aristocrats learned, when we believe in nothing, we'll believe anything. Closer to these times, the oft-overlooked "gland craze" of the post–World War I years illustrates our penchant for buying into stories that later eras can only marvel at.

In 1917, "Doc" John Romulus Brinkley opened a clinic in Milford, Kansas, population two hundred, where he grafted a goat testicle into the scrotum of a local farmer who hadn't had an erection for sixteen years. The operation (apparently) rejuvenated the patient instantly. Brinkley said he'd managed to transfer the goat's rugged, rug-chomping vitality into the Milford man. His infertility vanished. Nine months later, the patient's wife gave birth to a baby boy named Billy.

"You are only as old as your glands" became Brinkley's promotional catchphrase. A testicular overhaul, he claimed, could keep even the crustiest old-timer eternally young. In no time, he was performing as many as fifty goat-to-man operations each week. Brinkley amassed homes, automobiles, airplanes, yachts manned by full-time crews. Glandemonium broke out. Medical journals overflowed with gonad experiments. "Your Glands Wear Out," headlines trumpeted. If we could only prevent testicular deterioration, pitchmen insisted, old men would look and feel like twenty-five-year-olds again, forever.

Treatments were sold with taglines like "Why Grow Old?" The Vital-O-Gland Company, based in Denver, issued pamphlets announcing their discovery of "the secret of staying young"—*eating* dried animal testicles. Ad copy from Illinois's Youth Gland Chemical Laboratories swore that science had unlocked the secret to healthsome longevity: "It is based entirely on the principle of Feeding Actual Gland Substance Direct to the Glands, thereby renewing and rejuvenating them."

Transplants of ape testicles directly onto human testicles became widespread in the 1920s. The primary architect of monkey-gland transfers was Serge Voronoff, a tall, magnetic, dark-browed Russian émigré living in France. Voronoff got his start as the royal surgeon to Egypt's khedive Abbas II, where he often dealt with ailing eunuchs from the harem. He believed an absence of testicular hormones caused their frailty. Independently wealthy, careful and meticulous in his work, Voronoff wasn't a quack—he was a bona fide scientist who simply managed to convince himself of a hypothesis he couldn't prove. When it comes to immortality, we're all so willing to deceive ourselves.

Despite the ridicule he later endured, at the outset most scientists agreed with Voronoff. At the 1923 International Congress of Surgeons in London, reported *Time,* "700 of the world's leading surgeons applauded the success of his work in the 'rejuvenation' of old men." Voronoff provided medical journals with technical instructions for grafting slices of primate testicle onto patients' testes. The *New York Times* stated that monkey glands successfully correct "the defects of old age." Everyone agreed that aging would soon be relegated to the past. If we didn't grow old, why, we'd just stay young forever. Registered testicle-graft doctors sprouted up everywhere from Turin to Rio de Janeiro.

By 1927, over a thousand customers had been Voronoffed. His clinic on the Riviera used chimpanzee and baboon glands. Each ape testicle was cut into six narrow segments, then sutured with silk onto the exposed human testicles, where new blood vessels would ostensibly grow in to nourish the graft. (A number of patients contracted syphilis from the transfusion. There is also speculation that AIDS came about as a direct result of Voronoff's interventions.) While many scientists lauded his experiments, some cautioned against overoptimism. No one then knew how challenging it actually is to transplant animal tissue into human beings. Such procedures are almost always insuperably difficult and require medication to help them take. In reality, Voronoff's approach resulted primarily in activating the placebo effect.

His experiments were repeated by other scientists, yielding unfavorable results. Within a few years, commentators pointed out that the method didn't actually bestow any prolongation of lives, as "none of those on whom the transplants have been done seem to live beyond the normal period." Critics started fuming that a grafted monkey testicle was "nothing more or less than a piece of dead meat put in the wrong place." Public opinion turned. Animal glands fell from fashion. Disgraced, mocked, Voronoff died in obscurity, forgotten alongside the time when testicles were the fountain of youth.

We no longer have gland fever, but have we really changed all that much? Time will tell. Unlike singularitarians, most of us wouldn't claim to know what the future holds in store. But most people also don't realize how often we've been down this road before.

Ray Kurzweil's Bible-size books explore the intricacies of mind-uploading with almost messianic fervor. There's a passage in one of his books where he and Bill Gates speak about the world's need for a new religion. Gates is quoted as saying, "We need to get away from the ornate and strange stories in contemporary religions and concentrate on some simple messages." The book insinuates that Gates wants Kurzweil to lead this religion, but Kurzweil demurs: "A charismatic leader is part of the old model."

The rest of their conversation concerns the implications of physical immortality. Despite some nitpicking (will an intelligent universe be silicon intelligence or more like carbon nanotubes?), they seem to agree with each other on most points, at least until Gates makes an insightful remark about how everything of value is fleeting. This truism invalidates Kurzweil's central premise. For if our lives continued forever, how could they still be of value?

In immortalist thinking, paradoxes and rationalizations accumulate rapidly. Is eternity something more than an idea? The only way to find out is to achieve immortality. Even if we could live for only a hundred trillion bajillion years, we'd eventually have to contend with the sun's extinguishing or proton decay. If people stopped dying, wouldn't overpopulation become an insurmountable problem? Not at all, counters

Max More, extropian and member of the Immortality Institute. He brandishes figures suggesting that populations have maxed out and are on the decline. "We can expect population growth to continue slowing until it reaches a stable size," he adds, by which time aging will no longer be an issue.

Immortalists don't notice anything unreasonable in the desire to never die. The certainty of their belief is bolstered by its own justifications. Just as biblical literalists are positive that the world began only a few thousand years ago, radical life-extensionists disregard contradictory facts. They are so deeply passionate about technological progress that they are prepared to forsake their humanity, to become half machine, in order to live forever. They view themselves as spearheading the most important endeavor humanity has ever engaged in. Doesn't everybody else see that preventing death is a biological imperative?

The Immortality Institute's book on infinite life spans turned out to be one long pitch for the importance of artificial intelligence, mind uploading, and cryonics. Its contributors often veer into incoherence. Even if the first half of this sentence isn't something you consider ludicrous, the second half makes no sense at all: "The average age at death of those born in wealthy nations in the year 2100 will exceed 5,000 years, which is perhaps five times the value resulting from a permanent enjoyment of the mortality rate of young teenagers in such nations today." What sort of calculator can calculate the permanent enjoyment value of rich tweens' mortality rates? Or perhaps a better question is, what sort of brain comes up with that sort of equation? The answer: Aubrey de Grey. He's one of the most prolific of theoretical immortalists, so it's a given that he'd contribute an essay to the institute's anthology. Aging, he argues, simply shouldn't be allowed.

His chapter in the book describes, in gory detail, the island colonies of humanlike apes that would be vivisected and experimented upon to keep humans living forever. Apes are perfect not only because they're genetically similar to us and age even faster than we do, but, as de Grey writes, because "they don't talk, so if the biomedical imperative is sufficient society feels entitled to do more or less anything to them."

He may not want to hear any dissenting opinions, but those uncertain about becoming cyborgs deserve to know exactly what the Immortality

Institute's fellows propose. It isn't simply the colonies of apes, or the vagaries of cloned versions of ourselves, or even that the hybridization of man and machine will result in a new species. "The whole idea of a 'species' is a biological concept," clarifies Kurzweil, another of the institute's essayists. "What we are doing is transcending biology."

But aren't all living organisms by definition biological?

As an appendix to their book, the Immortality Institute published a list of reasons why living forever would be a good thing. To question the desirability of immortality is, in their view, "banal," but they acknowledge that some people may have trouble comprehending "why anybody would want to live beyond the currently fashionable limit of about four score." For those with limited imaginations, they offer the following answers:

- to find out what the future will be like
- to have more time to figure out the meaning of life, if there is one
- to have a chance to really grow up and find out what kind of wisdom and maturity might be attainable by a healthy eight-hundred-year-old
- to have more time to help others
- if you live, you can always change your mind about it later; death is irreversible
- to spend more time with friends and loved ones without a time bomb ticking quietly inside you all the while
- to explore exotic mental states
- to live happily ever after

These arguments, torn between the longing to know and a need to escape, are all ways of not dealing with the present reality. The list isn't likely to convert nonbelievers. Certainly it will appeal to anyone obsessed with the time bomb of death. But by the time most people make it to their twenties, they realize that the meaning of life, "if there is one," may have something to do with the fact that it ends. Even Bill Gates knows that ephemerality is what makes life precious. Keats, who died at twenty-five, didn't need eight hundred years to understand that we can't understand. All we know of life, all we *can* know and all we need to know, he wrote, is that "beauty is truth, truth beauty."

The most striking aspect of the institute's list is its "living happily ever after" conclusion, which reveals something significant about the fairy-tale nature of immortalism. Adherents seemed to me to be adults afflicted with the sort of aging denial typically found among teenagers. Their essays spoke of never suffering, of living pain-free lives devoid of all "corrosion by irritability, envy and depression." Some people, myself included, feel that life is made meaningful by struggle, but then again, most of us also consider death a fact—not an immoral and unacceptable "imposition."

21

It Was the Future

What an absurd creature man is,
Striving to reach divinity,
Yet stuck in his own image till infinity!

—Goethe, *Faust*

There is nothing more useless than an organ.

—Antonin Artaud, *To Have Done with the Judgment of God*

HUNTINGTON BEACH, in Orange County south of Los Angeles, bills itself as Surf City, USA. On the sun-drenched Sunday afternoon I arrived, the waterfront was crowded with teenagers in Ray-Bans playing volleyball or skateboarding on half-pipes. The pier teemed with athletic young people. Twentysomething street performers bounced around on pogo sticks. A pro-marijuana rally was under way. High-school-aged demonstrators chatted about "dank nuggs" and chanted, "It's my mind, it's my decision."

I struck up a conversation with a busker named Corey who was playing acoustic guitar on the esplanade. With his plaid shirt, Justin Bieber mop-top, and all-American good looks, he seemed to me as if he could have been a member of the Beach Boys—had he been born a half century earlier.

"Huntington Beach is chill," he told me, tossing his bangs in the wind. "There's a lot of bros here, guys who race dirt bikes, wear bandannas, and have spiderweb tattoos. You know, dudes who like to rep things."

"What does *repping* mean?" I asked.

"It means wearing shirts for a brand or company," he responded patiently, with a mellow SoCal accent. "Like *representing* them. Huntington Beach bros also love lifted trucks, like lifting the suspension on their trucks ten inches higher."

As we were speaking, an older man who resembled The Dude from *The Big Lebowski* came over to give us a flyer for his show. He was in a Led Zeppelin cover band.

"I think my parents have been to see you," Corey said.

The singer coughed uncomfortably. He wasn't quite sure whether Corey was joking or not. "You know, I'm also in a rap band," he added, "Terminally Ill Cartel."

"Sick." Corey nodded.

"Come check us out sometime."

"For sure. Can't wait to grip one of your sets."

After a few moments of silence, the older guy gave Corey a fist bump and then shuffled along his way.

"It's kind of pathetic how people try to stay young here," Corey confided. As he spoke, a couple of nubile forms sashayed by in bikinis. Corey let his tongue hang out and said, "But you can see why: this whole place is like a shoot from *Teen Vogue*. Swag."

It was true. Young people were everywhere, and everyone else was trying to look as young as possible. It's hard to age gracefully in a place such as this. Newport Beach, just down the road from Huntington Beach, is the plastic-surgery capital of Southern California. What a perfect place for immortalist ideas to thrive.

"We have our own commandment here," Corey said. "'Thou shalt not grow old.'"

"Do you have any examples of that?"

"That guy from Terminally Ill Cartel." Corey laughed. "Like, this is a place where girls get Botox when they're still in high school. And if dudes here are bros, the female version of a bro is called a *bro hoe*. For their eighteenth birthdays, they get boob jobs. When people excavate Southern California thousands of years from now, they'll find landscapes full of silicone implants. Their boobs will live forever."

Corey started playing a song on the theme of immortal injectables. When he was done, we said good-bye, and I headed in the direction of the party, enjoying the fading light and thinking about youth, so fleeting. Having reached my midthirties, I'd learned that a significant percentage of one's waking hours are spent dealing with challenges—such as aging.

Happiness comes in brief, rare bursts. Who would want to prolong the hassles of this life indefinitely?

Past the Sun and Sand Motel stood a large, four-story home overlooking the beach. I was a little early for the get-together, and my friends Clay and Jenna hadn't arrived yet with our costumes, but the front door, ajar, beckoned irresistibly.

It opened onto a marble-and-black-leather foyer. Two van Gogh reproductions hung next to the imposing fireplace. The room was empty, except for a lone turtle pacing between some potted plants. Voices drifted down the stairs.

On the second floor, a dozen people were hanging out. One man, standing by himself, had a little sticker on his shirt that said BOB 412. I introduced myself and asked how he'd ended up here. He said that he'd posted on the Immortality Institute forum asking whether there were any longevity fans in Laguna. Nobody had responded. "I have only one friend in Laguna who's into longevity and immortality—and he's really strange," Bob 412 said. "Then I noticed the posting for this party and decided to come check it out. I wanted to know what a roomful of immortalists would be like."

"Me, too!" I said. "In fact, I don't know anybody here."

"You don't even know Dave Kekich and Kat Cotter?" he said, pointing at a man in a wheelchair and a bubbly woman in her fifties explaining things to a couple of maids.

"Only from the invite."

In the course of our brief correspondence, I'd also read a profile of the couple in *Life Extension* magazine, which explained how Kekich had become paralyzed in 1978 when his spine was compacted during a weight-lifting mishap. He met Kat some years later, when she was the director of the Lifespan Anti-aging Center in Los Angeles. "Have you ever wished you could go back in time twenty years and get a second chance?" concluded the magazine profile. "Well, if Dave Kekich and Kat Cotter have anything to do with it, you probably will!"

"I've never met them either," Bob 412 said, "but they're well-known in the prolongevity community."

"So, Bob 412, what brought you here to the year 2068?" I asked, invoking the party's theme.

"Longing," he answered without hesitating. "When you get to be my age, you want to believe in 2068 and miracles coming down the line. It's gonna happen. In your life."

"What do you mean by *it*? What's going to happen?"

"Radically extended life," he said, deathly serious. "Indefinite life spans. The Singularity. C'mon, man, get into the groove!"

The year was 2068. On a gold-dust evening toward the end of May, fifty or so guests had gathered in Southern California to celebrate their friend David Kekich's 125th birthday. Some were standing on the lawn, admiring the ocean sunset and sipping resveratrol-astragaloside shakes. Others milled around the open-concept kitchen, picking at shrimp rings and supplement bowls while debating the merits of various telomerase activators. Blue and silver balloons hovered over the dance floor.

In the living room, a fifty-eight-year-old man with an XLR input port on his neck was talking with a Barbie-blond, middle-aged woman. The name tag on her low-cut black dress read HER MOST SERENE HIGHNESS, THE CELESTIAL EMPRESS. She had enhanced lips and introduced herself as the CEO of a stem-cell company. His sticker said BILL ANDREWS. He researched biological longevity and operated a website called www.cure-aging-or-die-trying.com. The two of them were chatting about an upcoming ultramarathon around the moon that Andrews was organizing. "According to my calculations of the moon's circumference," he explained, "if we start at sunrise, finish by dusk, and run six-minute miles the whole time, it's totally doable."

"What if someone can't keep up?" asked the woman.

"Well, normally you'd freeze to death." He smiled benevolently, touching her elbow. "But if anything slows you down, you'll just get stem cells injected into all your joints. That way, you'll be able to do it."

Andrews was an avid jogger on earth as well. En route to the party, his car ran out of gas, so he simply pulled onto the Pacific Coast Highway's shoulder and sprinted the rest of the way. He was wearing a full-body white outfit made from ultralight, UV-blocking fabric, including a modernized legionnaire hat. With his aerodynamic goggles and a salt-and-pepper Vandyck, he looked the part of a classic 2060s interstellar scientist.

Six decades earlier, a journalist had asked Andrews to describe his job. "If you were to meet me at a cocktail party and ask me what I did and I told you I was selling something to immortalize your cells, you would think I was saying I'd found the Fountain of Youth for cells," he responded. "You'd think I was a snake-oil salesman selling bullshit, and I wouldn't blame you." In reality, he never *sold* immortalizing products so much as formulated them. His "age-reversing" nutraceuticals were distributed through an Arizona-based, multilevel marketing company called Isagenix.

As several people at the party gathered around Andrews to hear more about the moon race, a beetle-browed, dark-haired man wearing a gaudy gold-sequined shirt strolled in.

"Joe! You look *so* futuristic," the Celestial Empress cooed, patting his sleeves.

"In the future, we'll all be shiny," Joe quipped. Everybody laughed. It was Joe Sugarman, the direct-marketing pioneer who'd made millions in the final quarter of the twentieth century selling BluBlockers sunglasses over 1-800 numbers. He'd brought samples of his latest blockbuster breakthrough: an antiaging product that could make anyone live to 120.

"What's that quarter-inch port for, Bill?" Joe asked, gingerly fingering Andrews's neck.

"Mind uploading, of course," Andrews answered.

The blond woman looked over Sugarman's Stem120 potions. "Your packaging is amazing, but people don't need this sort of thing anymore," she said. "We can live forever, remember?"

The disease of aging had been cured long before 2068. Everyone was immortal—except those who didn't bother to back themselves up.

Decades earlier, scientists figured out how to digitize the contents of human minds. Using microscopic semiconducting crystals, a technology called quantum dots, they were able to etch a mirror image of the brain's neuronal wetware into silicon substrates. Whole-brain emulation through magnetic resonance imaging was a paradigm-shattering breakthrough. For a reasonable fee, all our thoughts, memories, opinions, and experiences could be scanned, mapped, and cached onto computerized information systems. By duplicating synaptic interconnections and plugging them into the cybernetic matrix, humans could transition back and forth between flesh and data. Once noninvasive, static uploading was legalized, our bodies stopped being bodies the way bodies used to be, all messy and illness-prone.

First we replaced the blood in our veins with cell-size nanobots. The legalization of respirocytes—tiny robots that patrol, clean, and repair the body's four trillion tissue cells—led to the discovery that mechanized lungs were far more effective than the inflatable disease bags we'd initially been saddled with. Once we eliminated the need to breathe, we soon dispensed with the digestive system, which allowed us to eat as much as we wanted without gaining weight. Biotech lobbyists succeeded in making fiber-optic spines mandatory. After bionic white blood cells called microbivores became available, pollution-munching nanoscale filterbots were released into the atmosphere, solving all concerns about industrialism. Kidneys just caused stress, so once they were gone, nobody ever worried again. The liver, the pancreas, and the spleen turned out to be extraneous. "Although artificial hearts are beginning to work, a more effective approach will be to get rid of the heart altogether," prophesied a futurist named Ray Kurzweil in 2004. He was right: Human Body 2.0 was mainly nonbiological. We made ourselves into bodies without organs.

Our eyes were surgically enhanced with monitors that connected our retinas and minds directly to the Internet. Chips and processors implanted in our brains were synced to Google. We all came to know everything there is to know. Designers developed a new type of skin, made from color-morphing polymers, so we could choose whatever skin tone suited our mood. Soon, there was little distinction between virtual and analogue realities. What the inhabitants of the early twenty-first century considered physical products became information files—e-mail attachments. Pleasure devices were invented that allowed us to have sex with tangible replicas of our fantasies. Annual dechronification checkups continually reversed the body's age to an optimal twenty-four. After Permanent Rollback became a reality, nobody ever needed to die again. Each person was archived in various storage media (biweekly identity-reconstruction scans were recommended), and if one's body was damaged, broken, or killed, all one had to do was download oneself into a new clone-husk.

These artificial bodies were composed of subatomic, self-reconfiguring robotics components that swarmed together in malleable utility fogs. They'd rendered plane tickets obsolete—we could instantly transfer our "selves" across the globe into foglets (as bodies were now known) wherever we needed or wanted to be. Through identity diffusion we could even be in several places at the same time, using wireless links. The expression *I changed my mind* came to mean something else entirely. By midcentury, we evolved. We became living machines. Cyborgs. Immortals.

On the second floor of David Kekich's four-story home, the costume party was heating up. A small group danced jerkily to a CD of music generated by the human genome. Bleeping chimes and chirps issued from the speakers. It sounded like synthesizer bells playing space-station Muzak. "Is this music helping us age backwards?" asked Dr. Lord Lee-Brenner, a self-described pioneer in the use of human growth hormones. His date did a little moonwalk. She was wearing a beauty-pageant sash that said C. ELEGANS, the Latin name for the roundworm traditionally used in biological longevity experiments.

Next to them, in an I'M FROM THE FUTURE T-shirt, was the birthday boy himself, David Kekich. He was in his wheelchair, setting up a microphone stand, when two fiftysomething women in schoolgirl kilts and fuck-me heels walked over to greet him.

"Happy one hundred and twenty-fifth birthday!" one of them said, leaning in to kiss his cheek. "You don't look a day over sixty!"

"Thank you!" he replied. "I'm sorry—I know we've met, but what's your name again?"

"You don't remember?" her friend asked in mock indignation.

"No," Kekich responded, laughing. "Memory has been eradicated, right?"

On the ground floor, the topic was insulin-swing lag times. A lady in a black sweater emblazoned with the words THE SINGULARITY IS NEAR glided in on shoes with little wheels on the bottom. "Just like the kids!" she said, laughing.

"What's the Singularity?" asked a child whose clothes were covered in aluminum foil.

"It's the old-fashioned term for humans merging with computers to become immortal," answered the roller woman.

A copy of *Ending Aging* lay on the coffee table. When it was published in 2007, it claimed that people alive then would easily live to a thousand. Everybody at the party felt positive that would happen.

The lady with a HELLO MY NAME IS: KAT COTTER sticker affixed to her ample chest smiled at me. Interweaved within her waves of auburn hair were serpentlike tresses of glittering tinsel. She wore a black-and-gray robe with a futuristic, almost cultlike silver pendant. The large buckle on her belt was made of interlocking circles bedazzled with cubic zirconia.

I walked over to introduce myself as the writer who'd come from Canada to attend. "You must be Kat?" I said politely.

"No," she said, continuing to smile enigmatically.

I glanced down at the name tag.

"Kat couldn't come. I'm her wild clone who drinks too much."

"Aha: Kat's clone from the year 2068—of course." I shook her hand. "My costume is stuck in a wormhole, but it'll be here soon."

A woman wearing a plaid schoolgirl skirt came over and handed us each a cocktail. Her name was Jackie Silver. "I'm aging backwards," Jackie purred. "I'm actually underage-drinking right now."

Upon learning my profession, Jackie immediately started telling me about her antiwrinkle cream. It's called Replenish and she sells it through her website www.agingbackwards.com. "Replenish can quite possibly reverse the aging process," she explained. "I discovered this age-defying coconut-honey complex. The active component is called baicalin. It's a proven, risk-free cellular regenerator. It basically creates an invisible cushion of moisture that makes you look decades younger—like me!"

Even dressed like a porned-up version of a Catholic schoolgirl, she clearly looked like a lady of a certain age. "I get mistaken for a younger woman all the time," she writes on her website. "It happens when I'm out with my teenage son, Trent. People think I'm his sister! Sometimes, I'll even get asked out by men in their twenties."

Whoever thinks she's in her twenties would also have no trouble believing that the key ingredient in Replenish, baicalin, has miraculous properties. "Not only can it delay cellular senescence, it also helps your skin recover the youthful characteristics you had ten years ago," explains her promotional material. "It fixes the problem of telomeres shortening by creating the missing end piece. It's also been proven to increase skin firmness and elasticity, improve skin restructuring, and increase the number of fibroblast duplications." I wondered what the Nobel-winning scientists who discovered telomeres would say about her cream creating a "missing end piece."

My phone rang. Clay and Jenna were a few miles away. "Wait until you see your costume," Jenna said, laughing. "Clay is dressed like a mad scientist, and I'm a radical futurist-feminist. I've published several volumes of poetry on the vagina—now that it's no longer needed."

As we spoke, a lady in a bright red wig and mirrored ski goggles came into the kitchen. She put down her bag and removed her tie-dyed T-shirt and denim miniskirt. For a moment, I thought she had gone totally naked, but then I realized that she was wearing flesh-colored, head-to-toe pantyhose. She had no bra on, so the nipples on her small breasts were plainly visible. It was impossible to see what her face looked like as the wig and shades covered them so much. "Your timing is perfect," I gasped into the phone. "See you out front in ten."

As the quasi-naked redhead sashayed out of earshot, an elegant-looking woman walked over. She seemed to me like a conservative, churchgoing mother, and she was taken aback by what she'd just witnessed. "That lady in the body stocking has more nerve than anyone I've ever met," she muttered.

"Welcome to 2068!" I said.

Within minutes, it became clear that Barbara's prim attire did not reflect her ideas. She had joined the Methuselah Foundation 300, a group of three hundred philanthropists who agree to donate $1,000 per year for twenty-five years to help create "extended healthy life, enjoyed by all humanity . . . forever." The names of each donor, she explained, will be etched into a steel-and-marble monument for all time.

"We're even into going off-planet," she continued. "We're going to be off-planet by 2068, for sure. I would love to live off planet Earth. I don't care if we make fake planets. We'll find something new and make it better. I'm looking forward way past what we're doing now."

Two young men came over to join Barbara. They were brothers: one in high school, the other doing an undergraduate degree at Harvard in human evolutionary biology. I asked them if they were interested in living forever. "Of course!" said the younger one. They had been passionate about immortality since childhood. "When we were little, my brother and I decided neither of us wanted to die in any way ever," explained the older one. "Later in life that translated into a fascination with the purpose of life rather than a fear of death. There's too much we don't know to just leave it all unresolved when we die."

"Are we immortal?" I asked in the spirit of 2068.

"I hope so," he answered sincerely. "With the improvements we're making, it seems pretty certain to me. There's no way of predicting when the breakthrough will come, but when it does—*pow!* I have to believe yes, otherwise there's nothing to be working for or towards."

As we spoke, the two brothers and I couldn't help taking peeks at the woman in the body stocking.

"Is that what it's going to be like in 2068?" I asked.

"I don't know if it'll be like *that,*" the elder one said. "That's kind of gnarled out."

His brother just laughed.

"Would you like it to be like that?" I asked him.

"Selectively." He grinned slyly, stroking his chin.

Then a ripple of excitement spread through the room. A thin man walked in dressed up like the life-extension theorist Aubrey de Grey. He had the look down perfectly: the long beard, the pallid countenance, the matted ponytail, the skinny jeans, even the ratty tennis sneakers.

"Have you met Aubrey?" Barbara asked.

"Not yet," I said.

"He is the guru of nine-tenths of what we espouse," she said.

It wasn't someone dressed like Aubrey de Grey—it *was* Aubrey de Grey. His long, scraggly beard gave him an eerie and wizardly Rip Van Winkle vibe. He looked consumptive. The first thing out of his mouth was "Where's the beer?"

On June 25, 2000, almost a decade earlier, Aubrey de Grey and Dave Kekich, alongside an invitation-only group of other biogerontologists, spent the day debating the future of immortality research. They were in Manhattan Beach, a forty-five-minute drive up the coast. The round-table discussion was heated, and fruitless. Heading back to his room at the Marriott that night, de Grey felt frustrated at the gathering's inability to agree upon a concrete plan for combating aging. He wanted results, and he wanted them soon.

In bed, his mind kept racing with ponderings about the nature of metabolism. Running his hands through his abundant beard, he stood up and started to pace around the room. Until then, he and everyone else in the prolongevity community had seen aging as a hopelessly complex inevitability, a nightmare about which nothing could fundamentally be done. But was it a question to theorize and argue about, or was it one to answer? There must be a way of reengineering the body, he mused.

At around 4:00 a.m., de Grey had his eureka moment.

"I swept aside all that complexity," he recalls, "revealing a new simplicity in a complete redefinition of the problem." To end aging forever, he realized, we simply need to address what he calls the Seven Deadly Things. His "possibly comprehensive plan" entails fixing seven categories of cellular and molecular issues that we cannot currently fix. This includes curing cancer through WILT, eliminating the buildup of unwanted cells, rectifying chromosomal and mitochondrial mutations, flushing out lipofuscin, and wiping away other bulky detritus from our lysosomes.

If we could only activate an intracellular janitor, a broom of the system capable of cleaning out all the dust and scree that accumulates in our bodies as they age—why, then we wouldn't age at all. If each of the seven habits can be handled, we can all be highly effective people—forever. Calling his theory SENS, or Strategies for Engineered Negligible Senescence, de Grey declared a WOA (War On Aging).

The message was so simple it caught on. SENS became a media sensation. De Grey started foundations, wrote a book, and gave countless interviews to amazed journalists. He claimed that his acronyms would save "tens of thousands of lives a day, possibly including your own or your most dearly beloved." He started being compared to Jesus. He became the face (or the beard) of what he himself characterized as the most important mission humanity ever faced. The goal of preventing death never left his thoughts, he lamented. To cope with the responsibility, he drank many pints of beer every day.

What could the rest of us do to help? He recommended writing letters to Congress telling them to support his research. "But an even more powerful thing you can do is to donate to the Methuselah Foundation," he wrote in *Ending Aging*. It couldn't be easier: all we have to do is give him money and we'll never die. The donations are even tax deductible. Although his official title is chief science officer, it may be more accurate to call him a fund-raiser. "Five thousand dollars may not sound like much, but SENS Foundation is fantastic at spending money incredibly efficiently, on account of our fabulously committed workforce, so it'll make a genuine difference."

De Grey is emblematic of early twenty-first-century extreme scientific optimism—and its attendant ignorance. That a computer programmer taken for a professor from Cambridge could gain so much credibility testifies to this era's incapacity for separating truth and fantasy.

In 2005, more than two dozen of the world's most important biogerontologists published a peer-reviewed paper distancing themselves from de Grey's seven factors. "None of these approaches has ever been shown to extend the lifespan of any organism, let alone humans," they concluded. "To explain to a layman why de Grey's programme falls into the realm of fantasy rather than science requires time, attention and the presentation of detailed background information."

De Grey has lashed back with hundreds of pages of interminable, quasi-rebuttals. "If you look at my reasoning, how I get to those conclusions," he claims, "it becomes very much harder to actually identify anything that I'm saying that is unreasonable." But wading through his papers, as I did in preparation for the 2068 party, isn't something most people would subject themselves to in their free time, unless they wanted to be freaked-out by his dystopian vision of primate colonies.

Beyond the apes, there's a simple reason why SENS is unreasonable: it's a mythology. A belief system. It's not of the realm of reason; it's of the realm of faith. His basic argument is that we could eliminate death if we eliminate death. How can you argue with that? De Grey uses a farrago of scientific-souding gibberish to dazzle nonspecialists, invoking the powers of lyphamacains, dr2cs, IL-7 mediated thymopoiesis, glycosylation crosslinks cleaved through phenacyldimethylthiazolium chloride, etc. The technical terminology obfuscates his theory's lack of scientific evidence. No tests are undertaken to prove his speculative hypotheses before he publishes them. "Aubrey has made no original contributions to this field," writes Leonard Hayflick, of the Hayflick limit. "He is a fly geneticist who, without training in the field of the biology of aging, is in my view misguided in his belief that the aging process will be capable of manipulation in the next decade or two. This belief has been twenty years in the future for the last 3,500 years!"

The best way to understand de Grey is as a *punter,* writes Jonathan Weiner in *Long for This World*. A punt is a flat-bottomed boat used for scenic trips down the River Cam in Cambridge, where de Grey lives. A punter is a person who drives a punt, kind of like a British gondolier, except a punter uses a pole rather than an oar. De Grey was a proud punting helmsman in college. "When I was a student, I bought my own punt, a secondhand one for a few hundred pounds," he recalls. "And I used it in the summer to do what's called chauffeur punting."

What does a chauffeur punter do?

"People come along, tourists, and you tell them lies for money," he explains.

Jenna and Clay were on the front lawn, looking like two space cadets from the future. Jenna—svelte, high-cheekboned, in her late twenties—wore a powder-blue terry-cloth leisure suit, unzipped to her belly button. She had large gold elephant earrings and a shock of fuchsia makeup above each eye. Clay, roguishly handsome, sported a wig of curly black Lionel Richie hair and an undeniably futuristic V-shaped vest whose Technicolor fabric effected a kind of jagged electrical pattern. Large rings of green plastic encircled each shoulder.

"We decided you should be the minister of intergalactic affairs for Azerbaijan," Jenna said, handing me a bag of clothes, barely stifling her laughter. The top, a loosely puffy, silver concoction, looked like a repurposed sleeping bag. My emerald-and-gold, ankle-length skirt was matched by a large, circular fez-type hat. The outfit sparkled.

I quickly changed in the foyer, as Clay critiqued the crepe blinds, the imitation van Goghs, the marble fireplace, and the pleather furnishings. "This is the sort of style you always see with insecure millionaires," he sniffed. "The whole place smells like white leather and juice machines."

As we spoke, the naked-looking lady in the wig ambled by. "Oh, I wanted to come in the pantyhose bodysuit," whispered Jenna. "Good thing I didn't. It would've been awkward for both of us."

A man in a dojo came over and asked where we were from. When Jenna said Canada, his eyes lit up. "I take citric acid intermediates as a caloric-restriction mimic. It's a tiny pill from Canada—have you heard of it?"

"No, no, I'm sorry." She shook her head. "Does it work?"

"I sure hope so. But the problem all of us have is, how do you test for it? You can't measure their effect upon individual genes. *Yet!* In the meantime, we gobble pills and mix shakes. I also take methionine, which is great."

On our way upstairs to the second floor, Jenna's heel got caught in the cuff of her leisure suit and her glass went flying. Red wine splashed all over the wall, her outfit, and the marble stairs. Luckily the only person who noticed was a maid, who rushed over to clean it up and told us not to worry. Within a few moments, Jenna's wine had oxidized the marble and turned it blue. As it drizzled down the stairwell to the first floor, she couldn't help wondering whether the stain would last forever. "If the owners of this house were to attain immortality," she asked Clay, "do you think they would eventually embrace this stain like a special wrinkle, a sign of having lived a good life full of parties and adventures? Or will they just have the help clorox it away?"

In the living room, Joe Sugarman of BluBlockers fame was regaling a group of partygoers with tales of daily life in 2068. He spoke of how humanity had solved every disease by the year 2020, but that people still loved using his hair-restoration potions and libido enhancers. He started hyping an amazing "feminine sexual-enhancement cream" called Pure Fulfillment. He recommended rubbing it on any part of the body, internally and externally, to combat dryness.

Next to them, on the balcony, Aubrey de Grey was chugging a beer, checking the Internet, and attempting to ignore Sugarman. "In the year 2030," Sugarman said, raising his voice, "Aubrey figured out a way to make his beard grow very fast."

De Grey made a pained rictus, as if to say, *What am I doing here?*

Sugarman continued, "In the following year, Aubrey was taken to—"

"A lunatic asylum," de Grey interjected, breaking his silence, and making his annoyance evident.

"—a hospital," Sugarman continued, unperturbed. "For tripping over his own beard."

A couple of partygoers giggled nervously. De Grey shrugged, emptied his beer, opened another, and turned back to the computer. His assistant, an ultrathin blond woman, stepped up to the mic. Speaking quietly, she introduced herself as Maria. A few people turned to watch, but the rest didn't seem aware that a performance was under way.

"Shut *up,* everybody!" de Grey commanded.

"Thank you, Aubrey," Maria said, louder now, with a thick Spanish accent. Starting an instrumental CD backing track, she began to sing a terrible, maudlin number about how special it was to be a seventeen-year-old girl in the big city.

When the song ended, I started a conversation with Aubrey, who had busied himself on the balcony with e-mails. "Are you enjoying the party?" I asked.

"This is the first party I've been to in over a year," he said, his thick British accent muffled by the density of the beard engulfing his mouth. "I don't have any social time whatsoever I'm so busy working."

I nodded and said I understood completely—being a writer is such a full-time occupation it can be hard to find the time for anything but research. He affected a professorial impatience, hurrying me along as I spoke.

"I'd like—"

"M-hmm, m-hmm?" he prodded.

"—to—"

"Yes, go on!"

"—interview—"

"And?"

"—you."

"For what?" he said even before my last syllable made it out.

"I'm-writing-a-book-on-the-fountain-of-youth-and-the-quest-for-immortality," I intoned, at breakneck speed.

"It's been done before," he sneered.

"Many times." I nodded. "Even five thousand years ago in Sumeria."

He huffed and took a gulp of beer. From what I could tell, he'd condescend to an interview.

"Would this be a good time?" I ventured.

"Sure, I do interviews in my sleep." He was undeniably charismatic, in a circus-sideshow kind of way. *Mountebank* is a sixteenth-century word for a quack who would stand on a bench or pedestal or knoll selling elixirs. To me, Aubrey was an Internet Age version of the age-old immortality pitchman, a cyber-proselytizer spreading his gospel through chat rooms and forums.

"So it's 2068," I said, kicking off the interview. "What have the past fifty-nine years been like for you?"

"I retreated into glorious obscurity in 2035, having achieved my goals."

"And what did you do from that point on?"

"I started watching movies and reading books, doing all the things I couldn't do when I was busy." He waved his arms around. "I moved to Madagascar. I entered a state of indefinite leisure."

"Why don't you just watch a bit of those movies and read those books now as you're working?" I wondered out loud. "You wouldn't really want to just watch movies all the time, would you?"

"I'm too busy to watch movies now," he replied severely. "But trust me, I wouldn't get bored, that's for sure." As he spoke, he ran his hands along his bushy beard. It seemed to have the texture of angel-hair pasta—softer than I'd imagined, less wiry. Perhaps he shampooed and conditioned it.

We spoke about his theories—SENS, the seven steps, the War on Aging (WOA). He told me about his late-night breakthrough after spending the day with Dave Kekich a decade earlier. He was positive that the sunshine of perpetual youth would drive away "the dark specter of the age plague" well before 2068.

What he calls the post-WOA era will shift into effect when a portion of the baby boomers—in his words, "the richest ten percent of those in the richest ten nations"—are able to afford treatments that will extend their healthy life span by twenty years. In those two decades, what with all the vivisecting of apes we'll be doing, it's "virtually certain" that new therapies will be found that will extend those privileged lives for a further decade or two. From there, it'll be a cinch to reach 150. Once we are sesquicentenarians, we will start to experience what he calls a diminishing mortality risk, or "an increasing *remaining* life expectancy as time passes." The more years go by, in essence, the longer we'll live. Until we never die. This achievement is called "life extension escape velocity."

I asked him how he thought people would react to his idea of having these colonies of apes—tens of thousands of them—that we would be using in whatever ways we wanted in order to allow humans to become immortal.

"Well, the main difficulty with animal rights activists is that they don't communicate with logic," he said.

"Sure, but you must realize that there will be opposition to dissecting living primates," I countered.

"Primates?" he scowled, getting worked up, but trying to maintain his temper. "We want the primates to live for a long time because *we* want to live for a long time. We want to live forever. Those primates are not going to have a longer-than-normal life span unless we treat them really well."

Here he looked at me as though I were foolish to not realize that immortality scientists were going to be treating these apes well.

I looked dubious.

He didn't care. "There are going to be lots of colonies all over," he continued agitatedly. "Many, many colonies." As he spoke, becoming increasingly enthusiastic, he started hopping up and down. The hopping made it even harder to make out what he was saying. The crux of it was a roundabout explanation of how our performing in vivo experiments on apes would actually be something they would want us to do. He kept using modifiers intended to lend a scientific dimension to his statements, such as "experts claim" or "it is probable" or "almost certainly."

We were both relieved when our interview was cut short by the sound of hushing all around.

Dave Kekich lowered the microphone and announced, "Welcome, everyone, to the house I lived in fifty-nine years ago. It's interesting that you've all shown up with your avatars, because we're actually in cyberspace, and you did so with no prompting."

He started describing the events that had taken place in the six decades leading up to the present moment in 2068. It all began when he started a stem-cell-product company with Joe Sugarman, a peptide doctor, and some molecular specialists. "From 2011 to 2016, we worked to turn around the aging process. Our company generated millions and millions."

"Zillions!" said Her Most Serene Highness, the Celestial Empress.

"Trillions!" added Joe Sugarman. "Billions!"

"We redirected all the money we made into other biotech ventures," Kekich continued. "We were getting spectacular results. Billions poured into antiaging research. Unfortunately, in 2018 there was the total collapse of the US dollar and all other currencies—but in 2019, a new global currency was released called the Universal Dollar. It stabilized the world economy."

The audience clapped politely.

"In 2023," Kekich said excitedly, "Aubrey's SENS finally took place."

Aubrey rested his lager on the terrestrial globe in front of him. "I thought it would be sooner than that," he said, not smiling.

Kekich was unfazed. "In 2025, scientists did the first brain-to-computer transfer. Aubrey became an overnight billionaire."

Aubrey coolly sipped his beer. Someone's phone beeped and rang in the back of the room.

"In 2029, cell phones became obsolete," Kekich said, getting a laugh. "By the following year, we figured out how to turn old people young. Twenty thirty-two was the dawn of immortality. We learned how to fully upload ourselves to the computer, and so if you stepped in front of a bus, you were backed up. Five years later, cryonics companies successfully managed to resuscitate and reanimate frozen bodies. In 2038, my parents, after a year of being reorientated, went on a second honeymoon. It was a cruise in outer space. Then, in 2041, the Singularity happened. And you all know what happened after that."

"We all became immortal," said Jackie Silver, the woman in the schoolgirl kilt who was aging backward. "Except for those who didn't bother to back themselves up."

Everyone applauded. I turned to Clay and asked for his impressions. "I was just thinking that everyone here seems so rich and afraid of death," he answered. "And then I thought, so am I."

Kekich poured little plastic shot glasses of Hennessy X.O. As they were being handed out, Clay started rapping about sippin' Hennessy while poppin' Ecstasy with "a couple naked chicks sittin' next to me." Kekich asked if anyone at the party came from out of state.

"I came from outer space," said a short man with male-pattern baldness and long, stringy, gray hair.

"What's your name?" Kekich smiled benevolently.

"Narayan de Vera, MD." He was wearing a maroon tunic covered in beads and mirrors. "You may also address me as Chief Olomayana, the Blessed One."

"Okay, Blessed One. Are you from somewhere close by in space?"

"Five light-years away," Narayan boasted. "From the solar system Zaikunta. I had to precipitate into a physical body to attend the party."

"Do you do that often?" someone asked.

"Every twenty thousand years." He took the microphone and began speaking about the birth of planets and moons, about the nature of elemental particles, and about the earth's thousand-year-long period of weightlessness (which is how the pyramids were built). He'd discovered a plant, he said, called *Kalanchoe daigremontiana bryophyllum,* that could reduce mortality rates by 95 percent. Still, long life doesn't eliminate poverty, and poor people in 2068 were simply poor for a much longer time. "So money is important," he added, making an appeal for funding. He ended his monologue with an aside about how he was getting sentimental remembering the days when we actually had physical bodies. Then he poured some of his cognac onto the plant next to him.

Jenna and Clay were looking at him with their mouths open.

"Okay!" said Kekich, getting back on the microphone. "Anybody else from out of state?"

A couple wearing matching blue capes lifted their hands. Beatrix and Jose said they'd traveled five hundred billion warped light-years to come to the party.

"And where did you come from?" Kekich asked.

"We flew here from the Galaxy Final," Jose said.

"Using our solar capes," Beatrix added, twirling the fabric around. "We belong to a civilization that knows everything."

"What's it like over there?" Narayan inquired. "Has the Singularity happened?"

"Nothing is the way we thought it would be," Beatrix answered.

22

Refrigerator Heaven

Once you had gone so far, you might as well test the limits . . .
and all but sleep in the waves.

Annie Dillard, *The Writing Life*

I stay under glass
I look through my window so bright . . .
I see the bright and hollow sky
Over the city's ripped backsides

Iggy Pop, "The Passenger"

THE SINGULARITARIAN vision of the future is not so singular. In fact, it's eerily akin to one proposed in the 1960s by another proto-immortalist movement: cryonics. Although largely ignored today, the "freeze-wait-reanimate" cult also made a faith out of science.

They were concerned less with the feasibility of attaining physical immortality (a certain eventuality, in their view) than with developing the means of storing frozen bodies until science had figured out how to make people immortal—by which time it would also be a cinch to resuscitate properly preserved corpses and heal them of whatever caused their terminal condition.

Today, when a cryonicist dies, others in their community don't see them as dead. They've simply "deanimated." Once frozen, they rest in indefinite suspension, awaiting reanimation. Diehards consider the brains of those in metabolic arrest to be "alive." To them, dead bodies are actually "temporary incurables," patients to be stored in a deep freeze until eternal life arrives. Sooner or later, cryonicists are convinced, biomedical specialists will be capable of curing any disease or illness. This isn't a fantasy, cryo-activists say. It's progress. We're nearly there. Imagine if we could just be transported by ambulance to a point in time when all injuries can be repaired and undone. This fast-forwarding ambulance is precisely how the movement's pioneers view cryopreservation.

Many prominent figures in the contemporary immortalist scene—de Grey, Kurzweil, Minsky, the Mores—are signed up to enter biostasis when pronounced "legally dead." They see this as a fallback plan in case something happens before the Singularity occurs. But not all cryonicists see eye to eye with radical life extensionists.

In fact, Ben Best, the president of the Cryonics Institute from 2003 to 2012, has suggested immortalists reconsider the desire for *eternal* life. "Forever is not just a long time," he points out quite sensibly, "it is eternity and therefore beyond realistic conception." Rather than strive for the unrealistic goal of immortality, Best suggests aiming for one thousand years—a more reasonable and, in his view, altogether attainable ambition.

Before his deanimation in 2011, the movement's creator, Robert Ettinger, also criticized modern campaigns for ending aging. The primary message of Aubrey de Grey's book, Ettinger emphasized, is "send-me-money." That didn't necessarily bother Ettinger: "I hope the appeal is successful." But as a more seasoned futurist (and mad scientist), he did take issue with the lack of calculations to support de Grey's guesstimate that "there is a fifty-fifty chance of a large improvement in longevity within thirty years."

Despite these gripes, there isn't much of a distinction between the two denominations—at least when it comes to their abuses of verifiable science. For example, Ben Best states that reversible suspended animation of the brain "could happen in anything from ten to fifty years." Whatever calculations one uses to arrive at that formulation, it's still just a random, magical prediction.

There are other points of overlap. Like the 2068ers, cryonicists, too, dream of putting human brains into machine bodies. A robot twin will likely be needed in cases where a cadaver is too damaged to be reconstructed. And Ettinger, well before Kurzweil's time, also breezily brushed aside the complications of consciousness: long-term memories, Ettinger once argued, consist merely of "changes in protein molecules in the brain cells." But if it's that simple, then why can't we cure Alzheimer's yet?

The primary goal of Ettinger's books, like de Grey's, is to hasten the advent of a bizarro sci-fi future. In 1962, he claimed that, by the year 2000, we'd all have Dick Tracy–style wrist radios and shuttle around the globe on transcontinental supersonic subways. Cities would be covered with retractable roofs, and most homes would have voting machine attachments built into their TV sets, doing away with the hassles of dimpled or dangling chads. The thawed supermen of the new millennium would live lives of plentiful leisure alongside their resuscitated superwomen lovers, whose updated bodies would have "cleverly designed orifices of various kinds, like a wriggly Swiss cheese, but shapelier and more fragrant."

And just as the Singularity's message has been disseminated through the books of Kurzweil and de Grey,* the cryonics conception grew following the publication, in 1962, of two books: *The Prospect of Immortality* by Robert Ettinger, and *Immortality: Scientifically, Physically, Now* by Evan Cooper, using the nom de plume Nathan Duhring. (N. Duhring: say it out loud.) Cooper formed a group called the Life Extension Society (LES), but didn't get cryopreserved as he became a sailing enthusiast and ended up lost at sea (LAS).

Ettinger is better known, largely because he went beyond ideas and opened an actual cryonics institute in Detroit in 1978. The first person he preserved was his mother. The second inductee, a full decade later, was his first wife, Elaine. Bit by bit, more bodies came on board. His second wife, Mae, joined the fray in 2000.

What will happen when he and both his wives are revived and rejuvenated? "My standard answer is a reminder about the old saying: 'The rich have their problems and the poor have their problems, but the rich have a better class of problems,'" Ettinger explained, shortly before he, too, entered biostasis. "I would consider that a very high class problem." Would Elaine view it the same way? Perhaps, he conceded, neither she nor Mae will want him. Perhaps being reanimated will be an excruciating ordeal and they'll all wish they'd never been reborn. Then again, perhaps both wives will defrost into the aromatic sexual Emmentalers of his superdreams.

---

*Books! How telling that technoutopians still have to resort to something as archaic as *writing books,* even with eternal life around the corner.

When Lenin died in 1924, his body was initially frozen. The Bolsheviks charged with preserving their deceased leader believed they would soon be capable of regenerating cadavers. At that point, following the revolution, anything seemed possible. If the czar himself could be deposed, then why shouldn't we live forever? Early Soviet theorists spoke of engineering "the defeat of death." At the very least, Lenin could repose on ice until technology revived him. Alas, in the end, their proto-cryonic attempt failed when Lenin started decomposing. He ended up embalmed and remains on public display in Moscow's Red Square.

There are other cryo-precedents, both in history and in mythology. Dante's *Inferno* portrays some of hell's denizens as being stuck—forever—in a lake of ice. Eternal, immobile life. Francis Bacon, that pioneering longevity researcher, died of pneumonia seemingly due to complications stemming from his investigations into preserving meat in snow and ice. Unfortunately, despite his attempts to answer nature's most "vermiculate questions," he couldn't preserve himself. But ice is such a good agent of conservation that explorers who've stumbled upon woolly mammoths locked in icebergs for millennia have actually cooked the prehistoric flesh and eaten it.

The first American to be cryonically preserved, a Californian named James Bedford, died of kidney cancer on January 12, 1967. (Cryonicists celebrate the anniversary of his passing each year on "Bedford Day.") Although the precise details of his preservation aren't clear, he appears to have been injected with a dimethyl sulfoxide solution, then frozen with dry ice, packed into a minus-196-degree-centigrade capsule, and transported to Arizona, where he ended up in a makeshift subzero crypt facility.

Over the next two years, nine other cryonauts took the icy plunge. Until the 1980s most cryonic activity took place in California and Arizona. Throughout those years, an advertisement trumpeted "Life Extension Through Cryonic Suspension" on Spruce Street, in Berkeley. A wigmaker named Ed Hope set up Cryo-Care Capsules in Phoenix, where he concocted Bedford's liquid-nitrogen storage chamber. And the Cryonics Society of California, which worked with Hope to preserve Bedford, rose to national prominence following a glowing report in *Life* magazine.

The society's president, a Santa Monica television repairman calling himself Robert F. Nelson, published a tell-all memoir of the Bedford case, *We Froze the First Man* ("the startling true story of the first great step toward human immortality"). Others allege that much of the book is fabricated. His cryo-methods were suspicious at best.

Convinced by Ettinger's book that cryonics would change the world, Nelson became a vocal crusader for the movement. In 1969, he told *Cryonics Reports* magazine that he'd opened the "world's first cryotorium," a temperature-controlled warehouse where bodies were stored in individual pods with viewing windows. Up to twenty pods fit into large containers retrievable by stainless steel cables. The article depicted lab-coat-wearing technicians overseeing panels of gauges and dials.

In reality, Nelson surreptitiously preserved his first four patients in a cryo-tank designed to contain a single body. "We put this one in headfirst, that one in feetfirst," recalled a mortuary technician enlisted to help. "It didn't look like there was room, but they fit." The metal cylinder was stored in the mortician's garage, until Nelson built a concrete vault underneath a cemetery in Chatsworth, just north of Los Angeles.

Each week, he'd haul hundreds of pounds of dry ice out to the storage facility in his Porsche. ("It really ruined the upholstery.") Soon he had four more corpses, again packed into a single canister. Fresh though the cryonics concept seemed, its realization couldn't have been messier. Nelson shipped one body across the country without informing the moving company that they were transporting a cadaver. In the summer of 1979, the entire scheme fell apart when some relatives checked in on their investment. The bodies were disinterred and found to be decomposed. Investigative journalists broke into the mausoleum. "The stench near the crypt is disarming, strips away all defenses, spins the stomach into a thousand dizzying somersaults," wrote one reporter. Another witness observed that the bodies had "sludged down into what I can best describe as a kind of a black goo."

The deterioration wasn't his responsibility, Nelson protested. He hadn't promised anyone anything. "They were told they would be frozen for a period of time," he pleaded. "Five minutes is a period of time."

Today, America has two main cryonic facilities. One is the Cryonics Institute (CI), in Detroit, founded by Robert Ettinger. The other is

Alcor, based in Scottsdale, Arizona.* Each has slightly over one hundred cryopatients, as well as dozens of frozen pets. Alcor is the more expensive of the two, with preservation there costing more than double CI's approximately $30,000. Most people who complete cryonics paperwork are encouraged to take out life insurance policies with the facility of their choice as the beneficiary.

There's one main difference between the two businesses. The CI places whole bodies in cryonic suspension, whereas more than three-quarters of Alcor's clients are neuropatients—just heads. These are known colloquially as neuros or discorps. While the popular myth that Walt Disney's head is frozen isn't true, baseball star Ted Williams's head (and, separately, his body) did in fact enter Alcor's containers in 2002.**

The distinction is more nuanced than just severed heads versus whole bodies. The CI, too, only *cryoprotects* patients' heads, and not their entire bodies. In other words, they store the whole body, but they only take measures to prevent damage to the head. As Alcor's directors often point out, you can't just freeze a dead body without ramifications—when the liquid in human tissue dips into subzero temperatures, it expands. Spiky ice crystals form. Cell membranes decimate. Organs rupture.

A possible way around full-body frostbite is a technique called vitrification, in which bodily fluids are replaced with cryoprotectants. Through perfusion, one's blood vessels are filled with antifreeze, such as glycerin, dimethyl sulfoxide, or M22. Both main cryonics institutions perfuse the head; Alcor then (in 75 percent of cases) cremates the rest of the body; the CI keeps it attached and sinks the whole organism into liquid nitrogen. But, in Alcor's view, it's a "senseless waste of time to take along a hundred pounds of peripherals."

The head is emphasized because most cryonicists assume that properly preserved brains will eventually be deposited into surrogate bodies. Even though a phalanx of uncredentialed researchers are exploring means of avoiding crystallization, there's (obviously) no evidence sug-

---

*It launched in California in 1972 as the Alcor Society for Solid State Hypothermia, but switched states in 1994 due to regulatory constraints, legal trouble, and earthquake concerns.

**In the 1970s, Father Gervais met Ted Williams at an airport when they were in line for the same flight. "It's really you, isn't it?" Gervais asked him. "Yes," he said, laughing, "it's really me."

gesting that a brain shrunken down with windshield-wiper fluid will ever function again. And the idea of the brain's being the only body part we'd need to resuscitate is cartoonishly Cartesian. What of the endocrine system? Our organs? Our other peripherals?

Badly wounded as a soldier in World War II, Ettinger spent four years convalescing in hospital, where he read about biologists freezing frog semen with glycerol and then reviving it.* Didn't anyone else realize that the same must apply to human bodies? If stored in liquid nitrogen, he came to believe, human bodies could be preserved for centuries. By that time, naturally, immortality would be ours for the taking.

When the cryonics hypothesis first appeared, pundits likened the feasibility of reviving a frozen person to that of reconstituting a cow from hamburger. To this day, only a tiny number of people have chosen to pursue cryonics, causing insiders to bemoan their collective failure. Most outsiders would agree with that assessment. "It would be going too far to say that stranger things have happened, because they haven't," writes historian Jill Lepore, deeming the hope of reanimating dead bodies to be one of the "sorriest ideas of a godforsaken and alienated modernity."

I wouldn't go that far. Though it's an obvious target for ridicule, the cryonics narrative has value as an example of the ways humanity copes with death. Though its prospects may seem limited, people are still signing up. Larry King announced in 2011, "I want to be frozen on the hope that they'll find whatever I died of and bring me back." When *The Prospect of Immortality* came out, its arguments convinced a number of otherwise respectable people to do the same. Stanley Kubrick became a convert, for a spell, although he eventually opted to be buried under a tree in Hertfordshire. Such figures as Peter Sellers, Gore Vidal, and Muhammad Ali were possible cryo-candidates. Timothy Leary went so far as to sign up with an outfit called Biopreservation. (In the end, after being cremated, Leary's remains were launched into spatial orbit. Cryonicists lamented this as a sorry "capitulation to 'deathist' thinking.")

"Besides being definitely feasible, the freezer-centered society is highly desirable, and in any case nearly inevitable," wrote Ettinger, ral-

* The scientifically legitimate field of cryogenics—studying low temperatures in chemistry and engineering—has long been at pains to distance itself from cryonics. It's one thing to store germplasm; reviving deceased bodies is an altogether different aim.

lying adverbs for his utopian vision. "Since there is going to be a freezer program anyway, and since the frozen will share the immortality of their descendants, the rationale of opposition, if there ever was any, evaporates." Reading his words today, one can't help but marvel at his untrammeled conviction. "At first a few, and then mounting numbers will choose freezing, and before long only a few eccentrics will insist on their right to rot," he concluded triumphantly.

Ettinger felt that cryonics would bring about the end of warfare. Why would any enemy of American interests risk "a fabulous life of thousands of years (including personal wealth eventually exceeding the total assets of the world today) for a moth-eaten bag of slogans and a shabby empire?" Rather than some mere despot, he will, after being frozen and thawed, become the owner of eternal life. "Once he understands this, he dare not risk war," Ettinger portended.

Doubleday purchased the rights to *The Prospect of Immortality* after Isaac Asimov told the editors he found its premise reasonable. Asimov himself, however, wasn't a convert. He considered the concept "unnatural" and spoke out against it in later years. What's noteworthy here is that a *science-fiction* writer validated the manuscript—not a *scientist*. As learned as Asimov may have been, there's a difference between sci-fi and sci-ence, even though cryonicists still tout the fact that Ettinger's manuscript "passed scientific review" with Asimov.*

The blurriness befits the entire endeavor. Ettinger began his writing career contributing sci-fi tales** to pulpy magazines such as *Startling Stories* and *Thrilling Wonder Stories*. In one autobiographical sketch, Ettinger notes that he "grew up just assuming that, in the natural course of events, we would learn how to prevent and cure senescence and all other diseases." This assumption, he explains in the same sentence, derived from his reading of sci-fi tales in periodicals such as *Amazing Stories,* starting in the 1920s.

---

*Alcor offered sci-fi great Frederik Pohl a free cryopreservation in 1979. He told them he'd think about it, but then never responded to their attempts to contact him again. More recently, when one cryonicist approached Ray Bradbury at a public event, the writer made it clear he "was anxious to terminate the encounter as soon as it had begun."

**Scientology's founder, L. Ron Hubbard, also began as a sci-fi fantasy writer. His extrapolations were so persuasive, they attracted millions of followers around the world. Immortal life plays a role in Dianetics as well. Members of Scientology's Sea Org (the corps of insiders who oversee the church's functioning) sign billion-year contracts, as they are expected to pledge allegiance to the organization for at least that long.

He's not the only immortalist who came of age confusing science fiction with actual science. The *New York Times Magazine* recently profiled a Japanese jellyfish scientist named Shin Kubota who believes "it will be easy to solve the mystery of immortality and apply ultimate life to human beings." That he hasn't gotten anywhere near doing that hasn't dampened his desire to "become miracle immortal man." A quick glance at his education sheds some light on how he formed his ideas. In school, he confides to the magazine, "I didn't study. I only read science fiction."

One surefire way to speak with cryonics leaders is to attend the annual open house at either the CI in Detroit or Alcor in Arizona. Journalists, however, aren't welcome. Understandably, given the derision with which they're regularly treated, cryonicists have an antimedia policy. Another means of getting an interview is to approach them as a prospective customer, someone contemplating taking out a membership. This was Stanley Kubrick's method. He met with Ettinger to inquire about joining CI when he was designing the hibernation chambers for *2001: A Space Odyssey*.

At first, I thought I'd have to attempt passing myself off as merely cryo-curious as well. But then, after a chance meeting, it turned out I wouldn't need to cold-call the CI.

At an art opening, some friends introduced me to a filmmaker named Korbett Matthews. He'd directed a documentary about a cryonics advocate named Frank Cole. Cole, a Canadian filmmaker, had crossed the Sahara alone by camel in 1990, with a Bolex camera. After spending ten years editing the footage into a film he intended to call *Life Without Death,* Cole returned to the Sahara to shoot some B-roll. This time, Tuareg bandits murdered him. His remains were shipped home and ended up in the CI, as per his final wishes. Korbett's film interspersed Cole's unfinished footage with his own interviews with relatives and peers—and with Ben Best, president of the CI.

Korbett's connection to Cole had granted him unfettered access to the institute, and he said he'd be happy to help me however he could. In fact, he hadn't yet mailed them copies of his finished film. It would probably help if I dropped off some DVDs when I visited them.

Not long after our first encounter, Korbett and I met up again, and he showed me what he'd filmed at the CI. The images looked as if they'd been shot at dusk. "There's something creepy about the neon-nitrogen ambience," he explained. "They really don't want natural light in there."

Ben Best appeared lanky and gaunt. Bald on top, he had an unusual V-shape of baldness razing down the thin, dark hair on his cranium's backside. It seemed like a mark of rank from some other society. An older man, in his midsixties, Best spoke softly and precisely. "To triumph over death," he intoned, smiling lustily, his spine lengthening, "you need to get very close to death." His eyes widened and flitted about with a kind of canine excitement.

Korbett turned to me, mimicking Best, and straightened his mouth into a line, tautened his jowls in a Frankensteinian way, and shifted his close-slit, beady eyes around suspiciously. "These people are terrified of dying," he explained. "So they turn themselves into human Popsicles. I call them humansicles. It's all just a fear of death."

"Or a denial of death," I added.

"Yes! Denial. But denying death is like denying birth."

"They would say that we're denying the reality of immortality with our deathist thinking."

On the screen, Best shuffled around the grim facility he presided over. I'd read stories comparing him to Igor. While he did seem to have a bit of a stoop, he came across more like an overimaginative Asperger's kid who'd created his own world and, as he aged, surrounded himself with frozen corpses and a select cadre of like-minded outcasts.

He did have a slightly robotic quality in his intonation, as though already preparing for the eventuality of a machine body. His intention may have been to create the impression of an evolved being devoid of emotions, but the actual effect was of a solitary eccentric who spent much of his time at the CI. Not that he cared what we nonenlightened normies thought of him. As he once wrote, "Long before I heard of cryonics I had a rich fantasy life and I would imagine myself happily surviving alone after the rest of Mankind had passed from the scene and the planet Earth had been turned to volcanoes and fire." As different as he may have been, he was the perfect person to speak to.

"Something occurred to me when we filmed this," Korbett said. "If you were actually to live forever, these are the last people in the world you'd want to hang out with.* They're really not fun, they're socially

---

* The former president of Alcor Mike Darwin has complained about the movement's penchant for attracting "dysfunctional and sometimes frankly sociopathic personalities as members."

inept, and they look so freaky from all the caloric restriction. Seriously, look at the dude: he's got zero style."

Korbett didn't think it would be too hard for me to get an interview. He offered to e-mail Best for me. "What should I say?"

"Tell him the truth: that I'm curious about cryonics, and that I'm writing a book about people's relationships with the idea of immortality."

Korbett looked at me. "Are you personally interested in immortality?"

"No." I shook my head. "At least, I don't think so. But the idea of it interests me, in its many permutations, the way it helps us ignore and transform reality."

"How so?"

"You and I could both die at any point in the next five seconds. We all have a constant proximity to death, which we can't understand. I guess I'm exploring how people grapple with that by telling stories, by creating narratives. It's universal. Cryonics is a story about overcoming death. And any story about immortality is really a story about what dying means."

The next day, Korbett sent Ben an introductory note about my upcoming trip to Detroit, adding that I was writing a new book on the scientific quest for immortality. "He will stop by the CI to say hello and drop off the long overdue films. All the best from Montreal."

"Okay, Korbett, thanks for letting me know what to expect," Best responded.

We agreed to meet the following Monday.

I arrived in Michigan on a clement morning—nothing like that summer, when it got so hot in Detroit that steam poured from sewer tops because the streets were melting. Still, the city felt as apocalyptic as ever. Homeless zombies teetered around, swilling forties of malt liquor. Abandoned houses abounded, their roofs burned off.

Compared to those skeletal dwellings, the Cryonics Institute's building looked fully innocuous. I found it in a typical industrial zone in Clinton Township. The exterior had something of a Frank Lloyd Wright vibe. Heavy curtains shrouded the windows. Under the CI sign, smaller letters spelled out ERFURT RUNKEL BUILDING (in honor of two perfectly named donors). In short, nothing to make anyone suspect that over a hundred bodies rest within. But, then, abnormal things do have a way of appearing overtly normal.

I rang the bell.

Ben Best unlocked the door and blinked out at me, not returning my smile. "State your purpose," he intoned gruffly.

An abrupt greeting, but understandable, given the CI's marginality. It felt less full-on rude than majestically nerdy, textbook *Star Trek*. An interloper had approached the fortress; precautionary communications were required.

"It's twofold," I replied, unfazed. I, too, had been in chess club and could play the part easily.

He stared at me, bug-eyed.

"First, to deliver the DVDs Korbett mentioned in his letter." I handed over the package. "And second, to speak with you about cryonics, for the book I'm writing about immortality." I held up my reporter's notebook.

He considered the two purposes momentarily, then opened the door wider. "Okay," he muttered. "Enter."

It smelled like sweaters and brine in there, with a hint of onion powder added to the mix. As we moved through the dim light, Best's manner shifted from standoffish to his more usual emotionless neutrality.

He gave me a tour. We started in a boardroom lined with framed sepia-tone photographs of the inductees. Here was Ettinger's mother, his wives, a number of nebbish, bespectacled men. It felt like perusing a yearbook, not of high school students but of people who'd graduated to the beyond.

We moved into the main storage area, where a grove of white, ceiling-high thermoses stood huddled together under the fluorescent lights.

I looked at him and pointed at them. "So that's where . . ."

He nodded.

The white fiberglass canisters are called cryostats. Arizona's Alcor, favoring stainless-steel exteriors, uses the designation *big-foot dewar,* in honor of vacuum-flask inventor James Dewar. Essentially oversize thermoses, these large cylinders maintain a constant internal temperature. But instead of keeping coffee warm, these hold corpses—six per tank—at liquid-nitrogen levels hundreds of degrees below zero.

Bodies are fitted into each container headfirst. The technical term for this is *racking the body*. The rationale behind lowering them in isn't just extra weirdness for weirdness's sake: being stored upside down permits the brain's temperature to remain as steady as possible during the requisite refills. Feet handle the fluctuations better. As I made some notes, Best told me the single oddest thing I learned about cryonics: that each body is placed into a sleeping bag before being slotted into place. There they snooze, camped out around the frigid bonfire of a better tomorrow.

As Best described the intricacies of cryostats, I couldn't help but imagine a half dozen clowns trying to squeeze into one of those miniature smart cars. The CI's aesthetic came off like a budget, outdated version of futurism; less iPod sleekness and more Atomic Age plastic flimsiness. Clusters of crumbs and lint had gathered in the room's corners.

In front of the thermoses, on the floor, were small, white cubicles: places to keep loved ones' mementos. They were empty for the most part—here a dried rose, there a card. I felt wings of depression flapping around me.

Best led me deeper into the room, to the rear of the facility, where we paused next to a boxlike chamber. I climbed onto its side and took a glimpse inside. An army-green mattress lay on the floor. I wasn't too sure what I saw: spikes, meat hooks, a dark, deep stain. They prepare dead bodies here before putting them in the sleeping bags: "the customer's thorax is cut open, the lungs' blood vessels are separated from the heart, and the brain is connected directly to the heart. . . ."

Best suggested I climb in.

Did he have a leer on his face? I could just imagine it: the cryonicist's revenge. Perfusing the skeptical journalist—the precise media type who had mocked the movement so many times. Yet posterity would ultimately show it had been for the young man's benefit! When resuscitated as an immortal decades later, he would certainly thank Ben Best for having saved him.

Or maybe I was just being paranoid. Being in a room full of frozen dead bodies can have that effect. Either way, I declined his invitation.

He brought me to a smaller, waist-high cryostat. "For pets," he clarified. Dogs, cats, parakeets, and other animals were in there. Just under one hundred frozen pets are preserved at the CI. He lifted the container's lid, and clouds of dry-ice mist billowed out. He waved his hand over the smoke, a warlock tending his cauldron, and pointed out a freshly stored cat on top.

He said something about how, if it comes back to life, he'd wrap it properly.

I wasn't sure I heard correctly and couldn't get the quote exactly because, when I asked him to repeat it, he brushed me off with a rapid-fire, bravura display of liquid-nitrogen handling. He placed both hands into the broth and brought out a handful. White fog curled from his palms for a moment. When the chill became too much, he flung the contents onto the floor, where they sizzled and fizzled away before our eyes for a few moments. He wiped his hands on his pants. It had singed him, but not badly, and he seemed to relish the pain.

I felt a pressing urge to leave. I hadn't spent much time there—fifteen minutes, maybe. But I'd seen all I needed to see. As we turned to head out, he showed me a chart on a computer, some research he'd been developing regarding freezing levels. My heart went out to him. He was sharing his work with me, unworthy though I may have been. In that gesture, his vulnerability, his humanity, shone through. I could see how, as with the 2068ers, Best's greatest passion in life, the thing that makes life make sense, is immortality. How strange that those who want life to never end seem so removed from it already. Perhaps it's a sense of benevolent usefulness that keeps him going, the hope that, even if he's misunderstood today, the defrosted superpeople of the future will be grateful for his sacrifices and toil.

The CI's competitors claim that, until Best took over as president in 2003, "even the simplest and most basic parameters of patient care" were not monitored or documented. He changed that, but it didn't change cryonics. His graph's plot points may have appeared scientific-like to a believer, but so did those in that Islam book the Sunni librarian lent me. Defrosting cats to correlate prospects of humans' revival is not science: it's somewhere between a bizarre hobby and a ritual activity that helps lessen death's incomprehensible sting. Oughtn't we know by now, a half century into the cryonics experiment, whether dead beings can be resuscitated? The answer, at least among those who aren't cryonics faithful, remains resoundingly unaffirmative. Despite the veneer of similitude, a belief system—a form of magical thinking—does not science make.

The entire logic-stretching enterprise of cryonics requires multiple leaps of faith. The first stretch is believing that one day we'll be able to bring dead bodies back to life. The second is that the only type of dead bodies we will be able to reanimate are those that were vitrified (rather than those that were merely heaped into the earth and covered with clods of turf). One also has to buy into the notion that medical science will one day be able to cure all illnesses. This is where the immortalist thinking comes in. For in that medical utopia over the horizon, we will also know how to make people live forever, thereby justifying a return from the frigid hinterlands.

"Fact: at very low temperatures it is possible, *right now,* to preserve dead people with essentially no deterioration, indefinitely," wrote Ettinger, in *The Prospect of Immortality*. But postmortem deterioration isn't the point: the point is that dead people are dead people. They aren't "patients" tasting "the wine of centuries unborn." They're cadavers. How we dispose of them—or hold on to them—is our choice. To be capable of perfusing craniums in no way implies that we'll be able to regenerate them. Only believers could conflate cryopreservation with immortality. Storage of bodies is not eternal life.

Cryonics isn't that different from mummification or sky burial. Other American options include becoming fireworks or artificial diamonds. Cremated human ash can be transformed into lead and inserted into pencils. Our remains can be melted down, formed into bullets, and fired into the night. They can be inflated into balloons and floated over the Salton Sea. They can be mixed with bird food and dispersed by sparrows or tossed into concrete mixing bowls and paved into garden paths or sucked into airport ventilation systems and blown into the ether or shaped into man-made coral reefs and submerged beneath the ocean. Or we can choose to float upside down in vertical repose among our fellow ice dreamers.

While waiting for Best to write up a receipt for the souvenir CI baseball cap I picked up on my way out, I noticed a plaque on the wall of his office. It was their cemetery license. *That's it,* I realized, with a start, *a cemetery*.

Ettinger once spoke of the cryonic movement as a "mighty undertaking," a double entendre even more fitting now that I understood that the so-called "temporary incurables in suspended animation" are simply "dead people" buried on ice rather than in subterranean coffins. They'd deanimated, sure, but they're undisposed-of cadavers, nothing more. The CI is a long-term morgue. And Best isn't so much an Igor figure as a mortuary director.

When he handed me the invoice, he also offered me an autographed copy of Ettinger's final, self-published book, *YOUNIVERSE: Toward a*

Self-Centered Philosophy. The book describes itself as an attempt at "unriddling the universe, or at least the innerverse." Whether it is successful or not depends on one's attitude toward death, but it certainly attempts to tackle some weighty philosophical questions.

How did the universe originate? "Nobody has a clue. One guess is that *our* 'universe' is a simulation in some joker's computer."

What is life? "A disease of matter, a skin condition of the planet."

Is physical immortality possible? Absolutely. "Aging and death can be conquered—not only for the species, but for individuals now living. Duration of your life need have no limit—even if you die next month."

Regarding the works of Shakespeare, with their "puny" language and "weak" intellect, Ettinger argues that once we attain immortality, they will "interest us no more than the grunting of swine in a wallow." He saw himself as the movement's lyricist: "Blue is for violet, Red is for rose; After you croak, I hope you'll be froze."

One of the most disconcerting sections of the book describes his wife Mae's deanimation. They were living in Arizona at the time, and Ettinger himself pronounced death, mere seconds after she'd stopped breathing. He immediately set about operating on her to get her ready for the deep freeze. True cryonic love.

Different states have different rules, which is part of the reason Alcor moved from California to Arizona. In 1988, Alcor was subjected to a protracted legal battle after cryopreserving a woman named Dora Kent who hadn't been pronounced legally dead. The coroner's office claimed she'd been murdered. "How dead is dead enough?" asked the defendants. In the end, no charges were laid.

In Michigan, Ettinger wrote, "You are supposed to wait for the paramedics to pronounce death." The subtext to that "are supposed to" suggests "shouldn't have to." Indeed, until his own death, he had been a vocal opponent of what's called *legal* death, which is when authorized physicians pronounce a person dead. Ettinger called this "death by jabberwocky," meaning totally ridiculous. Waiting to get "snuffed-out by a fundamentalist physician," he fumed, can interfere with cryopreservation. The worst eventuality is having an autopsy, which eliminates or reduces chances for a successful perfusion.

The best time to intervene, cryonicists agree, is *before* death, particularly with terminally ill people. That way the disease can be paused until the body is pulled out, revived, and cured. Unfortunately, the meddlesome legal system doesn't allow for that. Soon things will change, Ettinger prophesied, for "who would not trade a few declining years in the present for a larger number of more active and rewarding years in the future?"

Even though *YOUNIVERSE* is addressed to a hypothetical "dear potential customer," its core teaching is summed up in its subtitle's aim: *Toward a Self-Centered Philosophy*. Self-centered. The dedication says it all: "I dedicate this book to myself or selves." It's pure narcissism, the mind-set in which we place ourselves above Shakespeare and tell people we've solved death and charge them tens of thousands of dollars to access immortality. Still, his vision offered others a novel means of contemplating death. Now that he has passed on, and life goes on, the most revealing thing he ever wrote can be laid to rest in the cryostat of history: "If the Tiger of Death eats you, that is the ultimate tragedy; that is when the world ends, when the cosmos disappears, when Everything becomes Nothing."

Best signed me up to a cryonics society e-mail list before I left Detroit. For the next few months, I looked in on their chats. The members, united by their interest in being frozen, displayed community spirit. Those with homes near cryonics facilities willingly let strangers spend the night so they could see the technology firsthand. They held potlucks together. They celebrated Bedford Day. They debated the ins and outs of vitrification. They recommended substrates made of carbon nanotubes from general nano assemblers. They spoke of how biologically based humans will become superfluous within decades.

At a certain point, I sent a "please unsubscribe" e-mail. The response came directly from Ben Best: "I have removed you."

23

Secret Santa Barbara

How will you find some madder adventure to cap this
coming down alive to Hades among the silly dead?

—Homer, *The Odyssey*

I have grown to love secrecy. It seems to be the one thing
that can make modern life mysterious or marvelous to us.
The commonest thing is delightful if one only hides it.

—Oscar Wilde, *The Picture of Dorian Gray*

I'D MET with physical immortalists. I'd been removed by cryonicists. Although they were all drawn to the fringes of science, it seemed to me that belief played a crucial role in everything they did. But I hadn't yet spent enough time interviewing scientists from Harvard and MIT. Surely they would be able to tell me clearly and honestly where we stand with regards to the possibilities of life extension and immortal life.

With that in mind, not long after my visit to the Cryonics Institute, I found myself driving up the coast to Santa Barbara on a curious tip from a woman who'd read my first book. Her name was Nicole Nicole (or so she identified herself in her e-mail message). She'd come across my research and knew of someone worth meeting: "He is the president of a foundation based in Carpinteria, CA, that spends millions on anti-aging research," Nicole wrote. "If you are looking for the fountain of youth, or for people looking for the fountain, he is your man. He must know something, because he looks forty and he's over sixty. He also has a habit of leaving weird letterpress cards with random information and instructions in strange places for people to find."

Mark Raymond Collins's website explains that he's hidden notes and images and $2 bills beneath tables in hotels and restaurants all over the world. He describes himself as a "consultant to experts." Alongside his longevity-research financing activities, he co-owns a vodka brand, RND Vodka ("five percent of profits go to science education"). The *Santa Barbara Independent* quoted him as saying he would love to make "vodka from stem cells." The *Montecito Journal* has written about his Mark Twain–inspired campaign to have the word *very* excised from the English language. He prefers other words. *Ergo,* for example. He considers *ergo* to be a "caviar of a word."

In a follow-up e-mail, with the subject heading "The fountain of youth=mark r collins," Nicole described him as "a very curious fellow" (a description that would have been just as apt without the adverbial emphasis). "He's a really trippy guy," she wrote, adding, "He is really into confessing things that some might find inappropriate." She provided examples, but I felt it more appropriate to let Collins himself decide what confessions he wanted to share.

Nicole arranged for the three of us to have dinner together. She and I met outside my hotel lobby and then headed to the restaurant. Nicole, in her twenties, gave a friendly, cultured, and stylish first impression. When asked about her double name, she confessed that her patronymic is actually Pierpont. I'd just driven down a street of that name. "Are you related to the Pierponts of Pierpont Avenue in Santa Barbara?" I inquired.

"No," she responded. "I'm related to the Pierponts who founded Yale and wrote 'Jingle Bells.'"

Nicole worked in film production, shooting what she called "death video diaries"—recordings that allow elderly, infirm, or dying people to communicate messages to their descendants. Previously, she'd been a barista in a nearby café, where she'd encountered Mark Collins. He came in for coffee often and chatted openly about his work administering antiaging funding. He'd also been a customer at her mom's antique store in nearby Summerland. "He bought a painting that looked like his son," she reminisced. "Then Oprah came in to buy some lion statues. She requested that we paint them black."

Mark Collins had booked us a table at Sly's, his favorite restaurant in Carpinteria. There was something sly about him as well. A healthy-looking, close-cropped gray-haired gentleman in a nice suit, Collins liked affecting a serious veneer that couldn't quite conceal his default amiability and bemused good cheer. It was immediately clear that Collins belonged to a different category of prolongevist than the immortalists I'd encountered at the 2068 party. Not only was he handsome, insouciant, and well adjusted—he also wasn't seeking physical immortality. "I reckon that I'm a skeptical optimist" was how he put it.

As president of the Glenn Foundation for Medical Research, Collins oversees a nine-figure endowment that fuels research into the biology of aging. They regularly give out five-year, $5 million grants to Harvard, MIT, Stanford, and the Salk Institute. The scientists working under their aegis are the most important biologists in America on aging. All four of these institutions' labs are named after Collins's boss, Paul F. Glenn, who amassed his fortune as an investor and venture capitalist. "Paul's successful," as Collins put it. "Suffice it to say, he's a wealthy man. He's active in his trading."

Glenn's consortium unites collaborative institutions working on advancing humanity's understanding of aging—not defeating aging. They work using the peer review process, Collins underlined, not with speculative theoreticians. The research funded thus far had led to breakthroughs in lengthening mice life spans. But, he cautioned, putting on a faux-pedantic air, "going from a mouse to a marmoset is a huge leap."

Nicole brought up his penchant for leaving scraps of interesting information in specific places around the world. Collins said that he'd recently hidden a photo of himself alongside Jonas Salk in the Salk Institute.

"What else do you hide?" I asked, hoping to find a scrap myself.

"You know, notes, ideas, pictures. It's another one of those ancient things that people do, like wanting to live forever. Whenever I stay in a hotel, I always write a little something in the Gideon Bibles.* Thirteen years ago in a castle that had thirteen rooms, I put a card in volume thirteen of a green leather-bound series of books. If you're interested, I'll send you clues to their whereabouts."

I said I'd love that, then steered the conversation back to immortality. He immediately interjected, "I want to reiterate that our foundation is not part of the immortalism movement. Wanting to live forever *is* an interesting psychological phenomenon, but I personally don't have

---

*Not long after our meeting, he started pasting QR-code stamps into the Bibles. They linked to his own website.

mortality anxiety. Vanity defeats itself. You start down the Botox slope and you never get off. You go Mach two." Here he pulled his face backward so his lips swelled into an imitation of a bad plastic-surgery job.

"Have you met people who want to live forever?" I queried.

"In my line of work?" he retorted ironically. "Our foundation runs antiaging research. I deal with everybody from run-of-the-mill, mainstream gerontologists to futurists who want to teleport themselves into machines. So, yes, immortalists solicit us all the time."

"What are they like?" Nicole demanded, wrinkling her nose slightly.

"They're the union of anxiety and narcissism. They're all terrified of dying, and they simultaneously can't hide their blatant self-motivation. They are unbelievably self-focused. It's all 'me me me, I want to live forever.'"

The waiter took our orders. Nicole opted for local abalone—a mollusk I'd never eaten before—which the three of us shared as an appetizer.

"I think denial is terrific," Collins remarked. "I'm a heavy user. Everybody uses denial until they have to deal with their health issues. That's when self-interest pops up. And I'm not against the selfish types—they *could* possibly come up with some solutions. It'll sort itself out in Darwinian fashion."

"Who are some of the selfish types?" I pressed.

"There's a bunch of them. And I have nothing bad to say about any of them. It's their message that bothers me. When Aubrey de Grey came to meet us, he started out by talking about death as an elective. I told him, 'Look, if you want to interest people in this area—whether you're just a *me me me* person or you genuinely want to benefit mankind—the first, most important thing is to have an unassailable mission statement. Something that no one is against."

"What would that be?"

"'Our goal is extending the healthful and productive years of life,'" he offered unhesitatingly. "What I always say is that we are trying to alleviate suffering and help public treasuries. That's unassailable. Nobody can contest those aims. Stopping death is another question entirely. When people are told they might be able to live forever, their eyebrows go up and then they go silent. That's where Aubrey and his gang go wrong: they take silence for approval."

On the topic of taking silence for approval, I relayed the contents of my conversation with de Grey about primate experimentation, including his highlighting of their inability to talk back.

"That ape colony idea is just really, really weird," Collins muttered. "What, are they going to be living coolers for organs? Or organ warmers? Or just used for experiments? It makes no sense. There'd for sure be some Ebola-level blowback, too."

"Every solution anyone comes up with always has unforeseen flaws and ramifications," I ventured.

Collins nodded as he took a sip of wine. "That's certainly the case in longevity research."

"Does anyone else find it strange that Aubrey is one of the oldest-looking people alive?" Nicole chimed in. "I mean, *the* face of staying young forever is a grizzled long-beard."

"He got attention 'cause he's fast on his feet and he looks like Rasputin," Collins affirmed. "He's a Roman candle who came out with six balls of flame. Plus even though he's surrounded by Raelians and other space people, he's very available for journalists. If you e-mail him, he'll answer you within five minutes, at any time of day or night." That corroborated de Grey's statement about doing interviews in his sleep.

The abalone appetizer arrived. As the waiter set the platter of pearlescent shells on the table, Collins grimaced. The dish was garnished with a sprig of mint. "I *hate* garnish," he snarled, removing it disdainfully. "No lettuce leaf, no orange twist, no little strawberry slices. I'm a no-bullshit guy. No garnish."

"You were speaking about immortalists," I reminded him.

"When you get to the hard-core nucleus of people obsessed with technical immortality," Collins went on, "it's a very small group. They're kind of culty."

I told them how attendees at the 2068 party had spoken about "it" happening, meaning attaining immortality.

"Yes!" chimed in Collins enthusiastically. "They always speak about 'it.' What *is* 'it'? They say things like, 'We are going to solve *it*.' That two-letter word is the shortest euphemism. Ha! If that isn't a fig leaf, I don't know what is. Does *it* mean not dying?"

"I think so, or finding a way to upload your mind onto a computer."

"Ah, yes, the Kurzweilians," Collins sighed.

"Is there no link between you and them?"

"There is a link," he allowed. "They're the distal end of the instrument."

"Meaning you are both part of the same general movement, but they're at a far remove?"

"I'm a living human." He pointed both hands toward himself, then extended them forward. "They are *transhumanists*. Kurzweil is intensely interested in an advanced form of pouring himself into the next version of himself, which could be a machine. He envisages making the transhumanist leap, whatever that is. I haven't set that as an objective, but who knows what's possible?"

"Would the Glenn Foundation get involved with Kurzweil?" I asked.

"Not on my watch," Collins declared curtly.

"Why not?"

"I don't know anyone doing scientific research into the field of transhumanism. Like maybe people really do know how to teleport and I just haven't seen the papers yet. Until then, I have no way of knowing whether they're onto something. There's a lot of greed and misplaced trust in the immortality world. Whenever people are vulnerable, it's a fertile territory for quackery. That's why we always bring up the unassailable mission statement. We're here to fight disease and help the treasury."

Our steaks arrived. By now, we'd had a couple of glasses of wine, and Collins seemed quite relaxed. He chided Nicole playfully for not calling him often enough. "Whenever you get the abalone urge, just give me a shout," he told her. It felt like an appropriate moment to bring up Paul F. Glenn, the mysterious head of the foundation.

The notion that human life spans are not fixed had fascinated Glenn since the early 1950s, Collins recounted. Glenn was inspired by an article in *Parade* magazine that suggested humans may one day be able to control aging through interventions on the molecular level. He's trying to bankroll that hope into a reality.

"Early on, people thought his interest in the field was odd," Collins conceded. "Any attempts to study aging were considered to be the pursuit of bio-alchemy. That changed as science started revealing how various aspects of aging work. Now, of course, there's widespread interest in longevity studies. Paul was ahead of his time."

"How is he perceived today?"

"As an augmenter. Until relatively recently, his vision wasn't seen as realizable. 'It' is now considered to some extent doable, and to some extent desirable. There's a spectrum of attitudes towards that."

I wondered what he meant by "it" in Paul's case. Was he speaking about prolonging life—or full immortality? What lay behind that unassailable mission statement? I tried broaching the question obliquely.

"If Paul could envision the best possible outcome of his financing of research, what would it be?"

"That's dicey to answer." There was a long pause. Then: "Death would be elective."

"Has he said that?" I asked casually.

"No. It's an inference. Paul doesn't say outrageous things."

"Do you think 'it' will happen?"

"What is 'it'?" Mark threw up his hands.

"What you just said: death as an elective. That you choose whether to die or not."

"I haven't any idea. Do I think it's possible? I don't even know. Some people are convinced it's probable, some think it's impossible *and* undesirable. There's no proof to be had. You're either running towards it, away from it, or standing still. I'm standing still."

"Do you think there will be increasingly effective interventions?"

"I'm standing still. We want to alleviate suffering and help public treasuries."

"Unassailable. But how will you help public treasuries? What does that mean?"

"It's very simple. Our foundation's aim is to help people live long illness-free lives. We want to counteract the slow descent into senility, to eliminate age-related diseases so you can live healthy and then die quickly. If you keep people out of old-age homes and hospitals, you save money. The country's aging population will result in treasury-breaking health-care costs unless something changes. We should be more worried about aging than dying. Growing old and infirm is the problem. What's to fear about being dead? It's the getting there that we're trying to affect."

After dinner, Nicole had an appointment in Los Angeles, so she suggested I meet up with some friends of hers that evening. A group of seven or eight of them were having drinks at a bar in Santa Barbara, and I decided to join them for a nightcap. Nicole's good friend Paris texted me with directions to a small country-western lounge.

When I arrived, Paris greeted me warmly and introduced me to the group, all in their midtwenties. Paris's Australian boyfriend, Nic (which he pronounced *Neeque*), looked a bit like Emilio Estevez, blond and beachy. Next to him sat another Aussie, Josh, whose raven locks were complemented by a psychoanalyst's mustache and goatee. His girlfriend's name was Sophie. A couple of others were with them as well, but I ended up spending most of my face time with that foursome. Paris explained to the group how Nicole had contacted me after reading my first book.

"Josh and I also work in books and film," said Nic.

When I asked in what capacity, Josh shot Nic a sharp look and deadpanned, "Sorry, we can't tell you."

I laughed, thinking they were joking. Neither of them smiled.

"Really?" I asked.

"Yes, really," insisted Josh solemnly.

"Hmm. Secretive," I suggested, not yet realizing how true that statement was. Nicole later explained that Nic and Josh were part of the creative team behind the self-help movement based on *The Secret*. The book, which solipsistically claims we just need to *want* things in order to get them, was written by an Australian woman, but her company works with a whole team, including Nic and Josh. Coincidentally, the two of them began as magicians doing street-magic performances in Australia.

"I used to do illusions for crowds," Nic told me. "It's all autosuggestion, sending thoughts elsewhere."

Magic, in its essence, is the basis of *The Secret*: before acting on something, we first desire it. If we don't believe it can come true, we rarely pursue it. But is it magical to believe that fate can be affected by our thoughts, or simply human hubris? What's the difference between destiny and the law of attraction? I tried to strike up a conversation with them about immortality.

"Im-mor-*tality*?" drawled Nic, nonchalantly feigning disinterest. "In Sufism, they spin to attain immortality. In fact, I'm pretty sure there's a certain direction that you have to spin in, in order to maximize the eternal benefits."

"Counterclockwise?" I offered, spinning my index finger.

"Depends on where you are," he reprimanded.

"And where you're measuring it from," I countered.

"It's all interrelated," Nic mused, clearly comfortable inhabiting enigmatic utterances. "It all corresponds, it all just is. It is all only one moment."

He asked if I'd come across the teachings of Mony Vital, PhD, a California-based breatharian who leads seminars on becoming a physical immortal. His book *Ageless Living: Freedom from the Culture of Death* contends that people just need to stop believing in death for death to become unreal. "To start with, you have to discard all death beliefs from your body. You have to deprogram your body: physically, mentally, and emotionally. You have to get the beliefs of death out of every cell."

I asked Nic if his "books and film" company had approached Mony Vital.

"It doesn't matter," Nic answered, gazing into the distance.

"But the company you work for is developing an immortality story?"

"No. Maybe. The company I work for is thinking of taking a TV-audience perspective on it."

"So maybe it's for television?" I said sarcastically.

"We don't do television." He folded his arms, relishing the confusion.

"But you just said—"

"It's an expression," he snapped. "I said a TV *audience*. See the difference? Not *TV*—for a TV *audience*."

"So whatever it is we're speaking about is aimed at a mass audience?"

"You could say that, sure."

"Moving pictures?" I attempted, not quite sure how this conversation had gotten so convoluted.

"Perhaps. Moving audio maybe."

At that point I lost patience. "Man, what a mindfuck."

Nic threw an arm around my shoulders and yelled out, *"Yes!"* as though someone else had finally understood. He then spoke about death as a belief. "The only reason I have to believe that I have an expiration date is that I've been told so. Other than that, how do we *know* that we'll die? We don't. We *believe* that we'll die."

"Does this relate to the work you do with your company?"

"No, my interest in immortality is personal. What we do at the company . . . It's about changing people's beliefs. Thoughts-to-feelings-to-beliefs is the underlying thing motivating us as people. And what's the ultimate dead belief? Death. That's the hardest one to uproot. Other than gravity."

He told me a story about how he knows immortality is real, a certainty stemming from in utero memories. "I remember being in the womb," as he put it. "It's like a memory or a reoccurring dream. That's the only way I can compare it to anything."

"What was it like in the womb?"

"Scaleless."

"Scaleless?"

"There was no scale for anything, no sensation of scale. There was no way of saying how to get from here to there. I remember a wheeling sensation. Everything moving: a scaleless, motionless motion."

He'd practiced meditation techniques that allow him to get back to that sensation. "Have you ever done Holosync?" he asked, pausing briefly for me to shake my head. "It's specially modulated drones of white noise. When you listen to it, it alters your brain waves to a subtle body frequency. Doing Holosync meditation made me feel the sensations of . . . *that*."

We drank more. Nic started doing coin tricks at the table. As the night drew to a close, I asked him if he believed that statement he'd made about death being a "dead belief."

"It was just something I said," he replied. "I don't know if it's a statement. I don't know if I believe it."

The following morning, I checked my e-mail and found a note from Mark Collins: "A coincidence, being introduced to you in a crowded intersection." He'd also sent a link to a press release about the Glenn Foundation's grant of $5 million to the Salk Institute. In the PDF, Collins himself spoke of wanting answers to "one of the most elusive questions in biology: is there a defined biological process of aging?" The scientists at the Salk would be studying stem cells' capacity to self-renew and differentiate into functioning cells, and why that activity decreases dramatically during aging. "The biology of aging underlies all the major human diseases," Collins was quoted as saying. "To understand the fundamental aging process and to intervene is to delay the onset of disease, to extend the healthful years of life and reduce costs to society."

"I think I see the unassailable parts in that press release," I responded, via e-mail. "PS Where's my clue?"

He replied with a photograph of the reading room at Thornbury Castle, a paneled chamber with two baronial chairs upholstered in red velvet and gold trim. Bound books lined the library shelves. "Look here," Collins wrote. "Volume XIII, p. 5931." Unfortunately, the castle was located in Gloucestershire, England. But he made an offer a little closer by: Was I available to meet Paul, at his home? "And if I can be of help with introductions to scientists seeking to understand the biology of aging, I'd be happy to."

Through him, a direct path led into the inner fields of scientific longevity research. I wrote that I'd gratefully take him up on Paul and those introductions.

I met Collins at his home, a large bungalow perched on a craggy hillside overlooking the ocean, so that we could drive to Paul Glenn's place together. The first thing I noticed upon entering was a sculpture of a torso in the living room. The house was full of flowers. Strains of classical music wafted through the air. "I leave the radio on to keep the flowers company," Collins said. Even though he'd just separated from his wife, he seemed in high spirits.

On the table lay a copy of *The Wine Advocate* magazine, with the Glenn Foundation–funded biology sensation David Sinclair on the cover. Sinclair's research into resveratrol at Harvard had led to the widely reported media story about red wine's being a source of longevity. The issue of *The Wine Advocate* had just come out that week, explained Collins, clicking through his iPhone to find me a text message from Sinclair. "Guess what's next?" Sinclair had written. "The *Sports Illustrated* Swimsuit Issue."

Collins's bookshelves were heavy with nonfiction hardcovers, as well as knickknacks, such as a leather hippopotamus and an abalone-shell bottle opener. He walked through the house, joking and openly discussing his own story as well as the world of longevity research. I'd never encountered such a peculiar blend of honesty and avuncularity in a subject before.

As we spoke about the previous night, it became apparent that he wasn't a fan of *The Secret*—"a lie on stilts," as he characterized it. "Marketing geniuses are sometimes rapacious cross-dressers, indistinguishable from altruists," he aphorized. It struck me then that Nicole had said something about his being a marketing person before he'd started working at the Glenn Foundation.

Collins then shared a quote from Mark Twain about the one serious purpose of writing: "the deriding of shams, the exposure of pretentious falsities, the laughing of stupid superstitions out of existence." He owned a copy of Kurt Vonnegut's high school yearbook (Shortridge High School, Indianapolis, class of 1940) and had even gotten it inscribed by the famous author: *Mark: charming rascal, sure to go far—Kurt Vonnegut.* The full name under Vonnegut's yearbook photo was Kurt Snarfield Vonnegut. "To snarf," Collins elucidated, "means to sniff a bicycle seat."

Framed on the wall was a business card from Collins's dad's butcher store, DON'S FINE MEATS AND ART GALLERY. The shop's motto read "Where meat cutting is an art." "My father was a nut and so am I," explained Collins. "He was always writing letters to the postmaster general, to utilities companies, to politicians. He was constantly doing odd things, so I think I must have gotten his 'do odd things' gene."

To illustrate this, he told me how one of the main activities of the Glenn Foundation was handing out unsolicited $60,000 grants to scientists investigating the biology of aging. "Applications are not accepted for the award," he declared imperiously. "We just give them to unsuspecting nominees doing what our committee considers to be important work."

"How do the scientists react?"

"They're always astounded and happy and grateful. It's totally unexpected and kind of magical to have sixty thousand dollars just arrive into your world. We've given out forty-two so far, and we plan to give out many more. How could we *not* do something so fun and useful? Beneficiaries can double down. Just because you've gotten one, we can still give you another. And Paul, much to my pleasure, loves breaking rules. Meaning I can bypass the committee with awards. I just call Paul and see what he thinks. And then he says, 'Sure, go for it.'"

"I don't suppose you support writers investigating the science of immortality?" I hinted, pointing a thumb at myself.

"Not unless your biology lab's results are peer-reviewed," he answered firmly. But he did offer to show me one of the awards, and we went upstairs to his study, which abutted the bedroom. The sole object on top of the chest of drawers next to his bed was a crystal ball. I peered into its fathoms. All it revealed was the room upside down: the ceiling's reflection flipped over the table, an interplay of windows and lamps, layers of curved glasswork folded into diaphanous opacity. "It's important for aesthetic reasons," he clarified. "Not magical ones."

In the study, Mark brought out a glass cube with the words THE GLENN AWARD FOR RESEARCH IN THE BIOLOGICAL MECHANISMS OF AGING etched into its base. A three-dimensional, laser-cut image of the Glenn logo—four horizontally stacked disks of varying thickness and diameter topped with curved lines and trails of dots—was imprinted within the transparent cube. "People look at it and say, 'Oh, a fountain of youth,'" confided Collins. "That's their first reaction." To me, it resembled a jester's hat on top of a stubby Grecian column.

He then showed me a photo of a pretty young blond woman pouring heart-shaped ice cubes into a cup. "That's Lauren," he said fondly. "She's twenty-one." Collins, recently separated from his wife,* specified that Lauren and he had become "intimate friends who can sleep together in a bed without making wooga-wooga. She asked me to go to the doctor's office with her." He smiled broadly, his left eye twitching ever so lightly. "I mean, it doesn't get any better than that."

He picked up a first edition of *The Picture of Dorian Gray*. "It's number one hundred of two hundred and fifty copies signed by Oscar Wilde. To be bequeathed to Lauren." She was in university, he added, where she was researching the nature of secrets. "Secrets are the uncirculated currency of gossip," Mark intoned, repeating a line he'd certainly already come up with. "Once circulated, their value drops to zero."

"What's up with Santa Barbara and secrets?" I puzzled.

"Lauren might know. She's been sending out e-mails asking people what secrets mean to them. Everybody ends up responding with their own secrets. Let's find out, shall we?" He dialed Lauren's number.

As they spoke about her gleanings into the secretive, I flipped through a book of Oscar Wilde's quotes and came across the section "The Secret of Life." In 1893, Wilde felt there was no such secret. "Life's aim, if it has one, is simply to be always looking for temptations." But then, a few years later, in prison, he revised his position: "The secret of life is suffering. It is what is hidden behind everything."

Collins put his hand over the mouthpiece and paraphrased Lauren's current research, which involved understanding the effects of writing secrets down on paper. "She says that looking at a secret from a distance allows people to separate it from themselves. Also, she's wondering if you, as a writer, would be open to answering a few questions by e-mail?"

"Of course," I assented, thinking of Nietzsche's aphorism about how when you gaze into the abyss, the abyss also gazes into you.

Heading to Paul Glenn's home, Collins drove us through serpentining streets lined with purple jacaranda in full bloom, their fallen blossoms dusting the pavement like periwinkle snow. Collins wore vintage mirrored shades, small and circular. The mood was chatty, the day sunny.

---

*His divorce counselor had suggested he undergo a rite of passage into manhood, an idea Collins said made him feel "smeary."

He pointed out establishments frequented in a prior career, when he did marketing. He said something about a cosmetic-surgery company that specialized in breast implants. He called breasts *djibobis,* his own Vonnegutesque, made-up word.

"Part of being human is the fantasy that we can improve ourselves beyond our limits," he asserted. "Whether it's bolt-on djibobis or the idea of immortality or any other utopian notions."

I asked him what he, Mark Collins the person—not the foundation's president—thought about the notion of death as an elective.

"It's a bad idea," he exclaimed.

"For example, does growing old bother you?" I pressed.

"Not really."

"How about Paul Glenn? How does he feel about it?"

"You can ask him yourself. And he'll decide how he wants to answer. Paul has earned the right to not do anything he doesn't want to."

The two of them got to know each other when Glenn owned shares in a company where Collins worked, he noted, as the vice president of marketing.* Their professional relationship began in 1986, when Collins mailed Glenn a postcard of a Tupperware party.

"What was the message?" I asked innocuously, although the similarity between bowls of plastic Tupperware and silicone implants did spring to mind.

"I wrote him something polite and funny," Collins replied, waving the question away. "I forget what exactly." Regardless of its contents, that fateful postcard spurred Glenn to offer Collins a consulting job. They'd been working together ever since. "I always try to figure out why he keeps me around, and I suspect it's my sense of humor," Mark admitted, peering at me through his spectacles. "There are so many people better suited to this job. I think he prefers funny over adequate."

Glenn isn't the biggest donor in the longevity field, although he's up there. The founder of Oracle software, Larry Ellison, has a foundation that gives out $30 million a year, six times more than Glenn's average of $5 million, a number certain to rise when Glenn's estate passes over to his foundation.

"What'll happen to you at that point?" I asked Collins.

---

*Curious whether it had something to do with that breast-implant company, I later asked Collins for details. He responded with characteristic brevity: "It was a company that failed to rise to the point of relevance to my story or yours."

"I hope to stay around. I want to keep doing what I'm doing for many more years. I don't mind being a public nutcase. I once left a huge sign on the lectern after introducing Paul saying, 'Your fly is wide-open.' He didn't miss a beat."

For a moment, then, Collins grew serious and brought up "certain sensitivities" and apprehensions he had about our meeting with Mr. Glenn. I assured him that if they didn't want me to write about his personal life, I wouldn't.

"It's not that," Collins attempted. "He'll say whatever he wants to say, but part of my job is to oversee his legacy. I want to ensure that you don't make him seem . . . *foolish*. Because he isn't. He's a very intelligent, discerning man. You can make me seem like as much of a nut job as you want as long as you don't make him look foolish."

"No problem." I nodded.

Collins then said something that I subsequently thought of frequently: "Each one of us is responsible for their actions and what they say."

Moments later, a text came in from Lauren: "Love is everything."

He burst out laughing. "This is absolute insanity!"

"Why?"

"*Why?* Because I'm sixty-five. And not yet divorced. And people here really disapprove of us being friends. Did I mention that we're going to Paris together?" Collins elaborated on the journey. They were going to fly first class on Air France. He'd booked an absurdly expensive suite at the Plaza Athénée. They'd conceived of the trip as a one-week wedding ceremony. "I'm going to be bringing bags of confetti and tiaras," he said, then paused. "I'm looking forward to a tragic ending."

"What?! Why a tragic ending?"

"It'll be tragic because I'll have to make her move on. We're on a schedule. It's necessarily brief because I'm so much older and she needs to have her life."

Collins said they were planning on getting identical tattoos while in Paris. Lauren had chosen the symbol: three straight lines. I asked whether they were going to be horizontal or vertical. "Horizontal seems natural to me," Collins answered, after considering it momentarily. "Anyways, this is our 'secret' that we're telling you. . . . You will be free to unwrap it for your tale, if you wish."

I looked out at the petal-strewn streets, wondering why he'd chosen to divulge all of this. Lauren's essay on secrets. *The Secret* of secrets. To conceal and to reveal, to hide and to want. Our innermost desires are for love and a way out of death. Collins clearly didn't care about living forever, yet I couldn't help acknowledging that his openness had something to do with a desire to be preserved, in book form, for posterity. And perhaps that was the part of the reason I, too, had landed in this car heading toward the home of Paul Glenn, whom Aubrey de Grey deemed "by far the world's oldest private philanthropic supporter of biogerontology research."

Glenn's home was tastefully and, given his magnificent wealth, not at all ostentatiously appointed. On the coffee table lay a book entitled *Reversing Human Aging*.

"How do you feel about the idea of reversing human aging?" I asked.

"I'm of the anything-is-possible school," Glenn answered jauntily, putting on a brown ivy cap. On the edge of eighty, with a gentle stoop, Glenn came across as sharp, jovial, keen witted, and not foolish in the least. He wore a large houndstooth shirt, corduroy pants, and loafers. The red turtleneck under his shirt unintentionally drew my attention to the wattle dangling off his chin. Given Collins's request to observe sensitivities, I felt merely noticing it was ungracious, yet the flappy neck was the most prominent detail of his physique. While I jotted uncertainly in my notebook, Glenn turned to me and said, "Would you like to see my *Euphorbia ingens*?"

I stared at him. "That's a plant," Collins pointed out jokingly.

We walked out to the immense backyard, which contained a collection of plants so vast and magnificent it could have been a botanical garden. Nasturtiums bloomed by the thousands, cycads burst with life, *Aloe ferox* plants were covered in bright red cones. A patch of coral-reef-type flowers resembled anemones and urchins. Flesh-colored brain plants looked like warty organs pulsating in the breeze. "I love blue-tinted plants," said Glenn, pointing out a blue cactus called a monstrose. "In fact, some of my monstroses have turned green and that ticketh me off."

The *Euphorbia ingens* he'd wanted me to see formed a group of surprisingly tall and spiny African cacti. "One of them nearly killed me while I was in the pool," he disclosed. "It came down diagonally across the lanes and narrowly missed impaling me."

As we walked into the Technicolor vegetation, mottled sunlight streamed through the leafy canopy. Glenn paused next to a Jacuzzi. "My exercise consists of climbing up these stairs, and then back down, after the session."*

"Session? Is that what you call them?" Collins teased.

Glenn pointed out that, even though he never married or had children, he'd been with the same girlfriend for twenty-three years. He then patted the foliage of a small bunya tree. "I won't be here to see this bunya mature," he pontificated. "But Collins might be."

"No, no," chuckled Collins tightly.

"Collins is young for his age," Glenn maintained, as we ambled along.

"It's a nice thought," Collins responded, "but I don't think so."

Glenn pointed at some seedpods and said they grew fast and died young. Collins joked about their having progeria. Coming to a pond, Glenn grumbled about raccoons eating his goldfish for canapés. We gazed into the pond, which didn't seem to have any fish in it at all. "I need to buy a hundred more tomorrow," he huffed.

I tried to steer the conversation toward immortality, asking Glenn about the foundation's mission. He started explaining how he'd been interested in longevity since childhood, with a major turning point being an article he'd read in 1951. The reminiscence trailed off as we came to a rare type of agave whose central flower stalk jutted fifteen feet into the sky like a gargantuan asparagus. "That stem is called a mast," he said, pointing at the magnificent totem. "When the flowers open, clouds of hummingbirds gather to drink the nectar."

Here he elbowed Mark and said, "Your spike was bigger than that in its prime, wasn't it?" They both laughed, and Mark winked at me.

We arrived at a tumescent, waist-high cactus tied to supporting props. "We call this one *the Mark,*" Glenn deadpanned. "It's so big you have to strap it down to keep it under control."

When I brought up some more questions about longevity, Glenn said they would be better directed toward the scientists they funded, particularly Leonard Guarente from MIT and David Sinclair from Harvard, both of whom researched sirtuins, which Glenn called "an unusual benediction." Guarente was in town for a conference, and Collins could arrange a preliminary conversation.

---

*Alex Comfort, author of *The Joy of Sex* as well as a noted gerontologist, once wrote, "Probably the nearest approach to the fountain of youth is the Jacuzzi, or the California hot tub. These small, sociable, whirlpool baths don't 'rejuvenate,' but they are stunning ice-breakers at any age. And they reveal the fact that many older people—since uncovered skin ages faster than covered skin—still have good bodies."

Paul Glenn ended our garden tour with an aside about how getting into ending aging had been psychologically logical to him. It all started with the terrible deterioration he'd seen overtake his grandmother when he was still a child. "She lived past ninety, and at the end she could barely walk." He sighed. "She was very ill and infirm. At that time, it struck me that aging was a terrible phenomenon. I thought, 'Life is so beautiful—why does it have to end?'"

The meeting with Leonard Guarente took place at La Super-Rica Taqueria, a turquoise-and-white taco shack reputed to have been one of Julia Child's favorite hangouts. I'd been instructed to await the end of a black-tie ceremony, at which point Collins and Guarente would skip out on the after-party to come meet me. While standing by, I flipped through Guarente's autobiography, *Ageless Quest: One Scientist's Search for Genes That Prolong Youth.*

The book described Guarente's discovery of genetic aging mechanisms within yeast. Investigating why certain genes expressed themselves in long-lived mutants, he realized that enzymes called sirtuins (from SIR2s, itself an acronym of *s*ilent *i*nformation *r*egulators) play a role in cellular pathways affecting longevity. These SIRs silence the expressing of certain genetic regions in chromosomes; but when the SIRs are inhibited, peculiar behaviors start expressing themselves. Some of the yeast strains he was studying lived up to 50 percent longer than average. But these long-lived yeasts were also sterile, as well as hermaphroditic. The SIR-controlled pathways that allowed them to live longer also rendered them incapable of mating.

As his research deepened, Guarente found that SIR2-like genes exist in every free-living organism known to biology, from bacteria to humans. He started suspecting that they might be "the master regulators" of the survival mechanism. Their biochemical activity has proliferated for over two billion years, which suggests that they perform some crucially important function. Guarente felt certain that they were part of the mechanism that turns on during caloric restriction. He came to suspect that sirtuins evolved as a means of surviving periods of food scarcity. An adaptation that became hereditary, sirtuins are best understood as evolutionary remnants from organisms that made it through famine, drought, or other extended periods of stress. "Imagine, as an example, I had a gene to slow down the aging process during a famine and you didn't. When the famine is over, and assuming I survived, I would likely remain young enough to reproduce and you would not. In that way, the gene which slowed down aging will predominate in subsequent generations." Guarente also started imagining what it would be like if there was a way of targeting those regulators, to make their antiaging effects express themselves.

Ageless Quest, published in 2003, covered the years in which that notion caught on—while remaining shackled to the speculative. The book began with his reminiscing about early breakthroughs alongside graduate students. Their work accelerated upon the arrival of an Australian whiz kid named David Sinclair. The book ended by noting that a major goal remained: "convincing big pharma that there is a new area of human health and product development that they will need to be a part of." That certainly became true five years later, when GlaxoSmithKline purchased Sirtris.

In that process, however, tensions mounted between Guarente and his protégé Sinclair. For a spell, they stopped talking, experiencing "a bitter dispute that crescendoed," as *Science* magazine put it, "when the pair published dueling papers." Guarente told media that their quarrel had run him "through so many emotions, some of which I didn't know I had." The mood settled subsequently, with the two of them ending up as cochairs of Sirtris's scientific advisory board.

Collins called my cell phone just after my order for rajas tacos emerged from the kitchen. I ate about half my plate before he and Guarente arrived. They pulled up in a black town car, wearing tuxedos. Guarente, tall and bald, said he liked "the vibe" of La Super-Rica, and we had a brief, interesting chat.

I started out by telling him that I liked the title of his book, *Ageless Quest,* because it had a whiff of the popular to it, which made him laugh. Although his interest lay in longevity, not immortality, he acknowledged that both urges are innate and intertwined.

"And for you it began with yeast," I said. "The same yeast that's in beer and bread?"

"Yes," he answered. "Yeast cultures actually consist of live, dividing cells, which makes them easy to study. The notion that understanding their life span would shed light onto human aging used to be deemed preposterous. Of course, that's no longer the case."

He described an evening when, leaving the lab, one of his grad students exclaimed, "You're gonna win the Nobel Prize for this!" Possibly a tad smug, Guarente was nonetheless clearly brilliant. He spoke technically, with precision. A classic egghead.

He considered Paul Glenn his patron, describing the ways in which the foundation had assisted him, particularly in providing the funding for his lab to start testing mice. "He helped me so much in chasing the dream of sirtuins, from worms to mammals," Guarente said. "What he's done for aging is deeply inspiring. Paul has also taught me so much about how best to spend this brief, little period we have on this earth."

When Lenny excused himself to go to the bathroom, I asked Collins if he thought I'd be able to visit Guarente's lab. "Sure," he answered. "If you really want. But all labs look the same. They all have the same glasses-wearing postgrad students tinkering on the same sorts of tests in the same nondescript buildings—labs, including his, are pretty uninteresting."

I wondered aloud if there was some better place I could visit to understand what was happening with sirtuins and other developments in scientific life extension. He suggested I come to Boston in June for the annual Glenn symposium on aging at Harvard. It would be a chance to meet all the most important biologists on aging alive today, he said, plus I'd get to sit in on presentations about where real longevity science is at.

24

The Harvard Symposium

Meryl Streep: "What is that?"

Isabella Rossellini: "What you came for. A touch of magic in this world obsessed with science. A tonic. A potion . . . It stops the aging process dead in its tracks and forces it into retreat. Drink that potion and you'll never grow even one day older. Don't drink it—and continue to watch yourself rot."

Meryl Streep: "How much is it?"

—*Death Becomes Her*

And the mind
trips, numbering waves; eyes, sore from sea horizons,
run; and the flesh of water stuffs the ears.

—Joseph Brodsky, "Odysseus to Telemachus"

Housing a faculty of the world's brightest doctors and biomedical scientists, Harvard Medical School's warren of buildings spills off from a central green quadrangle. While many of the original nineteenth-century buildings radiate Grecian notes—Ionic pillars; marble façades; Parthenonical neoclassicism—the colossal New Research Building just down Avenue Louis Pasteur is a hypermodern, half-million-square-foot behemoth in reflective glass and shimmering steel.

On the morning of Harvard's Paul F. Glenn Symposium on Aging, a gaggle of students stood smoking and gossiping about the age of their mice. Passing them, I walked into the auditorium and slid into a back-row seat. The symposium's program guide listed all the speakers, ten longevity scientists who'd converged here from institutions around the country. The brochure contained an introductory note by David Sinclair, the Harvard-based sirtuins researcher and director of the university's Paul F. Glenn Laboratories for the Biological Mechanisms of Aging. "Because chronic illness in the elderly is a major medical cost," Sinclair had written in his position statement, "enormous savings would be achieved if mortality and morbidity could be compressed within a shorter duration of time at the end of life." As Collins would've said, unassailable.

The dean of Harvard med school, Jeffrey Flier, stood up to give the opening address. He spoke of a graying US population living longer and, as a result, spending more time in the throes of age-related illnesses. Most of us want to live as long as we can, he noted, but a side effect to extended lives is the burden of disability, chronic disease, and increased cost. Fortunately, developments already under way at Harvard, and at MIT, would soon help forestall disease. The speakers today were all making discoveries on the verge of being translated into therapies. He spoke of helping cerebral functioning in aged people, of trying to learn why stem-cell capacity decreases during aging, of the tantalizing possibility of finding a universal set of genes that regulates aging. Their greatest hope lay in the sirtuin work of Sinclair and Guarente. Potent activators of sirtuin genes had already gone into clinical trials, Flier noted, so even though the work remained embryonic, optimism abounded. And none of it would be possible without Paul F. Glenn, whom they were honored to have in attendance. "With your help we will continue to push the limits of human knowledge," concluded the dean, thanking Glenn repeatedly for his generosity.

Glenn took to the podium, looking skinny in his suit and tie. "As I approach eighty," he said, coughing a few times, "the symposia get closer and closer. I'm trying to find a way to increase the distance of time. We're working on that right now."

He got a polite laugh, then began to emphasize the importance of sirtuins. "There are fifteen diseases that may benefit from sirtuin research. And I must say, just in time, as we're facing unprecedented numbers of the elderly from the baby boomer generation. Life expectancy appears to be growing by a year per decade. The work at Harvard is causing that to accelerate. Which means people will draw on Medicare for even longer. There's only $35 trillion in health-care money to assist the seventy-five million boomers out there. Their entitlements may cost up to $120 trillion. All the total assets in the US are $40 to $50 trillion. Meaning we have no way to pay for them. In 2017, Social Security will be running in the red. By 2030, it will be empty."

The symposium's presentations were impenetrable to nonspecialists. The Salk Institute's Andy Dillin explained how toxic jumbles of protein clumps are regulated by insulin/IGF-l signaling pathways. Harvard's Bruce Yankner shared his investigations into the role of the presenilin proteins Notch and Wnt in neuronal signaling. A biologist from the Salk Institute named Leanne Jones described mechanisms regulating the size and maintenance of a stem-cell niche. She had had success tweaking a gene known as PGC-1, also found in human DNA, in the intestinal stem cells of fruit flies. Doing this delayed the aging of their intestines and extended their life spans by as much as 50 percent.

At one point, a lovably nerdy jokester named Dan Gottschling, from the Fred Hutchinson Cancer Research Center in Seattle, took the stage. He started out by directing our gaze to the rectangular table onstage. Completely draped in black fabric, the table was barren except for a basket of flowers. "Thank you," he declared with faux solemnity, "for putting the end point on what we're all studying with this wonderful casket here." As everyone laughed, he discussed how he and his colleagues were aiming to research the opposite of that. He then spoke about cellular aging as well as genomic and chromosomal instability, explaining that he'd engineered mice to study promiscuous activity of telomerase. There were also new clues to be found in yeast.

The next biomolecular lecturer discussed ARFs and twixxits and other ways to fix life, something not broken yet apparently not quite right either. My attention lasted until he mentioned something called Mxi1-mSin3-HDAC1/2 and revealed that p53 suppresses tumorigenesis via activation of apoptosis in aberrantly cycling cells. He droned on, trying to persuade the audience of peers and financiers (the most important of whom, Paul Glenn, was by now snoring audibly a few rows over) that his variables were the ones to make all the colors in the universe come into focus. I snuck out and went for a walk.

When I returned, a woman in the lobby smiled at me wanly. "Yeah, I had to duck out, too," she offered. "If I didn't, I was going to be sick or explode."

I nodded a bit sheepishly.

"Why are you here?"

I told her I was researching longevity and immortality.

"Those are very powerful drives." She nodded. "Everyone in that conference hall has been bitten."

"Yes, and it seems like sirtuins are what they're most excited about."

"That's for sure," she half laughed. "I'm actually coauthoring a crash book about sirtuins with David Sinclair."

I told her I'd love to know more about what's happening within sirtuin research.

"What do you want to know?" she asked.

"To begin with, will it work?"

"Well, they sure hope it will. The only answer now is: no one knows for sure. They will once it completes human trials."

As we spoke, people started trickling out of the auditorium. The last speaker had finished.

"Do *you* think it's really going to work?" I asked.

She looked around, then pulled me by my shirt collar into a stairwell. "The FDA won't approve it for longevity," she whispered. "But it might have street value. It's a huge cognitive enhancer, and it gives you the most incredible endurance. Athletes are already using it. You can't write about sirtuins without mentioning doping."

What one scientific journal termed "the roller coaster ride of yeast aging research" began when David Sinclair revealed that resveratrol, a compound found in red wine and certain plants, activates sirtuins. Ever since, red wine has been touted as having life-extending attributes. What most journalists neglect to mention is that, to get the beneficient effects of resveratrol from wine, one would need to drink hundreds or thousands of bottles a night.

Since partnering with GlaxoSmithKline, Sirtris has been formulating concentrated, synthetic compounds that target sirtuin pathways more effectively than resveratrol. The hope is that one or more of these proprietary molecules will end up in a pill form, sold at pharmacies as a human therapy. It won't be labeled as an antiaging medication; aging is not classified as a disease, meaning the drug can't be regulated for aging. Glaxo's pill may, however, be packaged as a drug for other purposes, such as type 2 diabetes, psoriasis, or vascular disorders.

Even though they're already touting the drug's "potential to treat diseases associated with aging," they won't be allowed to market the drug as having antiaging benefits because advertising and promotional activities are limited to the drug's approved usage only. That does not stop doctors and others from mentioning that it also "may have" antiaging benefits.* Several of the compounds are in human trials, but are still a ways off from widespread availability. Regulatory approval usually takes close to a decade, and early results are mixed, at best.

On several occasions, Sinclair and Guarente's findings were disputed by scientists at Pfizer and Amgen, rivals of GlaxoSmithKline's. They found no evidence that resveratrol-mimics activate sirtuins. Then, one of GlaxoSmithKline's own human trials was halted after several patients developed kidney failure. The press release made it clear that the patients were already ill, but the drugs caused nausea, vomiting, and diarrhea, which may have "exacerbated the development of the acute renal failure." That particular compound, SRT501, which was plant-derived, has been shelved by the company, but other synthetic compounds including SRT2104 and SRT2379 are still being studied. In the summer of 2011, some hopeful findings were released confirming that the compounds do help obese mice live longer (meaning it may increase the health span of fat humans).

In the fall of 2011, *Nature* reported more negative news about sirtuins when scientists in London attempted to replicate Guarente and Sinclair's experiments. "We have re-examined the key experiments linking sirtuin with longevity in animals and none seem to stand up to close scrutiny. Sirtuins, far from being a key to longevity, appear to have nothing to do with extending life," concluded the report. The Bostoners took issue with the British takedown, and a transatlantic dispute opened up.

The *New York Times* published a number of articles following sirtuins, first with enthusiasm and then with increasing dubiousness. Headlines went from "Quest for a Long Life Gains Scientific Respect" to "Doubt on Anti-Aging Molecule as Resveratrol Trial Is Halted." By 2011, Nicholas Wade, who'd been reporting on the saga dutifully for a number of years, characterized the hope that human aging might somehow be decelerated as "flickering."

---

*Many drugs today are actually prescribed for off-label uses, regardless of evidence of effectiveness. Well-known examples include antihistamines for insomnia, propranolol (a blood-pressure medicine) for performance anxiety, and antipsychotics such as risperidone for eating disorders. According to the *Archives of Internal Medicine,* 20 percent of prescriptions in the United States are for non-FDA-approved uses.

In January 2012, a prolific resveratrol researcher at the University of Connecticut named Dipak K. Das was charged with 145 cases of fabrication and falsification of data in work he'd published in peer-reviewed journals. Dr. Das's work had been touted by a Las Vegas supplement maker called Longevinex as proof that resveratrol can turn a mortal heart attack in animals into a nonmortal event. Longevinex suggested that for this reason, people should take the company's formulation daily. Das himself dubbed resveratrol "the next aspirin."

In the fallout, it emerged that David Sinclair had himself at one time been a consultant for Longevinex. He also had a paid position on the Scientific Advisory Board of a company called Shaklee, who manufactured a syrup called the Vivix Cellular Anti-Aging Tonic, sold as "the world's best anti-aging supplement." The price for a one-month supply of the grape-flavored resveratrol concentrate was $100. Sinclair attended sales conferences for the tonic, in which he hyped salespeople. "Over a year ago, we set out together to do this," he enthused, "to make a product so that you could actually activate these genetic pathways that can slow down aging."

When reporters from the *Wall Street Journal* questioned him about the affiliation, Sinclair immediately stepped down from Shaklee's scientific board and told journalists that his name had been "misused in connection with Vivix." In their own statement, Shaklee declared, "Every implied product endorsement was in Dr. Sinclair's own words and every Shaklee use of his name—whether in print or video—was pre-approved by him in keeping with our agreement."

If that wasn't suspicious enough, in the summer of 2010 it was revealed that executives from Sirtris were covertly selling their own formulations of resveratrol on the side. An outfit called the Healthy LifeSpan Institute, run by Christoph Westphal (who cofounded Sirtris alongside Sinclair and was the company's CEO), was selling one-year supplies of resveratrol for $540. The day after the first reports surfaced, GlaxoSmithKline ordered Westphal to cease selling the dietary supplement.

During one of our many interviews, I asked Collins whether he'd take resveratrol.

"I already do take resveratrol," he answered. Initially, he told me he bought Longevinex, but then later mentioned he used a confidential source. "I don't see why I shouldn't be taking it," he confided. "If it does no harm, I think everybody will take it. For now, it's like Pascal's Wager. If I'm wrong, no problem. If I'm right, then great."

"So taking sirtuin activators is like believing in God?" I retorted, marveling at the comparison. "Now there's a declaration of faith, especially from a skeptical optimist."

Our conversation turned to the difference between immortality and longevity. Did he think the distinction had become blurred?

"I'm not going to let you set me up for a false dichotomy here," he replied tellingly. "One man's longevity is another man's immortality, perhaps. I don't know. I don't know."

At the time of my visit, the Harvard Medical School found itself in an ethics quandary, with hundreds of students and faculty protesting the influence of Big Pharma in classrooms and labs. Earlier that year, the school had received an F grade from the American Medical Student Association for conflicts of interest raised by its openness to drug industry money. Just under 20 percent of the medical school's professors and lecturers reported having "a financial interest in a business related to their teaching, research or clinical care." A Senate investigation of several medical professors was under way.

"If a school like Harvard can't behave itself, who can?" declared Dr. Marcia Angell, a Harvard faculty member and author deemed to be among the twenty-five most influential Americans by *Time* magazine. In another article, Dr. Angell wrote, "It is simply no longer possible to believe much of the clinical research that is published, or to rely on the judgment of trusted physicians or authoritative medical guidelines. I take no pleasure in this conclusion, which I reached slowly and reluctantly over my two decades as an editor of the *New England Journal of Medicine*."

Why was there so much corruption in the biomedical community? A twelve-story answer to that question hovered inches away from the symposium, in the form of a giant Merck research facility. It looms over the Harvard Medical School's New Research Building, two stories taller, seventy-five thousand square feet bigger, and all but rubbing up against its academic neighbor. The result of the Bayh-Dole Act, adopted in 1980 and intended to give American universities intellectual-property rights over their innovations, was the interpenetration of corporate and educational institutions. In Dr. Angell's words, "The big drug companies bribed and corrupted the medical establishment so that we no longer know which drugs are effective or why our doctors prescribe them."

The Glenn Symposium itself was being held in the Joseph B. Martin Conference Center, named after the former dean of the medical school, a booster for industry involvement in academia. During his tenure as dean, Martin concurrently sat on the board of Baxter International, a medical-products corporation, which paid him an extra $197,000 a year on top of his university salary. Martin appeared to have been aware of how such behavior would be perceived, having written papers about "the increasingly complex web of financial relationships between corporate sponsors of research and the investigators who perform laboratory research and clinical trials on their behalf." He also noted that eradicating conflicts of interest would be impossible, and that a wiser goal was finding productive, collaborative ways to work with industry.

His successor, Dean Flier (who'd spoken that morning at the symposium), followed in Martin's footsteps, noting (unassailably) that faculty collaboration with industry facilitates "scientific discoveries and clinical translation that will benefit the sick and suffering." According to the *New York Times,* he had also been the recipient of a $500,000 research grant from Bristol-Myers Squibb.

When the talks ended, speakers and other VIPs gathered for a "mingle." The hors d'oeuvres consisted of local Massachusetts cheeses, the sort a multimillion-dollar endowment buys. Strangely, given all the talk of resveratrol, there was no wine. One disappointed gray-haired attendee turned to another and said, "Maybe they'll just give out pills."

In a corner of the room, Dean Flier traded stock tips with Paul F. Glenn, who implored the dean to get rid of his US dollars. "Convert whatever you can, because they're going to lose all their value," Glenn intoned solemnly. The dean paid close attention, nodded, and said he'd be sure to call his broker.

Collins introduced me to a mustachioed, dark-haired man named Gary Ruvkun. He'd just authored an important new paper in *Nature* that said, "The secret of long life may lurk within the genetic activity profile of sex cells." I told him I'd seen reports about the paper. He was pleased to know that it was generating nonspecialist interest.

"I didn't get a chance to read the whole paper," I explained, "but the discovery had something to do with sex, right?"

"What?" he said, and looked at me as though I were joking. "Sex?"

I grew a little uncomfortable, as though I'd made an embarrassing faux pas. "Well, as far as I recall, the title mentioned the word *sex*."

He shook his head. "It's got nothing to do with sex. Not unless the press release got it wrong."

"Sex cells, then?"

"*Oh!* I see, the PR people called them sex cells. Germ cells, yes. The paper is about somatic cells in worms that have the ability to take on some of the attributes of germ cells—or sex cells."

"So that means regular body cells can become germinal cells?"

"Essentially, yes, if you start hitting insulin in all sorts of directions, you see this ability to turn into a germ line, which is called totipotency. The idea that somatic cells maintain the potential to reacquire pathways lost during differentiation, or the generation of new cells, is a tantalizing prospect."

When our conversation ended, I turned to David Finkelstein, a program director in the Division of Aging Biology at the National Institute on Aging, who told me that their recent studies suggested that rapamycin (a bacterium from Easter Island) appeared to have a greater effect upon aging than resveratrol. He also pointed out that no one has any idea how either of them delay aging. "The smartest longevity scientists in America are gathered here today," he explained, "and none of them know how aging works." As he spoke, he accidentally spat out a bit of couscous and then picked it off my jacket, apologizing.

When Mark Collins came over to schmooze with us, Finkelstein said, "Paul Glenn *is* the portrait of Dorian Gray—he hasn't changed in seventeen years."

Collins nodded significantly and enthusiastically. "How did you like the symposium?" he asked me. "Heavy stuff, isn't it?"

"I understood it all," I joked.

"Yes, it's pretty technical," he conceded.

"Do *you* even know what they're talking about?" I asked.

"No." He laughed. "Not at all, which is why I'm the perfect person to be helping fund such research."

Leonard Guarente overheard us and mentioned how even he'd been having trouble understanding some of the lectures, in particular one speech on stem cells. "Just by looking at the speaker, it's clear that he's saying something really profound, but I don't grasp the profundity of it," Guarente admitted.

As Collins and Guarente chatted, I spoke with John Furber, whom Collins had pointed out earlier as an acolyte of Aubrey de Grey's. I asked Furber if he thought it was possible we would live for five thousand years, as de Grey insists.

"Five thousand years *is* possible," Furber answered. "If you don't believe it's possible, you wouldn't work on it. But say it correctly: 'There's a fifty percent chance that—'"

"But what would you do if you could live for five thousand years?" I interrupted.

"I'd live one day at a time. I'm enjoying today. I'm sure I'll enjoy tomorrow. I'm in no hurry to die. What would you do? Would you plan it out or would you take it one day at a time?"

"I don't believe we'll be here for five thousand years. So those questions aren't really relevant in my case. I respect your beliefs, but in my belief system, reality means being here for a brief time. It also means dying."

"Well!" Collins cut in. "I do agree that the merits of death deserve to be discussed, but perhaps another time? Adam, I'd like to introduce you to David Sinclair."

Collins pulled both of us together so we could shake hands. "This is the guy I was telling you about," Collins told Sinclair, "the one writing a book on the fountain of youth."

"Ah, yes, we have to talk," Sinclair said with a thick Australian accent. He was shorter and more boyish than I'd anticipated, with a pointy nose, a broad smile, and sideways-parted hair. Something about him was Russian looking. Just as Guarente did at the taco stand, Sinclair started telling me how indebted he was to the Glenn Foundation. "Without Paul none of us could've gone from yeast to mice. He's an angel."

Perhaps expecting further discussion of Glenn's involvement in longevity research, he seemed taken aback when I asked how excited he'd been when he found out that GlaxoSmithKline had purchased Sirtris for $720 million.

His smile dropped immediately. "What do you mean?" he asked flatly.

"Well, were you excited? How did you react? Was it an exhilarating feeling?"

He had personally received over $8 million at the time Glaxo acquired Sirtris. As a consultant for the company, he also earned an annual salary of $297,000 (on top of his university salary). Without any expression whatsoever he said, "It's a dream."

I asked him where sirtuin research was at, and he said, "We'll know in twelve months or twenty-four months. In a few years. The earliest that the molecule will be available is in three years."

"Why is it that the media has been focusing on resveratrol and red wine rather than sirtuins?"

"I'm rather amused by that myself," he said, a modicum of joviality returning. "It would be as if Alexander Fleming had discovered penicillin and the media said, 'Drink mold juice.' Now they're saying, 'Drink red wine.' But the public is always five years behind discoveries."

"So based on current discoveries, where will we, the public, be five years from now?"

He pondered it for a moment. "In five years, there will be a few diseases that our pill is being prescribed for. In fifteen years, like Lipitor, it'll be a preventative. In a hundred years, it'll be at the drugstore for a cent a pill, which is what it should be. Ideally it could be free. I like to think about the molecule being added to the water supply like fluoride."

"Do you really mean that?"

"Well, I won't go as far as saying it *should be* in the water supply," he hastened to add.

"Would you *like* to go that far?" I pressed.

"No. Because I think people should have the choice."

The first modern gerontologist* was Charles-Édouard Brown-Séquard (1817–94). He taught at Harvard before becoming chair of physiology at the Collège de France. His work made the idea of comprehending—if not controlling—aging a respectable aim. The discipline's beginnings, however, were as suspicious as they were inauspicious. Brown-Séquard's late-period research occupies one of the more bizarre footnotes in medical history: toward the end of a distinguished career, he stunned the scientific community by announcing that he'd found a glandular elixir of eternal youth.

His speech on June 1, 1889, at the assembly of Paris's Société de Biologie, is widely considered to mark the commencement of gerontology. Most members of the society were in their seventies, as was the swarthy, six-foot-four, bushy-bearded gentleman onstage. In unscheduled introductory remarks, Brown-Séquard confessed that his natural vigor had declined considerably over the last decade.

---

* Gerontology, from *geron*, meaning "old man" in Greek, is the systematic study of aging.

At that time, many scientists felt that old age was not a natural phenomenon, so a murmur of commiseration rippled through the room. Those graying authorities knew full well what it meant to grow old, nodding as Brown-Séquard lamented his own chronic pain—the lassitude, the insomnia, and, most delicate of all, the decline of his manliness. He had a pretty young wife, he was rich, successful, accomplished—*et quand même*.

"I have always thought that the weakness of old men was partly due to the diminution of the function of their sexual glands," Brown-Séquard explained. Shouldn't there be a means of stimulating those tired old glands? To shock them back into action, he posited, would jump-start the entire system. (As we've seen, this concept was appropriated decades later during the "gland craze" of the interwar years.)

In Brown-Séquard's time, the functioning of internal secretions, hormones, and our glandular organs was even less understood than it is today. The endocrine system remained an undiscovered enigma. Testicles were thought to produce a sort of vitalizing substance that dissipated with age. Contemporary science recognizes testosterone as an essential component of bodily function. Brown-Séquard believed it to be the fountain of youth.

For the past few months, he told the assembly, he had been performing auto-experimentation with liquefied extracts from various glands. Several weeks earlier, he'd opened up the scrotum of a puppy and removed the testicles. Cutting them up, he blended them with an aqueous solution, ran the liquid through a filtration device, and filled a syringe. He then administered injections of the extract into his own thigh. Guinea-pig testes worked just as well, he declared, having inoculated himself with them as well. The physiological effects were, he testified, most surprising. He'd regained his youthful vitality—in *every* way. Just before the lecture, in fact, he'd managed an elusive feat offering empirical proof. He had been able to "pay a visit" to his young wife.

Brown-Séquard boasted in peer-reviewed publications of having cured impotence. His brand of organotherapy—*la méthode Séquardienne*—became a hit with consumers. He constructed a fantastic machine "with a belt pulley, tubes, alembic, aeration bladders, instrument dials: into it he fed bull testes pulped, filtered through sand, ascepticized with boric acid, drawn off as a liquor." Hundreds of elderly men started shooting up testicles. Their deafness diminished, they claimed, hair thickened and darkened, energy levels rose. Penises hardened. Unfortunately, as it turned out, they were only getting hits of the placebo effect.

Critics railed against Brown-Séquard's "senile aberrations." His contemporaries repeated the experiments, checking them against controls of plain-water injections. The results were conclusive: testicle-pulp inoculation had no noticeable effects. In fact, the saline serum Brown-Séquard had patented contained barely any hormones. It was just a turbid mélange of mashed bollocks in salt water. He was widely denounced. The pretty bride abandoned him. Prestigious journals no longer accepted his papers for publication. He fled to the seaside and died soon after.

•

After the Harvard symposium, Mark Collins invited me to stay for dinner in a circular ballroom on the fourth floor of the New Research Building. I was seated at a round table with Paul F. Glenn, Dean Jeffrey Flier and his wife, and sandwiched between Collins and Glenn's lawyer. David Sinclair and Leonard Guarente, whom everyone called Lenny, were hovering around, as were waiters, who informed us that the choice of entrée was steak, chicken, or salmon. Everyone at the table seemed to have smirks of wealth, their faces padded with the knowledge that they'd amassed massive amounts of money.

Once our orders were in, Glenn apologized for having fallen asleep during the lectures. "It seems I had a nap." He grinned. "Who did I miss?"

"A bit of each, Paul, a bit of each," said Collins, and we all laughed.

The discussion turned to which of the various synthetic molecules would best isolate and enhance sirtuin activators. I asked whether sirtuin drugs would ever be approved as antiaging medications by the FDA.

"There are no known biomarkers for aging, so how can you test it?" Collins explained. "There are putative markers, but you can't do a phase three for any drug on aging. But of course it doesn't need to be granted regulatory approval as an antiaging drug to be an antiaging drug, if you know what I mean."

Moving to the podium, David Sinclair picked up the microphone and said he just wanted to thank Paul Glenn for what he was doing for humanity. Sinclair admitted that the future of sirtuins was uncertain, but that it undoubtedly carried immense promise. Even if they failed at extending people's lives, they might cure diabetes in the process.

Glenn stood up and took the microphone for another speech. "We see through the glass darkly at the Glenn Foundation," he announced. "We aren't scientists. We're just a living checkbook. Your job is to explain what you're working on so that we understand whatever the hell it is you are talking about. . . . If you are a researcher who hasn't been paid yet, come and beat on Mark Collins over here. I believe he has learned to live with pain. He's rather masochistic. If you hit him hard, you might have a fast friend."

"So you're a masochist?" the dean asked Collins.

Mark shot us both a wry look. "I prefer romantic."

At the end of Glenn's speech, which went on way too long, he said, "If I don't recognize you next year, remind me who you are—and who I am." This got a laugh. "So stay in touch, beat on Collins, and hope to see you here again." Glenn then suggested someone design a new poster for the event, adding that the same worm had been on the symposium flyer for the past four years. He called it "Andy's worm."

Whoever Andy was, he piped up from somewhere in the room and said, "Hey! Let's not talk about my worm." A number of people groaned audibly. The dean's wife grimaced. Collins assuaged the situation by joking that the worm has a grin because it's immortal.

"I thought people here were into longevity, not immortality," I whispered to him.

"They are." He frowned. "This is a serious place—this isn't Dave Kekich's pad in Huntington Beach."

I turned to Glenn's lawyer and asked whether he wanted to live forever. "No, not at all," he said calmly. "People at this event want to stay healthy longer and then fall asleep and die peacefully, rather than die in drawn-out suffering and incoherency."

He drew a little graph on a napkin of how he wanted to die. The drawing consisted of a straight horizontal line and then a ninety-degree plunge downward. A lifeline of health punctuated by an instant downturn of painless demise. He then illustrated how most people die, his ink forming a slow, descending curve into the final, pain-glazed years of age-related illness.

Conclusion: If ______ Is Possible

We are all deeply accustomed to seeing science as the one enterprise that draws constantly nearer to some goal set by nature in advance. But need there be such a goal?

—Thomas S. Kuhn, *The Structure of Scientific Revolutions*

Paradoxically, the best arguments produced by any believing community, including perhaps the scientific community, have always led to mystery rather than demystification, expanding our sense of awe and wonder instead of explaining it away.

—Eugene Fontinell, *Self, God, and Immortality*

A YEAR LATER, I checked in on Mark Collins, who'd moved into a new Santa Barbara home with an "age-appropriate" (as he put it) forty-seven-year-old graphic designer named Lily. Sirtuin research was mired in contradictory results. Had the Glenn Foundation's researchers found anything of note not yet reported in the media? "Just a few more peptides in the haystack," he said. "Not enough to hang your hat on."

I asked him again why he thought the incremental scientific advances being made in the study of aging were being construed by some as leading to immortality. "Because it just happens to dovetail nicely with those people's anxiety," he answered, shrugging.

"But what's your position on it?"

"The lines have always been blurred between extending lives and making death an elective. That's why we have an unassailable mission statement: alleviate suffering; lose the slope; help public treasuries."

We spoke of how Aubrey de Grey seemed to have clued into that idea as well. He'd started saying he no longer wants to use the word *immortality* when describing his aims. "I work on health," de Grey clarified. "I am interested in ensuring that people will stay completely youthful, like young adults, for as long as they live."

As Mark and I chatted, Lily joined us for tea in the kitchen. He told her that I'd been writing about people who want to never die, an idea she found abhorrent. "Imagine being eternally imprisoned in your body?" She shuddered. "Life without parole."

"I've always wanted to know where that sweet spot is that makes people want to keep doing that special something forever," Collins said, nodding. "Is it when they're in the Jacuzzi looking at the stars? Is it at breakfast? Is it when they're having incredible sex? Even if they had the *ideal* day—the best poached eggs *ever,* the best OJ *ever,* the most robust bowel movement *ever*—it would still involve all those daily human activities. Every day—*forever*!"

"The whole thing comes down to 'Why?'" Lily lifted her hands.

"The pursuit of health is understandable, the pursuit of a long, fun, productive life is understandable," Collins continued. "But immortality would be like insomnia—*forever*! How cool would that be? Imagine an eternity of insomnia? I would not want to feel what I'm feeling for the next twenty badjillion years."

"Whoever those people may be, they are denying our condition," Lily replied. "There is a circle of life."

I nodded, but her conclusion made Collins bristle. "Just because things have been a certain way doesn't mean that they always will be."

"So you think you can change nature?" she charged.

Before Collins could reply, I jumped in. "Well, that's what these people think, the immortalists that we're speaking about."

"It's just crazy." She shook her head. "No cell lives forever."

"Nature gives you the default position," Collins said, growing serious. "Humans have the imagination and the ability to tamper with the natural order. I think it's fine to tamper with nature. And I don't fault immortalists for trying to see what'll happen."

"I don't have a problem with *tampering* with nature," Lily explained. "What I have a problem with is saying anything can live forever. That's crazy."

"It may be crazy, but if people felt that way about flying, they never would have gotten off the ground," Collins countered. "Are immortalists that different from Lindbergh? Where's the line of immutability? I'm not in a position to say they're wrong."

"Flying a plane isn't ending aging," she said, levelly.

"Nobody knows," he responded.

The mood had grown a bit tense. Lily pursed her lips and started doodling circles on a pad of paper in front of her.

"You've fallen into a trap of drawing a metaphor that works for you," he went on. "'The circle of life.' 'The balance of nature.' They're nice phrases, and there's probably some truth to them—"

"Probably?" she snapped.

"Yes, probably." He smiled, as though reminding her they were simply debating, not arguing.

"Well, there's probably some truth to everything which leaves us stuck," she conceded. "I love the world of absolutes, though."

"There's no such thing, though," he said, putting his arms around her shoulders.

Before Charles Lindbergh set off across the Atlantic Ocean, newspapers described the flight as a guaranteed "rendez-vous with death." While the *Spirit of St. Louis* hummed toward France, human-formed phantoms and vaporlike spirits materialized before Lindbergh's eyes. These "inhabitants of a universe closed to mortal men" spoke to him, reassuring him and helping him find his way. This inner experience, he wrote, seemed to penetrate beyond the finite. It was an epiphany that guided the rest of his life.

After his flight, he received millions of letters, thousands of poems, countless accolades. Three hundred thousand people attended a single parade in his honor. Wing-walking skywriters spelled HAIL LINDY high in the air. Former secretary of state and later US Supreme Court chief justice Charles Evans Hughes gave a speech in New York heralding "science victorious."

In the euphoria's wake, having managed one impossibility, Lindbergh wondered if he mightn't help solve another. Working alongside Nobel Prize–winning cell biologist Alexis Carrel (who claimed, erroneously, that cells divide endlessly and are therefore naturally immortal), Lindbergh came to question whether death is "an inevitable portion of life's cycle," musing that perhaps scientific methods could hasten the arrival of bodily immortality.

Lindbergh had been raised to believe that "the key to all mystery is science." The idea that science will allow men to become gods was instilled in him by his grandfather, a well-known surgical dentist. For postflight Lindbergh, solving the basic mystery of death seemed only as challenging as flying across the sea. It just meant doing what people said couldn't be done. Yet as he aged, and as his experiments didn't yield the hoped-for results, he began questioning his desire for immortality. He became an environmentalist, spending time in the wilderness, where his observations of nature made him start gravitating toward cycles of life and death rather than the linearity of unremitting progress toward eternal life.

In his later years, he characterized himself as a former disciple of science, someone who'd mistakenly enthroned knowledge as his idol. "I felt the godlike power man derives from [it]," he wrote. "I worshipped science." He publicly acknowledged how mistaken he had been, adding that "physical immortality would be undesirable even if it could be achieved." In his final years, he became convinced of the necessity of dying. In death, he concluded, "is the eternal life which men have sought so blindly for centuries not realizing they had it as a birthright . . . Only by dying can we continue living."

When Neil Armstrong landed on the moon in 1969, we felt sure immortality was next. After all, if we were able to reach other worlds, how could we *not* live forever? "Transcendence is no longer a metaphysical concept," trumpeted sociologist F. M. Esfandiary. "It has become reality." His book *Up-Wingers* argued that our ability to send rockets into space meant we would soon never die. By the end of the twentieth century, he predicted, we'd be ageless, interstellar denizens of orbital communities, easily space-hopping in and out of planets and moons. Robot servants would take care of our every need, armies of clones would fight alien enemies, teenagers would soar over clogged urban centers in winged cars. We'd also be able to alter our bodies' genetic scripture at will.

"Nobody in this generation has to die, unless they want to," Timothy Leary assured disciples in the 1970s. "They can become immortal and go to the stars." He himself did end up going to the stars—when he died. He didn't become immortal (or get cryonicized), but seven grams of his cremated ashes ended up in orbit aboard a Pegasus rocket. A different end from the endlessness he'd envisioned, but a fine one nonetheless.

The idea that scientific breakthroughs will help us escape the human condition is a common delusion. Whenever technology attains a new feat, we start imagining we'll live forever as a result. When Craig Venter created the first synthetic genome in 2010, newspapers claimed science had "officially replaced God." After CERN's particle accelerator apparently established the existence of the Higgs boson in 2012, journalists announced that we'd soon be traveling around the galaxy at light speed.

We've always been like this. As soon as frozen food became standard in grocery stores, cryonicists began staging demonstrations in front of funeral parlors, carrying placards saying DEATH IS A DISEASE AND CAN BE CURED and WHY DIE? YOU CAN BE IMMORTAL. Our ability to program computers filled us with more of the same overblown fantasias. "If we can make computers," we assured ourselves, "why, surely we can program ourselves to attain immortality!" It's a hallowed narrative. We found telomeres; we can live forever! We discovered a jellyfish that regenerates itself; we can live forever! We located sirtuins; we can live forever! Thanks to science, we can do more now than ever before. We may even become cyborgs one day. But we'll never be gods.

In 1974, Charles Lindbergh succumbed to lymphoma. As he lay on his deathbed, more and more claims emerged promising eternal youth. In February 1973, Michigan State University announced that they were close to releasing a pill that would extend life by two hundred years. That same year, others said an immortality pill would be available by the year "2000 or so." By that point, newswire outlets reported, aging would be fully reversible. Polls showed that the majority of Americans believed that there would be artificial eyesight for the blind, drugs to permanently increase intelligence, and chemical control of aging within three decades. Forty years have passed.

Cyberneticists of that time spoke of coding personality so that it could be filed into electronic circuits and reanimated at will. Knowing that DNA was simply an information system, we'd soon find a way to program and reprogram it. "The ultimate reason that immortality is possible is that we are not the stuff we're made of," they informed awestruck reporters. "You might say that the formula for Immortality is Cybernetics + DNA." We might have developed motherboards, but we still can't download eternal life.

A 1970s organization called the Abolish Death Committee issued press releases explaining that the end's end was nigh. "Because more than one science is hard at work on this problem, an early solution is forecast." It would take only "15 years maximum" before we'd succeeded in extending human life to an average of four hundred to five hundred years. Computer whizzes went on to change our lives, but as Steve Jobs's death proves, they haven't yet succeeded in trashing mortality. "We're born, we live for a brief instant, and we die," Jobs said. "Technology is not changing it much—if at all." Understanding computers—just like understanding how to fly, how to land on the moon, how the structure of DNA works—doesn't mean we can understand death. "I don't understand how someone can be here, then not be here," admits Larry Ellison, CEO of Oracle software and immortality financier. "It's incomprehensible."

Death is something we cannot rationally comprehend. The only way to contend with it is by attaching ourselves to stories. Some religions assure followers that reaching the hereafter is the way we understand what life was all about. Physical immortalists, on the other hand, believe that the inevitability of dying is itself just a belief. For them, what was once extramundane has become very much of this world. The symbolic notion of rising from the dead and being reborn has been transposed onto Internet-age computing systems.

Forget symbols. Immortalists don't see their viewpoint as a modern appropriation of the everlasting need to believe. They're engineering "the scientific conquest of death." To them, death is unconstitutional, an infringement on our fundamental liberties. Their paradise is physical, here for now and forever. It's a magically mechanistic place, endlessly stocked with enough new gadgets and products to sate even the most atrophied attention span.

Thinking about death, says Ray Kurzweil, "is such a profoundly sad, lonely feeling that I really can't bear it. So I go back to thinking about how I'm not going to die." One of the ways he goes back to thinking about not dying is by fantasizing about the nature of progress. Believing that history's terminal will be a point in which we're all perfect and immortal is, if nothing else, a useful way of alleviating the fear of mortality.

The notion of progress as the general mechanism whereby time and history function is only a few centuries old. Until the Enlightenment, we couldn't have conceived of improvement as a fundamental law: all anyone had to do was look back to ancient Rome or Greece to be confronted with reminders of present-day shortcomings. Architecture, philosophy, culture, politics, agriculture, civilization—life, in short—was better then. We didn't dream of things getting better; we hoped for a return to prelapsarian perfection.

Throughout the Middle Ages, the dominating outlook in the West was that our lives are ruled by providence. The fates determined everything. People believed in their Islamo-Judeo-Christian God. Not doing so was punishable by death. Europeans didn't believe in progress: sometimes things got better, sometimes worse, but in general, the world, like us, simply got older. Time moved in cycles. We followed the seasons. In that era's "circle of life" worldview, everything came from the ground—and eventually returned to it.

The idea of progress emerged as knowledge was becoming its own end and not something that depended on or led to godliness. Our brains were the way forward. Europeans had found the New World—shouldn't eternal life be around the corner, too? Doctors started thinking about ways of extending lives. Medical authorities assured us that human life expectancy is malleable. By understanding it, we'd be able to improve on it. We'd just get healthier and healthier until we found ourselves living forever.

Our belief in cycles fell from fashion. The Enlightenment led us to start seeing history as a continual advance into the brightness of perfection. Da Vinci's straight lines pointed toward the vanishing point of life getting better and better, and longer and longer. Intoxicated by the possibilities, thinkers took up the argument that humankind's history consists of irreversible betterment. Political theorists spoke of creating a society without inequality. In no time, democracy would secure freedom for all future ages.

Science was on its way to replacing God. All our problems seemed soluble. The end point of evolution would be the advent of perfected humans—read *immortals*—in a perfect world: "The ultimate development of the ideal man is logically certain," wrote Herbert Spencer. "Always towards perfection is the mighty movement."

Education seemed as if it could elevate us above the human condition. If we studied something enough, so the reasoning went, we could know it fully. And the perpetual and unlimited augmentation of our mental faculties would inevitably yield heaven on earth. Our minds promised us nothing short of a full comprehension of the complete workings of the universe. Knowledge became boundless in its possibilities: it could stamp out all misery, save the world, and make us live forever. We just needed to start believing in science.

We never suspected we were under a magic spell.

Of all the wars that have taken place since then, none has endured so long as the conflict between knowledge and belief. For centuries now, knowledge has attempted, unsuccessfully, to supersede belief. But the entire clash stems from a misapprehension of the nature of belief. We can't not believe; and we won't ever know everything. We know this much: knowledge remains an endless advance toward an end point that endlessly recedes. Like the forms on Keats's Grecian urn, we're forever panting, forever pining, forever in pursuit of something we can't ever possess: Truth and Understanding.

We haven't yet found certainty. We can uncertainly state that we likely never will. All definitions are by definition incompletable. We can either believe that we're ascending an escalator of increasing rationality that will somehow deposit us in the realm of Absolute Knowledge, or we can try to accept the inevitability of unresolvable conflict.

The twentieth century prided itself on invalidating the metaphysical. Doubts about the afterlife arose even as so-called nonbelievers attempted to locate surrogates for the loss of meaning atheism occasioned. Enraptured with progress, we deepened our collective worship of science. As incredulity to metanarratives rose, so did a massive publicity campaign to convince us of the omnipotence of science. Research institutions and organizations enlisted PR companies to spread the message that their findings—if amply funded, of course—would rid society of disease and even death. Much-talked-about books such as *The Conquest of Disease* (1925) promised readers that all illness would soon be eliminated. The bestseller *Men Against Death* (1932) profiled scientists working toward immortality. The undertow of progress grew fiercer. If we could create an atom bomb, why, then, shouldn't we be able to live forever? In the boom years after World War II, Westerners came to see themselves as invincible, an "affluent society" forever young, forever healthy.

Every technological leap—from refrigerators and rockets to tablets and unmanned drones—intensified our unconscious faith in progress. Most of us still take it for granted that it's only a matter of time before we eliminate poverty, illness, and, eventually, mortality. Without even realizing it, our civilization remains in thrall to the obsolete ideologies of Hegel, Voltaire, and Locke. We still elevate certain people to the level of an expert—spiritual leader, techno-pundit, atheist bestseller writer—but the problem with venerating proselytizers is that they are as error-prone, unknowing, and ultimately as human as the rest of us.

The Enlightenment touted the arrival of omnipotent leadership that would dispense global equality; but the faith that governments can solve problems is by no means universal today. That epoch's rationality promised to be emancipatory; instead it led to strip-mining and nuclear warfare. It's how our oceans ended up peppered with plastic. Particles of petrochemical waste aren't just clogging up landfills and the craws of pelicans—they're also coursing through our bloodstream. To support an unsupportable lifestyle we ravage our surroundings.

Progress has not brought about universal happiness; realizing that simply requires opening a newspaper. Are we better off than hunter-gatherers? We'd like to think we are, but there's no way of knowing. Affluence isn't a guarantor, or even an indicator, of contentment. We aren't yet drivin' Buicks to the moon. The more we've surrounded ourselves with devices ostensibly intended to free our time, the less time we have to pursue our interests. A continual stream of gadgets reminds us that continual improvement is the way life works. Car manufacturers, credit-card companies, and smartphone makers all use taglines like "progress is great" or "the world is getting better." But progress isn't simply great; it's other, more problematic things as well. To deny this is to inhabit a make-believe land in which technology is humankind's savior.

In the past, the religious notion of salvation helped us find meaning when dealing with the universe's immense imponderables. To this day, we have no idea what time is or how it functions. In a century or two, we'll have a very different perspective. In the meantime, we're all ineluctably caught in the tension between who we are and what we might become. And so belief persists.

We can believe whatever we want to believe, which means we often don't realize it when we're believing. Our faith in science and progress is part of the reason people believe in physical immortality today. Mythologizing the unknown gives a storybook sense to our lives—and their termination. In mythology, science can make us live forever. In reality, however, science's aims are much humbler.

The scientific program consists of finding interrelations linking perceptible phenomena. Its purpose is to establish rules explaining reciprocal connections between objects. Successfully detecting patterns in nature always brings with it an opportunity for self-deception, of falling prey to illusions. While we *can* perceive traces of reasonableness in the structure of reality, the fact that certain events appear ordered doesn't mean that all of existence is open to logical apprehension.

There are limits to the purely rational conception of existence. Crossing those limits brings us into belief. To think there's no problem science cannot solve is a form of faith. "The scientific method can teach us nothing else," Einstein pointed out, "beyond how facts are related to, and conditioned by, each other." But not everything is a fact. Science can attempt to determine what is factual and what isn't, but then it cannot approach those aspects of experience beyond facts. Nature's profoundest chasms are off-limits to intelligence, inaccessible to the thinking mind. They give themselves not to demonstration but to revelation.

By refusing to acknowledge the vastness of all that cannot be proven or disproven, the uncertain mind shifts into a place of certainty, of conviction, of belief. In countless ways, phenomena have a uniformity and regularity that suggests they will keep on happening the same way over and over. This reliability has taught us how to move megatons of water through city pipes, how to make plasma screens receive hi-res signals from outer space, how to treat ovarian cancer with platinum—but there are still vast swaths of reality that we don't understand. One's attitude toward those unknowables reveals one's belief system: Will all of them one day be understood, or will complete understanding necessarily elude our grasping mind? Do we worship at the altar of science, or do we stick to the facts and calculate its limitations?

Materialism is predicated on the belief that only things that can be proven with evidence are real. "We shall end up understanding literally everything," declares radical atheist Richard Dawkins, a position statement triumphantly lacking any evidence. The scientific method is incredibly effective at dealing with material concerns, but can it deal with things that cannot be known? Are double-blind, quantitative, randomized, clinical experiments the only way of accessing reality? Reason distinguishes and defines; ultimate reality, being indivisible and irreducible, is impervious to such activities.

The benefits of science are undeniable, but those who task science with providing an answer to every question are simply subverting religious belief into scientism. Mystery can be dismissed as inconsequential or revered as an inexhaustible source of wonder, but it's hard to test the belief that untestable things aren't real. In the end, only faith can address unprovables. All science begins with the possibility of a world comprehensible to reason, a knowable cosmos, a harmony pervading the universe. "This source of feeling, however, springs from the sphere of religion," Einstein wrote. "I cannot conceive of a genuine scientist without that profound faith."

In this sense, the otherwise "non-overlapping magisteria" of belief and knowledge can coexist. Einstein called his belief system "cosmic religion." At its heart is a sense of awe, wonder, and reverence before the inner workings of nature. "The ancients knew something which we seem to have forgotten," he wrote, in *Out of My Later Years*. "All means prove but a blunt instrument, if they have not behind them a living spirit." Even though he derided the notion of a punishing or rewarding Creator concerned with human actions, he spoke openly of wanting "to know how God created this world." Einstein's solutions demonstrated links between energy and matter, but he himself acknowledged the impossibility of scientifically proving the existence of Truth. Scientific hypotheses don't lead to Truth—they lead to experiments that can then be replicated.

We can't explain why life *is* rather than *is not*. Science handles the provable facets of reality, but we can't scientifically prove or disprove reality itself—it's simply there (or here) and all tests are mere by-products of its unverifiable yet ubiquitous presence. We can either operate under the illusion that an accumulation of facts necessarily leads to universal truths or side with Oscar Wilde, who believed that "Truth is independent of facts always." Science is a clue in the darkness, a glowing password that opens many gates in the treasury of nature. But we shouldn't overestimate its capacities.

Scientific results don't yield infallible proofs; they offer data from which conclusions may or may not be drawn. These purported truths tend to erode under scrutiny. The results of replicability often decrease over time, a phenomenon known as the decline effect. As the British philosopher John Gray explains, "If we know anything, it is that most of the theories that prevail at any one time are false."

We're inevitably afflicted by some form of magical thinking. Self-deception is an integral component of mental activity. We often identify so deeply with our core beliefs that we only really notice whatever is in accordance with our ideology. "Publication bias" refers to the well-documented tendency for scientific researchers to favor evidence that confirms preexisting hypotheses. As William James argued in his 1896 lecture "The Will to Believe," science is a practice in which "a fact cannot come at all unless a preliminary faith exists in its coming. And where faith in a fact can help create the fact."

By verifying suppositions, science allows thinkers to investigate problems within a framework of belief. As philosopher of science Thomas Kuhn has argued, no scientific paradigm can bring us closer to the Truth, or to so-called objective reality. A particular paradigm of problem solving lasts only until enough irreconcilable anomalies lead to major breakthroughs, followed by the establishment of a new paradigm. This new paradigm isn't necessarily an improvement over the last—it's just a different way of looking at reality, a different shared set of assumptions that can then be investigated through experiments.

In choosing a new paradigm, scientists invariably gain in some ways while also losing in others. Institutional scientists are resistant to new discoveries and theories, primarily because they impinge on a way of making sense of the world. It's for this same reason "extreme" scientists denigrate faith, just as zealots dismiss any evidence not in keeping with their mythologies.

In truth, truths change and evolve, as do we, as do our attitudes to certain beliefs. There are countless things we don't know, that we *can't* know. Some aren't necessarily only true or false; they can be both true and false. There are questions we can't even ask, let alone answer, or prove with 100 percent certainty.

Beliefs masquerade as knowledge even in the scientific community, as Dawkins continually reminds us. "Science can absolutely be a religion," an eminent scientist at McGill University told me when informed that I was researching the belief in scientific immortality. "In fact, it's *my* religion."

It isn't hard to find examples of scientific myths. Daily newspapers are filled with pop-science fables. The Big Bang theory, to pick just one, is a hypothesis put forward by a Jesuit priest, Georges-Henri Lemaître, in the 1920s. Since then, particle physics has become predicated on the belief that the entire cosmos used to be subatomic in scale. A dimensionless, timeless universe smaller than an atom remains a concept impervious to testing. Can any equation ever replicate, let alone prove, what caused that primordial explosion? Can we observe what came before it? (Gas? Nothingness? The question of whether a something can come out of nothing has remained unresolved since Parmenides's time, despite recent atheist contentions to the contrary.)

The Big Bang is a lovely metaphor, simple and elegant, useful in astronomical calculations and abstract experimentation, but so riddled with inconsistencies at its physical core that it's a creation myth. "Myths are simplifications of reality," wrote William Irwin Thompson, "but so are scientific laws, for they magnetize the infinite information of the universe into the fields of their own formulaic descriptions." Our minds cannot wrap themselves around the immense endlessness of an ex-speck-size, now-ever-expanding universe. No matter how we try, we can't make disorder orderly. We can call ourselves chaos theoreticians, but we're not entitled to know how life began or why water came to be or what makes atoms move. Even the idea that nature is governed by laws is simply a belief. Randomness remains a guiding principle of reality. And there's still no way to measure chance, at least not on a cosmic scale.

All of this to reiterate that scientific immortalism isn't scientific; it's belief clothed in scientific garb. Eternal life, whether in this world or the next, is always a story about our inability to comprehend death.

As my research wound down, I often found myself thinking about selfhood: The self; the consciousness thinking these thoughts. Is it the same thing as a soul? Is it our personality? Impossible to say. I came across scientific attempts to explain it as a "distributed neuronal process," but science cannot prove that the individual self is real.

There's no consensus on the nature of the self, no adequate explanation of it. We can never fully understand this self, ourselves. Not only can we not answer "Who am I?"—we can't even answer "Am I?" No self-help book can help locate the self. If it exists, it is nonphysical. It has both conscious and unconscious qualities. Beneath an accessible surface lie the subconscious fathoms.

We have dark sides. We have good days. Are we our moods? Our reactions to unexpected obstacles? Whatever we may be, we're all much too multifaceted to be fastened down. We contain multitudes. Parts of our self can be found; capturing the whole is like grabbing a handful of penumbra. Just as a globule magnified by microscope squirms with squiggles, each of us is a world crammed into a body, a divided society, a parliament of selves as fallible as any government.

The so-called self, akin to an atom, is constantly in motion. The only way to understand it is, as Niels Bohr once said of atoms, by changing our definition of the word *understand*. The self is like happiness, or perfection: something pursued, possibly perceived for moments. A quest for ever after, an infinitely receding goal, a turning point in a tale. Still, we speak of finding our self, or being true to our self, of expressing our self. But a self isn't something we can pinpoint or grasp.

Once we realize the impossibility of knowing ourselves, we can begin to unknow ourselves, which paradoxically may be a way of getting closer to selfhood. The practice of meditation turns consciousness in on itself, preparing the mind for a form of nonrational enlightenment radically different from any reason-based approach. "To know yourself or study yourself is to forget yourself, and if you forget yourself, then you become enlightened by all things," explained the Zen sage Dōgen.

Concepts such as *nirvana, turiya, satori, chaturtha,* and *samadhi* all describe the possibility of entering into "consciousness itself" through self-abandonment. This version of enlightenment is the realization that our body and mind are just an outpouring of the infinite source, that even though we will one day die, we will return to that source. As remote as enlightenment may feel, mindfulness practices such as yoga and meditation are a step toward calming the vortices of unquenchable thought.

Eastern philosophies speak of developing new forms of consciousness, of observing thought, of tapping into a headspace beyond all comprehension and time; but the notion of consciousness is not even recognized as something real in the scientific worldview. Forget knowledge of the knowable; Eastern enlightenment is about the wisdom of the unknowable. As the *Mahabharata* explains, "The greater aim of all is to know that Soul is one, uniform, perfected, not subject to birth, existing everywhere at once, undecaying, stable, eternally unspoiled, of true knowledge, and dissociated with illusions and false realities." Not exactly Richard Dawkins's selfish gene.

Western Enlightenment has to do with everyone everywhere soon being able to understand everything rationally. Its secular paradisification of life here on earth ascribes to the doctrines of progress, freedom, and scientific perfectability. If medicine hasn't figured out how to overcome death by now, it will imminently. Or so many of us think.

In Eastern traditions, a perfected one is "a consummate logician and expert in all the sciences." But this version of logic is very different from what most Westerners think of as logic. Eastern logicians use breathing and stretching exercises as a way to *stop* thinking, to get from thought to thoughtlessness. Yogic immortality, or *jivan-mukti,* means living in the eternal present. In *Yoga: Immortality and Freedom,* Mircea Eliade speaks of "instasis" as opposed to ecstasy, of something within rather than without, of immanence rather than transcendence. To learn more about being on the verge of something that can never be understood, I started going to meditate at a local Buddhist center in Montreal.

Pouring herself another cup of coffee, Ani Lodrö Palmo ventured a joke about the importance of being awake. Even shaved-headed Buddhist nuns of the Shambhala monastic order can use a caffeine jolt, especially on such a busy day. The predawn practice had earlier marked the official start to the Year of the Iron Hare. Preparations were now under way for the annual Elixir of Life ceremony, set to begin in half an hour. Palmo adjusted her maroon robe while rereading that morning's verses of aspiration. The V-like collar on her saffron vest represented the jaws of death, a reminder that she, like all of us, could die anytime.

Celebrants started trickling into the center, a loftlike space on the fifth floor of an immense old downtown building. Bowing gently to greet Palmo, they helped themselves to croissants and coffee. In the small kitchen, a man in a tunic stood looking out the window. "Look at this weather," he huffed, jovially, with a vaguely Eastern accent. "Snow-ice-rain-mud-sun! Like a Russian salad all mixed up, the whole year in one, summer-winter-spring, every season at the same time boom-bang!"

I'd been to the meditation center on several occasions to learn about Buddhism, but I hadn't been in the kitchen before and couldn't tell if newbies such as myself had permission to partake in the orange pekoe. Putting down the kettle, I noticed a military-looking young man in a uniform watching me. This being my first time attending such a ceremonial event, I hadn't yet encountered a Dorje Kasung—as these guards are known—so didn't realize his role as protector of the dharma and bringer of wakefulness. In fact, I thought he was a security officer about to bust me for copping a tea bag.

"Good morning," he intoned, powerfully.

"Hello," I replied, nervously wrapping my hands around the mug.

A moment of silence.

"Are you the security here?" I ventured.

"You could put it that way." He looked into my eyes, neutrally.

"Does this gathering really need protection?"

"Actually, I'm posted here with a specific mission: to make sure you all remember to be awake and conscious."

"Well, this should help." I smiled, tapping my cup.

"Namaste," he said, mindfully bringing his hands together.

Shambhala is often described as Buddhism for Westerners. Its main aim is teaching meditation. Although Buddhism is certainly a religion, it can also be described as a philosophy or, as adherents prefer, a science of mind. Sitting in meditation calms the thinking brain, they say, allowing it to contact things as they truly are (rather than filtered through the ego). All Buddhists believe in impermanence—the idea that everything in life is fleeting. Whether it be the seasons or our bodies, everything changes. Buddhists see the desire for physical immortality as just another manifestation of our ego's mistaken need for things not to change. Alas, permanence is not the nature of nature. Death, however, is.

Rather than turning away from the unpleasantness of death, Tibetan Buddhists try to focus on that unpleasantness, to meditate upon it, to lean into it. The Dalai Lama visualizes his own death every day. A basic precept of their teaching is that our attempts to avoid suffering do not actually eliminate suffering. The Buddha's first noble truth, *dukkha,* is that suffering and discomfort are a real part of human life. And most suffering stems from our cravings and desires. The very things we do to escape suffering amplify our suffering.

Pema Chödrön, an esteemed American nun who is among the more recognized names in contemporary Buddhism, has often written about reframing the idea of hardship and struggle. Her book *The Wisdom of No Escape* explains that, by implementing the Buddha's teachings and recognizing the first noble truth, we can become sensitized to an underlying energy behind all of existence. It is always circulating in and around us, coursing through everything. She speaks about it in the same terms Father Gervais did. "If something lives, it has life force, the quality of which is energy, a sense of spiritedness," she writes. "Without that, we can't lift our arms or open our mouths or open and shut our eyes." This dynamo of impermanence is a basic creative essence we can come into contact with through the practice of meditation. But our daily life can also block out the life force's light. "Why do we resist the life force that flows through us?" Chödrön wonders. "If you can realize the life energy that makes everything change and move and grow and die, then you won't have any resentment or resistance."

The life force can be enjoyable or challenging. It manifests in myriad oppositional ways: easy or hard, fun or excruciating, brilliance or despair. An aim of Buddhism is to realize that our sufferings and our discomforts are simply ways in which we feel the life force. Shambhala's founder, the Tibetan monk Chögyam Trungpa Rinpoche (1939–1987), described obstacles as "adornments." The goal is to attain a level of sensitivity, through meditation, in which both pain and contentment become ornaments we hang upon the tree of our lives. At a certain point, mortality becomes just another bit of tinsel.

In Shambhala, enlightenment is a youthful, playful state of being. Self-realization is like a second birth, Trungpa Rinpoche used to say, in which we rediscover our innocent, childlike curiosity with all things. This exploration of life situations is connected with a sense of eternity, he writes, with an understanding of how the impermanence of life is connected to the eternity of life: "the constant changing process of death and birth taking place all the time." Trungpa Rinpoche called this the wisdom of eternity. In that state, "nothing threatens us at all; everything is an ornament. The greater the chaos, the more everything becomes an ornament."

The embodiment of this ideal is Padmasambhava, the Indian sage who brought Buddhism to Tibet. Also known as the Lotus-Born, he is associated with eternal youth, inquisitiveness, and childish wonder. In images, he holds a vase of longevity filled with the nectar of deathless wisdom. "Padmasambhava still lives, literally," Trungpa Rinpoche once declared. "The fact of physical bodies dissolving back into nature is not regarded as a big deal. So if we search for him, we might find him."

Padmasambhava hid spiritual treasures so that later generations would discover them. These leavings, called *termas,* became revered sources of Buddhist transmission. The scrolls and secret teachings buried away within caves, forests, and lakes concealed lessons that only reveal themselves at the proper time—and to the right person. Trungpa Rinpoche himself stumbled across a number of them, adding to his mystique.

He uncovered most of his treasures at the holy Tibetan mountain of Kyere Shelkar. He found his first scroll there as a teenager. Aged nineteen, he came across a block of marbled red stone studded with gemstones. According to eyewitnesses, as soon as he dislodged it from the cliff, "thunder clapped, it started raining, and a pleasant fragrance came from the sky. Everyone there was amazed and started crying uncontrollably." They put the stone in a nearby temple and meditated around it for seven straight days. At that point, the casket opened and a small scroll unfurled, containing instructions, prophecies, and sadhanas composed over a thousand years earlier.

The most extraordinary of Padmasambhava's *termas* is an eighth-century CE text found four or five centuries after he composed it, in the Gampo Hills of Tibet. The *Bardo Thodol,* or *Tibetan Book of the Dead,* as it came to be known, is an instructional manual for dead souls, and one of the key texts of Tibetan Buddhism.

Bardo means "interval," or "liminal space." It's an in-between place, an intermediate state. Life itself, the period between birth and death, is considered to be a bardo. Meditation is a bardo, as is dreaming. Three bardos come after we die. Being in them is bewildering, something like being shipwrecked. "You've left the shore, but you haven't arrived anywhere yet," writes Pema Chödrön. "You don't know where you're going, and you've been out there at sea long enough that you only have a vague memory of where you came from. You've left home, you've become homeless, you long to go back, but there's no way to go back."

Jumping into the sea, at the moment of death, we are granted a dazzling vision of clear light. This illumination is an opportunity to realize that our consciousness *is* the radiance. The voidness of our dead mind turns out to be the shining reality of Buddhahood itself. Unfortunately, most of us cannot reconnect with the immutable light that both is and is not us, so we pass into the second bardo, which entails visions of deities who judge and tempt. Just as a dream can be so exhilarating we feel we're living it, or so nightmarish it freaks us out well after we've woken, these karmic illusions are vividly real. They can be wondrous and peaceful or wrathful and terrifying.

We encounter the Five Dhyani Buddhas, each of which is a manifestation of Buddha. They proffer complicated ways of attaining liberation. We meet Akshobhya, the Buddha of seeing things as they truly are, of consciousness as an indestructible reality. As blue as the oceanic depths and as white as sunlight glinting on the sea, he embodies the ability to distinguish reality from illusion. He shows us the pure water of liberation, a radiant light that, again, is us. If we're not worthy of the Dhyanis' invitations, we end up confronted by "the fifty-eight blazing, blood-drinking wrathful ones." For the rest of the interregnum, we move through a tormenting series of hallucinations back into reincarnation and the bardo of life starts all over again.

Before the Elixir of Life ceremony, when I reached out to interview Ani Lodrö Palmo about the bardo and other Buddhist views on immortality, she suggested I speak with Esther Rochon, a local science-fiction writer. "She would be a better resource than me," Palmo informed me, by e-mail. "She has been studying this topic for many years."

Esther and I met at the center on a quiet Friday afternoon. A soft-spoken, Joan Didion–esque woman with gray hair and a certain 1960s quality, Esther radiated calmness. She chose her words carefully, speaking with a precision befitting her university training in advanced mathematics.

We started our discussion on the topic of the soul—a conceit that most Buddhists reject while simultaneously accepting the notion that some part of us gets reborn into a new body after death. "We do not speak of a permanent soul that survives forever after physical death," Esther said. "But we do speak of mental processes that are ever-changing, and which continue after the body decomposes."

Because the law of impermanence is so central to Buddhist thought, they have trouble with eternalism, the belief that a part of us lives forever in some static way. "Most of us are so into eternalism we need a cold shower of emptiness," Esther explained. "Meditating on the present moment is an essential step to undoing that. It's an antidote."

"If you aren't eternalists, then why does Buddhism speak of the bardos?" I asked.

"What might be called a 'thought lineage' or 'mindstream' does survive the death of the body. It's not an eternal soul, but rather a chain of mental events."

"That chain is what goes into the bardo realms?"

"Yes. Our bardo self, which is a voidness rather than a soul—a kind of astral body—is what goes there. It can get quite skittish because it doesn't have a body and it wishes it were in a body, but its fear makes no sense because it can't even have a body."

"So it's our tendencies that are reincarnated, our mental activity?"

"In a way, yes. But it's not what you might call the 'personality' or 'self' that lives on. It's a state of mind that we train ourselves to develop. A compassionate 'part' of the mind survives, but it's not attached to us. It's something that's in all of us."

"But essentially part of the mind goes on beyond death?"

"It depends on how you define *part*. The mind of enlightenment—the mind which is totally awakened—that may last. That potential is within us. It does not change. It is primordial and eternal. But it is not something you can touch with your finger."

"So it's a part of us that's not a part of us? Sounds paradoxical."

"It's subtle but not contradictory. The teachings speak of it as a water. Your essential mind is like water in a glass. When you die, it pours out and returns to the source. It retains no memories of what it had been before. A Buddha is someone who is always in contact with the essential nature of their mind."

As we were nearing the end of our conversation, I thanked Esther for taking the time to speak with me about Buddhist beliefs. She immediately corrected me: "Belief isn't something important in Buddhism. Buddha said you have to test every teaching I've given. Rather than just *believe* in it, you have to work with it to see what your experience of it is, to see how it works for you."

"But there's a human tendency to not see our beliefs as beliefs."

"Yes, but Buddhism is a nontheistic faith, unlike religions which believe in a God. The practice of meditation doesn't have anything to do with believing or not believing in God. It's often said that Buddha discovered the truth precisely because he didn't believe in God."

"Yes, but isn't the bardo a matter of belief? How can we really know what happens after death?"

"There are provisional teachings and then there are essential truths. Our stories and myths put people in a frame of mind where they can get at something really essential. The provisional stuff engages people's intellects so they are able to understand. But the real heart of the matter is beyond logic."

"So you can't discuss the essential stuff."

"Not at all. But the teachings allow people to see whether they are relevant for them or not."

"Is Shambhala the same as other denominations of Buddhism?"

"Different schools have different teachings. The pedagogical approach may be different, but the main ideas are ultimately the same. The Four Noble Truths aren't beliefs so much as categories of experience. Anyone can try them out. That's the point."

There's a famous old Buddhist aphorism: "If you meet the Buddha on the road, kill him." Many of their key texts caution against holding on to beliefs. As Pema Chödrön notes, in this regard, "You want something to hold on to, you want to say, 'Finally I have found it. This is it, and now I feel confirmed and secure and righteous.' Buddhism is not free of it either. This is a human thing." It's easy to judge others according to how their beliefs differ from our own. It's much harder to question one's beliefs, to let go of them, to transcend them. Still, we all interpret reality, and our beliefs help us do so. But Buddhism teaches that our interpretation of the noninterpretable is just that: an opinion rather than the Truth.

The very first mention of logic occurs in the writings of a fifth-century BCE Greek philosopher named Parmenides. His thought survives in an extraordinary poem, "On Nature," which recounts his journey to the heart of the universe. There he meets a goddess who introduces him to something not of this world: logic. She tells him that logic proves that nothingness does not exist: "That which is cannot not be." In other words, if you exist, you cannot not exist. Therefore, the human soul will necessarily endure indefinitely.

Logic begins as demonstrative proof of immortality. Thousands of years later, in contemporary science, the same set of variables filtered through a different belief system becomes demonstrative proof that there is no God.

We *can't* know; we can only choose what we want to believe.

Today we think of rationality as sensible, sound, evidence-based. Parmenides's version of logic is what we would classify as completely illogical. We arrive at it not through reason-based thought, deductive science, or sense perception (all of which prevent us from accessing things as they truly are, he said) but through mysticism.

Like Zen teachers, early Greek logicians wanted to make us aware of a realm beyond intellect, of something greater than materiality. Parmenides's disciple Zeno posited a series of paradoxes that logically disprove the possibility of motion. They were intended as metaphysical exercises: thinking about them launches the mind beyond duality into a place where the mind game actually makes sense. Paradox, in the pre-Socratic sense, is an arrow that connects us to the unseen. Movement is a shadow of true reality, Parmenides and Zeno argued, which is unchanging, static, and indestructible. They believed metaphysical contemplation helps us "be immortal, to the extent possible."

Classical Western philosophers claimed that we are all always able to reconnect with our inherent divinity. Human minds, Socrates argued, stemming from the Demiurge's mind, contain traces of their original omniscience. Our active intelligence, Aristotle added, is like a spark of God within the brain that returns to God when our bodies die. "It is this alone that is immortal and eternal," he wrote in *De Anima*. The part of our mind connected to that ulterior dimension has limitless capacities. It can make anything real. Simply by thinking something, it can bring that thing about. It reaches a point where "that which thinks is the same as that which is thought." When a thought becomes the object of thought, Aristotle concluded, our consciousness unconsciously realizes that it both emanates from the eternal and is itself eternal.

Such thinking may seem Eastern at heart, but such is the nature of early Western logic and science. At the start, math wasn't the rote, rational, and reliable discipline we think it is today. Back then, math was more like a preparatory branch of theology, a means of becoming familiar with immaterial nature. According to the early Greek mathematician Euclid, geometric shapes exist only in another dimension independent of present reality. We cannot see in nature any examples of pure geometry, he argued, only shapes that approximate our ideals. Perfect squares and circles exist in our thoughts—but not in real life. This epiphany laid the foundation for Plato's doctrine of Forms. Math linked this material world of divisible reality to the higher realm of indivisible Forms.

For ancient Greek thinkers, math brought the sacred into focus. Each soul, they affirmed, is made of mathematical ratios. Delving into geometry, explained Pythagoras, leads to the eternal. Empedocles described God as a rounded sphere equal to itself in every direction, without any beginning or end. It's for this reason that Plato hung a sign over the door to his academy that said "Let no one devoid of geometry enter here." For the Athenians, knowing geometry meant knowing that your soul is immortal.

Speculative mathematics today remains a realm of beauty and poetry, with thinkers working on turning orbs inside out, or calculating the square root of negative one, or measuring spheres with a plurality of centers and no circumference. Such science is fully aestheticized, above material needs, less about practical application than it is about art. Realizing that pi can be calculated until the end of time has nothing to do with the promise of solutions; on the contrary, the transcendental and irrational number relating a circle's diameter to its circumference demonstrates, again, our inability to attain certainty. This world remains a world of conjectures. We believe that dimensions exist yet can't quite define or comprehend them. And what on earth is time, the fourth dimension? Newton argued that time is absolute; Einstein deemed it relative; and then Gödel proved that it does not even exist.

Buddhists would agree. For them, this whole situation is illusory.

Around forty or so people had gathered to observe the Elixir of Life ceremony at Montreal's Shambhala center. We all sat down on our meditation cushions. Volunteers handed out a photocopied instructional page and a vase of water perfumed with saffron.

Ani Lodrö Palmo welcomed us, leading us through a silent meditation. I sat watching thoughts floating through my mind: *If it isn't belief, then how do we explain the idea that the current leader of Shambhala is a reincarnation of an eighteenth-century Tibetan teacher? Is that not belief? How can that be tested? Isn't Buddhism just another flawed system of dealing with the human experience? But aren't the benefits of meditation undeniable? Maybe I'm doubting too much, questioning things I shouldn't be?*

Palmo then asked us to gather saffron water in our hands and splash it onto our faces. Doing so, we repeated lines from a photocopied handout: "This body that I possess has come together due to inconceivable circumstances. It is aging moment by moment. This is simply the nature of the human condition and I should celebrate it."

Having moved into the second half of my thirties, I felt that to my inconceivable core.

We then purified our throats and uttered a prayer about speaking with compassion. Next, we sprinkled water over our hearts. Palmo described how all pleasures change to suffering. Within the shifting sands of impermanence, however, part of our mind is always connected to a greater infinite source: "From beginning-less time, this radiant, luminous mind has played in the space of uninhibited wisdom."

We then spent fifteen minutes contemplating a passage about the futility of hanging on to the past. We cannot overcome the process of aging, the ceremony's handout explained, urging us to realize that all beings suffer this plight. Rather than hide from it, we should celebrate it. One paragraph stood out: "All the logic in the world will not save me from the simple truth that I age. Sickness is my companion; it follows me everywhere. I try to avoid the truth, but the painful conversation with sickness never ends. Death is my friend, the truest of friends, a true friend that never abandons me. Death is always waiting for me."

So this was the Elixir of Life? Illness as our one true friend. Death as soul mate. It was hard. As hard as reality. I sat there, shifting on my cushion, and tried to focus on the ceremony's meaning while half listening to the sounds in the street below. There on my seat, I could feel it, life, as a person now alive who knows, as we all do, even if we try not to, that one day I must die.

Epilogue

Springs Eternal

The explanation must be that I have been filled from some external source, like a jar from a spring.

—Socrates, *Phaedrus*

I write. I write that I am writing. Mentally I see myself writing that I'm writing and I can also see myself seeing that I am writing. I remember writing and also seeing myself writing. And I see myself remembering that I see myself writing and I remember seeing myself remembering that I was writing and I write seeing myself write that I remember having seen myself write that I saw myself writing that I was writing.

—Salvador Elizondo, *The Graphographer*

ALTHOUGH I couldn't attend Auntie Tiny's funeral in Budapest, I thought of her when my aunt Kati passed away as this book neared completion. Kati, my father's older sister, was a paragon of elegance and sophistication who succumbed to pancreatic cancer on Christmas Eve 2012. The funeral service took place at the Mount Royal Cemetery. Her sons, my cousins, arranged for a memorial mass in Hungarian. The organist sang Gregorian alleluias about never dying. The priest sprinkled holy water upon her urn with a golden aspergillum. He spoke of "God's sacred secrets," the promise of eternal life, the way the cross teaches us that we live on, even though we die.

If we believe.

When the time came to take Communion, I wasn't sure whether to join the queue or remain seated. "Should we go eat the wafer?" I asked my brother, next to me. He'd never taken Communion in his life, so he shook his head. I decided to stay in the pew with my family. My dad told us he'd dreamed of Kati the night she died. She told him she was setting out on a new adventure in a new land. He felt sure she was still out there, somewhere, elsewhere, among the stars. I could see the consolation of immortality helping him in his grief.

As we put our coats on to leave the chapel, I noticed a Christmas tree covered in heart-shaped cards. Each had been inscribed with a message to a dead loved one: "Missing you, Suzanne." "Always thinking of you." "Grampa, I would've liked to have spent more time with you." "Here we are, seven years later: I still love you with each passing day, and I continue to raise our girls the way you would have wanted me to."

I couldn't help crying as I read one of them, in a child's handwriting: "Mom, I never knew it was gonna hurt so much. Love, Lori."

Next to the tree stood a stack of brochures advertising the cemetery's services. "The dictionary defines 'perpetuity' as eternity or the rest of time. At Mount Royal Cemetery, burial rights are offered in perpetuity." As I read, I remembered Father Gervais, his funeral, every priest I'd heard speak about living forever and ever.

As much as my time with Father Gervais had affected me, I hadn't become a believer, at least not in the monotheistic sense. But something *had* changed for me. It wasn't just that I'd come to believe so firmly in the need to believe, in the fact that belief is implicit in everything we do. It was something more nebulous. When I came across a newspaper report about thieves taking stained-glass windows out of an old country church, it felt extremely wrong to me. It wasn't just stealing; it was stealing from *the Lord's house*. Yet even if I believed in God, I didn't just believe in *one* God. I also believed in the gods of ancient Greece, in Haida divinities, in Hindu deities.

I liked Einstein's self-portrayal as "a deeply religious non-believer," but I felt that I believed in all religions—without following any of them. Perhaps I had what Byron called "faithless faith." He wrote of trying to wrench something out of death that might confirm or shake or make a faith, but all he found was mystery: "Here we are, and there we go—but *where*?" I certainly couldn't answer, even after speaking with all those spiritual leaders. I'd learned so much from each of them, but respecting a theology doesn't require converting. We can connect with political others simply by connecting with their belief systems, by appreciating their myths, by entering their stories. Even though I didn't go regularly to the Sufi center, I was pretty sure I believed in the Sufic idea of Ultimate Reality. I definitely loved the Hasidic notion that God is in everything. And I'd never seen the anthropomorphic, white-bearded God of Christianity, but I'd definitely experienced the life force.

Leaving Aunt Kati's funeral, it occurred to me that I didn't even believe in the life force anymore: I knew it to be real. If someone were to ask me whether I believed in it, I would answer like the man who, asked whether he believed in the possibility of being reborn, replied, "Believe in it? I seen it done."

I felt the life force on Copperfield's private island; at Florida's Fountain of Youth; in Marilyn Rossner's message circle; at the 2068 party; with Rabbi Haim Sherrf's family at Shabbat dinner. I felt it each time I saw Father Gervais. I felt it in Auntie Tiny's apartment, and I felt it after Aunt Kati's death. I felt it falling in love and I felt it while sleeping. That dream of the fountain had been a vision of the life force, the energy that sustained me throughout the years I spent writing this book.

You cannot end a never-ending quest, but you *can* finish the manuscript. I undertook several more research trips, but nothing brought me closer to the truth than the Buddhist Elixir of Life ceremony, with its message that aging isn't something we need to recover from. It's our condition, and we deserve to celebrate it. Death, too, is inescapable. It awaits all of us always. It's our one true friend, our constant companion, our soul mate.

Still, I kept hunting. I took a simulated submersible ride to the ocean floor, where volcanic vents are presumed to be the source of life on earth. I flew to Crete to see what the Minoans knew about immortality. (Answer: mysteries in caves.) I took mud baths, visited Turkish bathhouses, booked a sensory-deprivation appointment at Montreal's Ovarium, a womblike flotation chamber. I researched the rejuvenatory and life-preserving effects of conjugation and autogamy on the fission rates of unicellular organisms. I didn't get to the holy waters of Lourdes, but I did visit the ruins of a mineral-spring asklepieion in Athens. The stones glared silently.

Trying to find a new way of thinking about progress, I spent days sifting through the archives of Joel Hedgpeth, marine biologist and founder of the Society for the Prevention of Progress. I found unpublished essays about oceanic life, poems, and love letters—but no resolution. Whether

I believed in progress or not, I'd learned the First Noble Truth: suffering and cruelty won't ever end. But there's meaning in the struggle.

I spent a month investigating hot springs across Italy. I went to the ancient waters of Bagno Vignoni in Tuscany. On a tip from Martha Morano, Copperfield's publicist, I checked into a splendid thermal wellness resort called Adler Thermae. I headed up to sacred founts in the mountainous valleys of Alto Adige. I trekked into the crater of Pozzuoli, outside Naples, thought to be the entrance to Hades. All I found was mephitic vapors.

I took a ferry to the volcanic island of Ischia, famed for its healing waters, where I spent time in Apollo's Temple of the Sun. It consisted of three rooms dedicated to "the eternal cycle of life." In the first, water poured from a holy vaginal relic symbolizing the origin of life. I tried to figure out why the mossy orifice was crowning a stone pyramid, but I couldn't get too close without getting splashed. The second chamber celebrated earthly pleasures, with erotic bas-reliefs depicting various sexual positions. There were bacchanalian figures blowing into horns, flying toward the sun, serenading goats. It was bestial and lurid and full of strange vitality. Pink water lilies with erect yellow stamens smiled on knowingly. Apollo presided over it all in a carriage drawn by swans.

The third room was the darkest. Here the theme was death: the end of mortal life. That stream from the sacred opening trickled across the floor before vanishing into a green grotto-vortex above which a statue of the Cumaean Sibyl gazed on pensively, finger to cheek. She knew not whence we came nor where we go. On the way out, I came upon a rock engraved with a line from a seventeenth-century poem: "All bliss consists in this: to do as Adam did."

What this Adam did next was indeed blissful: I traveled to Capri and jogged up a cliffside path that ended at a breathtaking sea-view lookout. The blue-green immensity lay beneath me, concealing infinity. Its cresting waves resembled a bedcover. Beneath the sheets, morphing slabs of obsidian and jade writhed together. Light fell off in sheaths.

A fragment of Heraclitus came to mind: "As water springs from earth, so from water does the soul." Springs. How curious that the word spring, whether referring to the season or to the stream of flowing water or to the upward movement, has in all cases a subtext of rejuvenation. To often be dead, and often return to life. Some birds hung in the air, black against the sun. Circling, their silhouettes burst into whiteness for a moment.

Just like that, I was out of the forest. I turned around and looked upon the beauty of the trees. I looked back at the sea. Some themes will remain forever mysteries.

A few days later, I found myself walking along the seashore in Positano. The tide offered up surging ringlets of foam. I stopped to pick up a few emerald-colored pieces of glass, bottle shards smoothed and rounded by the buffeting sea. I'd collected a jarful of these rough gems as a fourteen-year-old, even then fascinated by the ocean's ability to make art of man's debris. This is what I learned: nature transforms. Whatever happens next, it'll be different from what we imagine.

Acknowledgments

MY HEARTFELT THANKS to the many people who helped this book become real, especially Mark Raymond Collins, Martha Morano, Sharry Flett, Warren Auld, Esther Rochon, Radwan Ghazi Moumneh, Himo Martin, Cheskie Lebowitz, Jonathan Freedman, Leslie Feist, Nicole Pierpont, Sophie Leddick, Taras Grescoe, Jocelyn Zuckerman, Bill Sertl, Peter Würth, Tyler Graham, Sarah Amelar, Korbett Mathews, Charles Levin, Ian Jackson, David Tobias, and Janice Kerfoot. Thank you to the St. Augustine Historical Society, the Rene Goupil House, the Cryonics Institute, the Esalen Institute, the Internet, and the libraries of the world. Thank you to the QWF, Lori Schubert, the Canada Council, the Public Lending Right Program, and Access Copyright. To Warren and the Hedgpeth family: I still hope to write about the Society for the Prevention of Progress one day. Much thanks to John, Vanessa, Dexter, and all at the Long Haul. Thank you to Demetra and the Zoubris gang for putting up with me (aka Larry). *Efharisto* George and Jimmy Vitoroullis. Oncle Jax Andison: "It's been a service pleasuring you."

For the conversations and insights: Melanie Sifton, Donald Antrim, Lorin Stein, Jenna Wright, Clay Weiner, Tracy Martin, Billy Mavreas, Doctor Oz, Michelle Sterling, Ithamar Silver, Daniel K. Seligman, Jessica Wee, Miguel Syjuco, Edith Werbel, Peter Meehan, Roger Tellier-Craig, Sabrina Ratté, Danny, Jesse, Nat, Tim, Cathy, and John Riviere, the Sanchez clan—Carlos and Anny, Jason and Elena, Louise and Joe, Suroosh Alvi, Tim Hecker, Brett Stabler, Yaniya Lee, Matt Brown, Robbie Dillon, Fred Morin, Cassady Sniatowsky, Michael "Guru" Felber, Elliot Jacobson, Arjun Basu, Philippe Tremblay-Berberi, Bartek Komorowski, Tim Fletcher, Theo Diamantis, Nathan Curry, Zoe

Mowat, Michelle Marek, Anthony Kinik, Thea Metcalfe, Robin Simpson, Anna Phelan, Sarah Louise Musgrave, Hart Snider, Aisling Chin-Yee, Mark Slutsky, Susannah Heath-Eaves, Mila Aung-Thwin, Bob Moore, Yung Chang, Allan Moyle, and Kurt Ossenfort.

The Tessler Agency played a key role in the story of how this book became a book. I'm thrilled to now be working with Tracy Bohan of the Wylie Agency.

My eternal gratitude to everyone at Scribner, especially Nan Graham, Susan Moldow, Kelsey Smith, Leah Sikora, and Paul Whitlatch. I'm indebted to Steve Boldt for his excellent copyediting work, and to Dan Cuddy and the entire production team. My editor and copilot, Alexis Gargagliano, encouraged me to go deeper and never wavered in her faith, patience, and understanding. Thank you, Alexis, for guiding me, for collaborating with me, for being so generous and so brilliant.

Much thanks to the fantastic team at Doubleday/Random House Canada: Lynn Henry, Amy Black, Kristin Cochrane, Christopher Frey, and Adria Iwasutiak.

Loving thanks to Annie Briard, David Gawley, Kicsi Néni, Kati Néni, Antal, Ricsi, Zsuzsi, Paul, Peter, Brian, Mandy, Ian, Sheelagh, Willie, Sam, and Laura.

Thank you to the incomparable Liane Balaban for the wisdom, poetry, and inspiration: *"Toujours nous irons plus loin sans avancer jamais."*

To Natasha Li Pickowicz, thank you for being in the life force with me. You know more than anyone what this entailed.

My deepest thanks to my family—my brothers, Miska and Julian, my father, András, and my mother, Linda.

Sources

Est enim benignum (ut arbitror) et plenum ingenui pudoris, fateri per quos profeceris.

—Pliny, *Naturalis Historia*

EVERYTHING IN THIS book actually happened or can be traced back to verifiable sources. The people, the places, the events, the ideas, the dreams: all are real. The sequence has been modified for narrative flow and structural continuity. Innumerable documents informed my research; those to whom I am most indebted are listed below. This is not intended as a work of academic scholarship; it is beholden only "to the sacred majesty of Truth." In that aim, I invite anyone curious about the derivation of a certain fact or passage to contact me (through adamgollner.com) should clarification be desired. Corrections, as well, are welcome; I alone assume responsibility for any errors or misinterpretations in the text. My thanks to all the authors listed below.

General Sources

Karen Armstrong, *In the Beginning; A Short History of Myth*
Margaret Atwood, *Negotiating with the Dead*
W. H. Auden, *The Sea and the Mirror*
Mikhail Bakhtin, *Rabelais and His World*
Julian Barnes, *Nothing to Be Frightened Of*
Graham Bell, *Sex and Death in Protozoa*
Harold Bloom, *The Visionary Company*
Lucian Boia, *Forever Young*
Jorge Luis Borges, *Collected Fictions; Selected Non-Fictions; History of Eternity*
John Burrow, *A History of Histories*

J. B. Bury, *The Idea of Progress*
Joseph Campbell, *The Hero with a Thousand Faces*
Anne Carson, *Eros the Bittersweet*
Pema Chödrön, *The Wisdom of No Escape*
Lorraine J. Daston and Peter Galison, *Objectivity*
Lorraine J. Daston and Katharine Park, *Wonders and the Order of Nature, 1150–1750*
John Dewey, *The Quest for Certainty*
Annie Dillard, *For the Time Being*
Wendy Doniger, *The Implied Spider; Other Peoples' Myths; Tales of Sex and Violence; The Hindus: An Alternative History*
David Eagleman, *Incognito*
Mircea Eliade, *Myths, Rites, Symbols; Cosmos and History: The Myth of the Eternal Return*
J. G. Frazer, *The Golden Bough; The Fear of the Dead in Primitive Religion; The Belief in Immortality and the Worship of the Dead*
Johann Wolfgang von Goethe, *Faust 1 and 2,* tr. David Luke
Stephen Jay Gould, *Wonderful Life*
John Gray, *Black Mass*; *The Immortalization Commission*
Béla Hamvas, *The Philosophy of Wine*
Thierry Hentsch, *Truth or Death: The Quest for Immortality in the Western Narrative Tradition,* tr. Fred A. Reed
Heraclitus, *Fragments*
William James, *Human Immortality; The Will to Believe; The Varieties of Religious Experience*
Carl Gustav Jung, *The Collected Works*
Károly Kerényi, *Dionysos: Archetypal Image of Indestructible Life; Asklepios: Archetypal Image of the Physician's Existence; Eleusis: Archetypal Image of Mother and Daughter*
Thomas S. Kuhn, *The Structure of Scientific Revolutions*
Georg Christoph Lichtenberg, *The Waste Books*
Thomas Mann, *Death in Venice*
Michel Eyquem de Montaigne, *Essays*
S. Jay Olshansky and Bruce A. Carnes, *The Quest for Immortality: Science at the Frontiers of Aging*
Parmenides, *On Nature*
Adam Phillips, *Houdini's Box; Terrors and Experts; On Balance; Side Effects*
Plato, *Symposium; Phaedrus; Phaedo; Republic; Euthyphro; Parmenides*
Porphyry of Tyre, *On the Cave of the Nymphs*

Robert Prehoda, *Extended Youth*
Kathleen Raine, *Blake and Tradition*
Mary Roach, *Spook; Stiff; Bonk*
Richard Rorty, *Contingency, Irony, and Solidarity*
José Saramago, *Death with Interruptions*
Susan Sontag, *Reborn*
Enid Starkie, *Baudelaire*
Charles Taylor, *Sources of the Self*
William Irwin Thompson, *The Time Falling Bodies Take to Light*
Ivan Turgenev, *A Sportsman's Sketches*
Evelyn Underhill, *Mysticism*
Eliot Weinberger, *An Elemental Thing*
Oscar Wilde, *The Truth of Masks*
Ludwig Wittgenstein, *Lectures and Conversations on Aesthetics, Psychology, and Religious Belief*
William Wordsworth, *Ode: Intimations of Immortality from Recollections of Early Childhood*

Introduction

Simon Blackburn, *Truth: A Guide*
Eugene Fontinell, *Self, God, and Immortality*
John A. Mann, *Secrets of Life Extension*
Albert Rosenfeld, *Prolongevity*
Edith Sitwell, *The English Eccentrics*
Lyall Watson, *The Romeo Error*

Belief, Grief

Ernest Becker, *The Denial of Death*
Ronald Britton, *Belief and Imagination*
Robert Buckman, *Can We Be Good Without God?*
Alex Comfort, *The Process of Ageing*
E. M. Forster, *Howards End*
John Hick, *Death and Eternal Life*
Howard I. Kushner, *Self-Destruction in the Promised Land*
Friedrich Wilhelm Nietzsche, *The Birth of Tragedy*
Meghan O'Rourke, "Good Grief," *The New Yorker,* February 1, 2010
Michael Shermer, *The Believing Brain*

Islam and Judaism

The Zohar
The Torah
Farīd al-Dīn Attar, *The Conference of the Birds*
Rav Berg, *Immortality*
O. M. Burke, *Among the Dervishes*
Nikolai Gogol, *Dead Souls*
Mahmud Shabistari, *The Secret Rose Garden*
Idries Shah, *The Sufis*

Christianity and the Life Force

Aristotle, *De Anima*
Henri Bergson, *Creative Evolution; The Creative Mind; The Two Sources of Morality and Religion*
Marc Gervais, *Ingmar Bergman: Magician and Prophet*
Ignatius of Loyola, *The Spiritual Exercises*
Diarmaid MacCulloch, *Christianity: The First Three Thousand Years*
Plotinus, *Collected Works*
George Bernard Shaw, *Back to Methuselah; Saint Joan; Man and Superman;* essays and lectures on the life force
W. S. Smith, *Bishop of Everywhere: Bernard Shaw and the Life Force*

Water

Apuleius, *Metamorphoses (The Golden Ass)*
Loren Eiseley, *The Immense Journey; The Night Country*
Richard Broxton Onians, *The Origins of European Thought*
Walter F. Otto, *Dionysus, Myth and Cult,* tr. Robert B. Palmer
G. Cope Schellhorn, *Man's Quest for Immortality, from Ancient Times to the Present*
Eric Trimmer, *Rejuvenation*

The Fountain of Youth

Pseudo-Callisthenes, *The Alexander Romance*
Luella Day, *Tragedy of the Klondike*
Gerald Gruman, *A History of Ideas About the Prolongation of Life*
E. W. Hopkins, "The Fountain of Youth," *Journal of the American Oriental Society,* 1905
Jill Lepore, "Just the Facts, Ma'am," *The New Yorker,* March 24, 2008

Lael Morgan, *Good Time Girls of the Alaska-Yukon Gold Rush*
Leonardo Olschki, "Ponce de León's Fountain of Youth: History of a Geographical Myth," *Hispanic American Historical Review,* August 1941
Richard Stoneman, *Alexander the Great: A Life in Legend*
Aleksandra Szalc, *In Search of Water of Life: Alexander Romance and Indian Mythology*
Various documents from the St. Augustine Historical Society

Esalen

Burkhard Bilger, "The Possibilian," *The New Yorker,* April 25, 2011
Robert Bly, *Iron John*
Richard Jefferies, *The Story of My Heart*
Jeffrey J. Kripal, *Esalen: America and the Religion of No Religion*
Michael Meade, *The Water of Life*
Géza Róheim, *The Eternal Ones of the Dream; Animism, Magic, and the Divine King; The Gates of the Dream*
Rudolf Steiner, *First Steps in Inner Development*
William Irwin Thompson, *At the Edge of History*
Victor Turner, "Betwixt and Between: The Liminal Period in *Rites de Passage*"
Arnold Van Gennep, *The Rites of Passage*
Walt Whitman, *Leaves of Grass*

Magic

Paul Christian, *The History and Practice of Magic*
David Copperfield, *Tales of the Impossible; Beyond Imagination*
Simon During, *Modern Enchantments*
Erik Hornung, *Conceptions of God in Ancient Egypt*
Walter Kaufmann and Forrest E. Baird, *Ancient Philosophy*
Eliphas Lévi, *The History of Magic; Transcendental Magic*
Thomas Mann, *Mario and the Magician*
Karol Myśliwiec, *Eros on the Nile*
David Silverman, Leonard H. Lesko, John Baines, and Byron E. Shafer, *Religion in Ancient Egypt*
Bill Zehme, *Intimate Strangers*

Eros and Symbolism

Charles Baudelaire, *Paris Spleen*
J. E. Cirlot, *A Dictionary of Symbols*

Samuel Taylor Coleridge, *The Statesman's Manual*
Sándor Ferenczi, *Thalassa*
Sigmund Freud, *Beyond the Pleasure Principle*
Melanie Klein, "The Importance of Symbol Formation in the Development of the Ego," *International Journal of Psychoanalysis,* 1930
Hanna Segal, "Notes on Symbol Formation," *International Journal of Psychoanalysis,* 1957
August Strindberg, *A Dream Play*
Emanuel Swedenborg, *Arcana Cœlestia*

Alchemy

Obed Simon Johnson, *A Study of Chinese Alchemy*
Carl Gustav Jung, *Alchemical Studies*
Joseph Needham, *Science and Civilisation in China*
Edward H. Schafer, *The Golden Peaches of Samarkand*
Eliot Weinberger, "China's Golden Age," *New York Review of Books,* November 6, 2008
David Gordon White, *The Alchemical Body*
Douglas Wile, *Art of the Bedchamber*

Antiaging Science

Marcia Angell, "Big Pharma, Bad Medicine: How Corporate Dollars Corrupt Research and Education," *Boston Review,* May/June 2010
Leonard Guarente, *Ageless Quest*
Stephen S. Hall, *Merchants of Immortality*
Siddhartha Mukherjee, *The Emperor of All Maladies*
David A. Sinclair and Leonard Guarente, "Unlocking the Secrets of Longevity Genes," *Scientific American,* February 20, 2006
Rebecca Skloot, *The Immortal Life of Henrietta Lacks*
Lewis D. Solomon, *The Quest for Human Longevity*
Judith Thurman, "Face It," *The New Yorker,* March 29, 2010
Nicholas Wade, Numerous articles on antiaging in the *New York Times*
Duff Wilson, "Harvard Medical School in Ethics Quandary," *New York Times,* March 2, 2009
Keith J. Winstein, "Harvard Anti-Aging Researcher Quits Shaklee Advisory Board," *Wall Street Journal,* December 26, 2008

Radical Life Extension

Anonymous, "Strength from Within," *Life Extension* magazine, April 2002
Various authors, *The Scientific Conquest of Death: Essays on Infinite Lifespans*
Aubrey de Grey, *Ending Aging*
David Hamilton, *The Monkey Gland Affair*
Ray Kurzweil, *The Age of Spiritual Machines; The Singularity Is Near*
Nathaniel Rich, "Can a Jellyfish Unlock the Secret of Immortality?" *New York Times Magazine,* November 28, 2012
Jonathan Weiner, *Long for This World*

Cryonics

Mike Darwin, *Cryonics: An Historical Failure Analysis*
Nathan Duhring, *Immortality: Scientifically, Physically, Now*
Robert Ettinger, *The Prospect of Immortality*; *YOUNIVERSE: Toward a Self-Centered Philosophy*
Jill Lepore, *The Mansion of Happiness: A History of Life and Death*
Robert F. Nelson, *We Froze the First Man*

Conclusion

Albert Einstein, *Out of My Later Years*
Mircea Eliade, *Yoga: Immortality and Freedom*
F. M. Esfandiary, *Up-Wingers*
David M. Friedman, *The Immortalists: Charles Lindbergh, Dr. Alexis Carrel, and Their Daring Quest to Live Forever*
Peter Kingsley, *Reality*
Jill Lepore, "It's Spreading," *The New Yorker,* June 1, 2009
Proclus, *A Commentary on the First Book of Euclid's Elements*
Sakyong Mipham Rinpoche, *Turning the Mind into an Ally*; *Ruling Your World*; "The Elixir of Life Ceremony" handout
Chögyam Trungpa Rinpoche, *Crazy Wisdom*
Robert Anton Wilson, *Cosmic Trigger*
Ronald Wright, *A Short History of Progress*
Yevgeny Zamyatin, *We*

Index

About the Author

Adam Leith Gollner is the author of *The Fruit Hunters.* The former editor of *Vice* magazine, he has written for the *New York Times,* the *Wall Street Journal,* the *Guardian,* the *Globe and Mail,* and *Lucky Peach*. He lives in Montreal.